A First Science Dictionary

4/50

David J. Lucas
Senior Administrative Assistant, University of Leice[illegible]

Harold I. James
Formerly Senior Division Head, Bosworth College

John Simpson
Head of Chemistry Department, Wanstead High School, London

Edward Arnold

976
ward Arnold (Publishers) Ltd.,
25 Hill Street,
London WIX 8LL

ISBN: 0 7131 0027 3

Text set in 9/10 pt. IBM Press Roman, printed by photolithography and bound in Great Britain at The Pitman Press, Bath

PREFACE

In this dictionary we have aimed to provide accurate explanations of the words and phrases of Biology, Chemistry and Physics that a student may meet in any course up to and including those in preparation for the ordinary level of the G.C.E. or an equivalent standard.

It has been our aim to use simple words in our definitions and explanations as far as this is possible and, as a further aid to the student, we have included many cross-references between definitions. All such cross-references have been indicated by the use of *italics.*

However we would ask all students to note:

1. Where, for example, the word 'polarized' is used in a definition and we wish to indicate a cross-reference to the dictionary entry which happens to be 'polarization' rather than the word 'polarized' itself, we have shown the cross-reference as ....polarized (*polarization*)..........

2. In the case of an entry for a phrase like 'alpha particle' where the words are always used in that order, we have *not* included an entry for the reverse order, i.e. 'particle, alpha'.

3. The very common words of science, e.g. electric current, have not been put in italics every time they have been used.

4. Diagrams can be located in the text by referring to their main subject.

5. Throughout the dictionary, the following abbreviations have been used: *At. No.* for *Atomic Number* and *R.A.M.* for *Relative Atomic Mass.*

D. J. L.
H. I. J.
O. J. S.

abdomen That part of the body which contains most of the internal *organs.* It is separated from the *thorax* by the *diaphragm.* See *alimentary canal* (Fig. 5).

aberration Faults of colour and, or shape in an *image* produced by a lens or mirror, when compared to the original object.

absolute alcohol *Ethanol* which contains not more than 1% of water by weight.

absolute temperature Temperature, of great importance in theoretical work, measured from the *absolute zero* in steps which are the same size as the *celsius* degree. Such temperatures are given in *kelvins.* As absolute zero is −273.15°C, absolute temperatures are equal to celsius temperatures plus 273.15 (or, in simple work, plus 273). Thus the temperature of melting ice is 273.15 K (or in simple terms 273 K). See also *Charles' Law, Law of Pressures.*

absolute zero The zero of the *kelvin* (*absolute temperature*) scale. According to theory it is the lowest temperature that there is, but it cannot be reached in practice, although temperatures very close to it can be. At this temperature it is thought that *molecules* no longer possess any heat energy and all movement stops.

absorption (1) A physical process in which one (or more than one) substance is taken up by another, e.g. gases by liquids (air by water) or solids (air by hot iron on cooling) and liquids by solids. Penetration is much deeper than with *adsorption.* (2) Applied to the taking in of *soluble* substances, e.g. *sugars* are absorbed into the *blood. Mineral salts* are absorbed into the roots.

absorption of radiation Dull black surfaces absorb (take in) *radiation* well, polished metal surfaces reflect (*reflection*) most of the radiation and therefore do not absorb well. The radiation which is absorbed makes the surface hotter. Good absorbers of radiation are also good emitters of radiation (*emission of radiation*).

absorption spectrum When *white light* is passed through a substance (e.g. a gas or a solution) some of the light of certain colours may be absorbed (taken in) by the substance. So when this light is used to form a *spectrum* some colours are missing. The spectrum looks like a *continuous spectrum* crossed by dark lines or bands. It may be possible to name the substance through which the light has passed by studying the pattern and position of the dark lines.

acceleration The rate at which the *velocity* of an object increases with time. The acceleration equals the change in velocity that has happened divided by the time taken for the change. The *SI unit* is metres per second squared and the symbol is *a.* See also *deceleration, retardation.*

acceleration due to gravity The *acceleration* of a freely falling object, in a vacuum, is the same for all objects. It is given the symbol *g* and has a value of about 9.8 metres per second squared. The exact value depends on one's position on earth, being, for example, greater at the poles than at the equator.

accommodation The eye is able to see objects at different distances by changing the shape of the *eye* lens; this is called accommodation.

accumulator An electric *cell,* six of which are used in many car *batteries.* It consists of two sets of lead plates in sulphuric acid. The positive plate has a coat of lead dioxide when ready for use (charged). When electricity is taken from the cell (discharge) both plates become coated with lead sulphate and the acid becomes weaker. It can be returned to its original condition (recharged) by passing an electric current through it in the opposite direction. Other chemical systems can be used, e.g. iron and nickel oxide plates in potassium hydroxide solution (*Nife cell*). (Fig. 1).

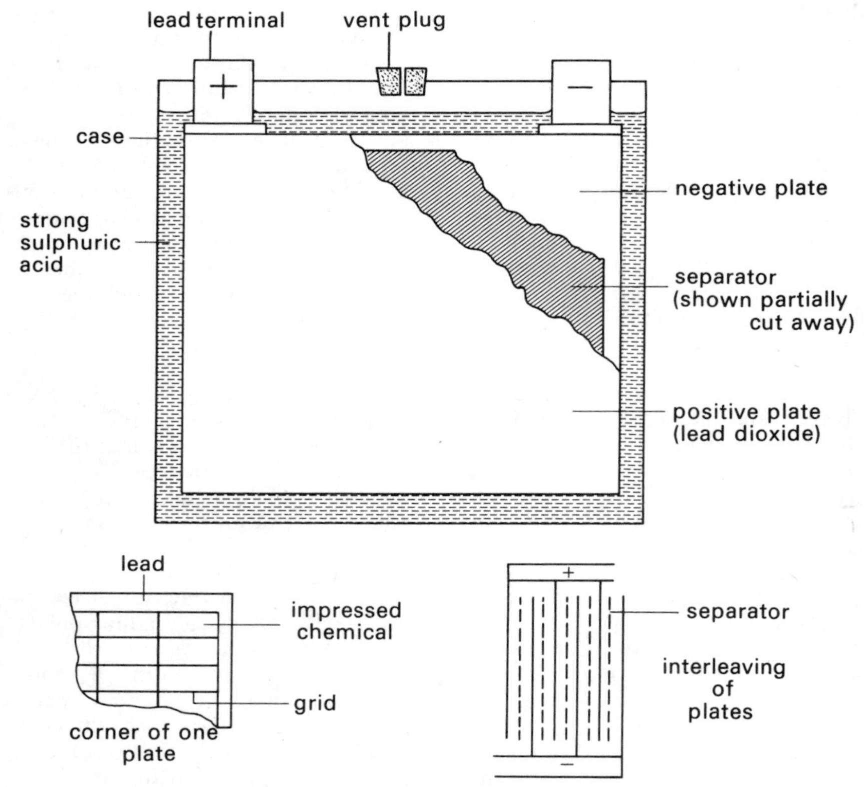

Fig. 1 The Structure of a Commercial Lead Accumulator

acetabulum Part of the hip *joint* (Fig. 97ii). A cup-like hollow in the *pelvis* into which the top of the thigh bone fits.

acetic acid CH_3COOH See *ethanoic acid.*

acetone CH_3COCH_3 A *mobile flammable*, colourless liquid with an agreeable *characteristic* smell. It is made industrially by *oxidation* of propanol $CH_3CH(OH)CH_3$ or by suitable treatment of ethanoic acid. It is used for *dissolving fats, resins* and greases, and *ethyne* is stored in solution in acetone. It is also called dimethyl ketone.

acetylcholine A substance made at *nerve* endings. It helps to carry the *nerve impulse* from one *neuron* to another where two nerves meet, or from the neuron to a *muscle.*

acetylene See *ethyne.*

achene A kind of *fruit* with one *seed* covered by a tough skin which does not split open, e.g. sunflower.

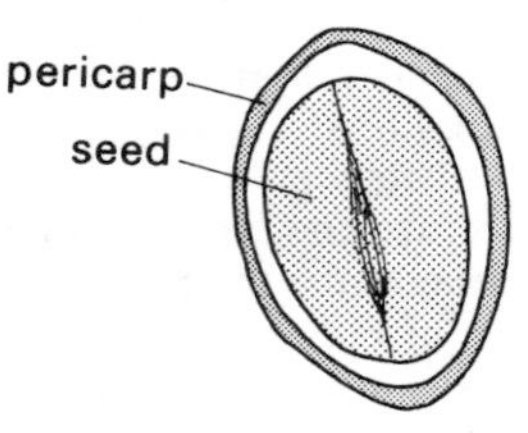

Fig. 2 Achene in longitudinal section

achilles tendon The string-like structure which joins the lower leg *muscle* to the back of the foot. If it is damaged it can prevent running.

achromatic lens A simple *lens* does not bring light of different colours to a *focus* at exactly the same point, and an image produced by such a lens will be falsely coloured. An achromatic lens is specially made to reduce this faulty colouring. See *abberation.*

acidic Behaving like *acids* in turning *litmus* red, and having a *pH* less than 7.

acid radical The *element* or group of elements present in *acids* and *salts* as a negatively-charged *ion*, which persists unchanged during many *chemical reactions,* e.g. *sulphate* radical or ion, SO_4^{2-}; *phosphate* ion, PO_4^{3-}; *chloride* ion, Cl^-; *ethanoate* ion, CH_3COO^-; *hydrogencarbonate* ion, HCO_3^-.

acid salt A *salt* formed by the partial *neutralization* of a *dibasic* or tribasic *acid* by a *base* and which still contains replaceable hydrogen in the *acid radical.* Such salts in solution do not necessarily have an *acidic* reaction although they possess some acid properties, e.g. sodium hydrogencarbonate, $NaHCO_3$, has a *pH* of about 8 and is slightly *alkaline,* whereas sodium hydrogensulphate has a pH of 5–6 and is acidic.

acids Substances which, in solution, have all the following *properties*:
(a) produce hydrogen ions H^+,
(b) have a sour taste,
(c) *react* with certain *metals* to liberate hydrogen,
(d) react with *bases* to produce a salt plus water,
(e) turn *neutral litmus* red, i.e. have a pH of less than seven.

At room temperature acids may be solids (oxalic, citric), liquids (sulphuric, methanoic (formic)) or gases (hydrogen chloride). In most cases they only behave like acids when dissolved in water.

acid-soda fire extinguisher A steel container, about three-quarters filled with a *solution* of *sodium hydrogencarbonate* and containing a bottle of *sulphuric acid.* By inversion or impact the two are caused to mix and the *carbon dioxide* given off forces a jet of the liquid (mainly water) through a nozzle. This can be directed on to a fire which is put out, mostly by the cooling effect of the water.

acoustics The study of sound.

acromegaly A disease caused by something wrong with the *pituitary gland.* The bones of the head, hands and feet become larger than usual.

actinomorphic Applied to flowers where the *petals* are all of the same shape and size.

activated carbon *Charcoal* obtained by heating vegetable carbon (nut shells, etc.) either in a vacuum or in *superheated steam* at about 300–1000°C to remove *hydrocarbons.* This process increases very considerably the power of the charcoal to adsorb (*adsorption*) gases and colouring matter. Activated charcoal is used in gas masks and for removing the brown colour from raw sugar.

activity series A list of *metals* arranged in order of decreasing chemical reactivity. This may be measured by the effect of exposing them to air or by the ease with which they liberate *hydrogen* from water, *steam* or dilute *acids* such as hydrochloric acid. Hydrogen is usually included in the table in such a position that the metals above will liberate it from dilute acid but those below will not. See also *electrochemical series.*

addition compounds *Saturated organic compounds* (e.g. *alkanes*) formed from *unsaturated* ones (e.g. *alkenes* or *alkynes*) by breaking the *double* or *triple bond* and adding on other *atoms* or groups. E.g. $C_2H_4 + H_2 \rightarrow C_2H_6$, and $C_2H_2 + 2Cl_2 \rightarrow C_2H_2Cl_4$. No further additions can be made.

adenosine triphosphate (abbr. ATP) A substance present in all living *cells.* It stores the *energy* in the cells until that energy is needed to do some work, e.g. muscular (*muscle*) effort.

adhesion The force of attraction between unlike molecules, for example, between molecules of a solid and of a liquid which touches it.

adipose tissue *Cells* which store fat or oil, e.g. under the *skin* of a man or woman.

adipyl chloride $ClOC(CH_2)_4COCl$ A *reagent* used in the preparation of *nylon.*

adrenal body The name of the *gland* which makes *adrenalin.* It is positioned above the kidney.

adrenalin A chemical made in the *gland* usually joined to the *kidney.* Adrenalin, made in the inner part of the gland, has an effect upon the *nervous system,* causing changes in various parts of the body when the animal is angry or afraid. The substance helps an animal to fight harder or run away more quickly.

adsorption The taking up by a *solid* of a *liquid* or *gas* (*vapour*) which concentrates on the surface. It differs from *absorption* in being a surface effect only and penetration may not be more than one *molecule* or one *atom* deep. It is shown best by substances with a large surface area for a comparatively small volume, e.g. *activated carbon* used in gas masks to adsorb poisonous gases. Palladium metal will adsorb large amounts of hydrogen. Many metals, such as platinum, which act as *catalysts* in reactions do so by adsorbing gaseous molecules on to the metal surface where they *react* together.

adventitious Used for roots which grow out of stems, e.g. the roots at the bottom of a corn stem or the roots on an Ivy stem which help it to climb.

aerobic respiration The release of *energy* in the *cell* using oxygen.

aerosol 1. Very small drops of liquid or particles of a solid suspended in air or in a gas (mist, fog, smoke). 2. A container holding a deodorant or insecticide along with a gas under pressure. When a valve is opened the liquid is released in the form of a very fine spray. The gas most often used is one called freon-12.

afferent Used for *nerves* which carry *nerve impulses* towards the *brain* or *spinal cord.*

affinity Chemical attraction: the tendency a substance may have to unite with one particular *element* rather than with another, e.g. *chlorine* has a strong affinity for *hydrogen* and will readily remove it from *compounds* in which it is present to form hydrogen chloride. Chlorine reacts with turpentine, $C_{10}H_{16}$, to produce carbon and hydrogen chloride. A solution of chlorine in water decomposes to give oxygen and hydrochloric acid.

after birth The *membranes* around the young baby when it is still in the *womb* and which come away soon after the baby is born.

agar agar A powder made from seaweed, used for growing *bacteria.*

agglutination The sticking together of *red blood cells* when the wrong kind of blood is mixed with them. It is used to find out whether two bloods can mix in *transfusion.*

aggregate fruit A *fruit* made up of a number of little fruits, e.g. raspberry.

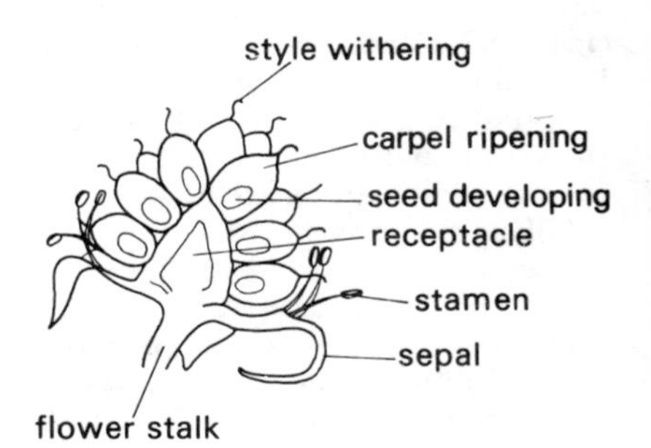

Fig. 3 Aggregate Fruit (Raspberry) in longitudinal section

agonic line The imaginary line round the earth which joins places having zero *declination,* i.e. where the magnetic compass points true north.

air bladder An air-filled space in most fish which allows them to float

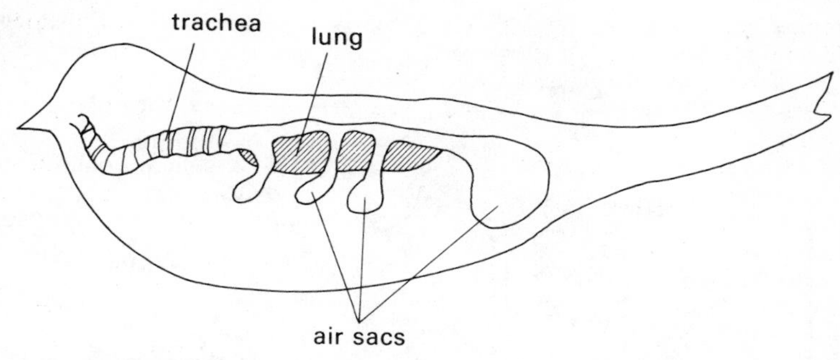

Fig. 4 Air Sacs of Bird

when they are not swimming. The amount of air can be controlled according to the depth of the water. It is also called a swim bladder.

air sac The extra lungs of birds. They are in pairs inside the bird. They hold air which passes in and out through the true lungs where *oxygen absorption* takes place. Small air sacs are also found in true lungs. See *alveolus.*

albumen The white of the *egg* (Fig. 61). A *protein* also found in seeds. Albumen is used as food.

alchemist One who practised *alchemy:* i.e. one who tried to turn common or *base metals* into *gold,* to find a liquid which would be a *solvent* for every known substance (the alcahest), or a substance which would give eternal youth to those who drank it (the elixir of life).

alchemy The name given to chemistry as studied from about 500 to 1400 AD when it was a mixture of astrology, trickery and some practical chemistry.

Alcohols R—OH General term for a class of *compounds* in which one or more *hydrogen atoms* in a *hydrocarbon* is replaced by *hydroxyl* groups, e.g. methanol, CH_3OH, glycol, $C_2H_4(OH)_2$. The word alcohol is often used for *ethanol,* C_2H_5OH. This is used as a drink by man and has an effect on the nervous system.

aleurone grain A form of *protein* used as a store of food in seeds.

algae Simple plants which contain *chlorophyll* and sometimes other coloured substances. An alga may be so small that it can be seen only with a microscope, e.g. a *diatom,* or very large, e.g. some seaweeds.

algin A substance found in seaweeds. Used by man in making icecream.

alimentary canal The *food canal* which starts at the mouth and goes through the body ending at the *anus.* Different parts are used in different ways to *digest* food. (Fig. 5.)

aliphatic compounds Carbon *compounds* composed of *straight* or *branched chains* of *atoms* or *groups,* but not ring structures of the type present in *benzene,* C_6H_6. Examples are ethane, C_2H_6; propanamide, $CH_3CH_2CONH_2$; ethene glycol, $CH_2OH—CH_2OH$. ethane.1.2.diol.

alkali A substance which will dissolve in water to give an *alkaline* solution, e.g. the *hydroxides* of sodium, potassium and calcium and also ammonium hydroxide. All alkalis are *bases* and will neutralize (*neutralization*) *acids* to form *salts.* Their solutions turn *litmus* blue and *phenolphthalein* pink.

alkali metals The *metals* of Group I of the *periodic table, lithium, sodium, potassium*, rubidium and caesium. Their *hydroxides* readily dissolve in water to form strongly *alkaline* solutions. All the metals attack water with the liberation of hydrogen and all form *positive ions*

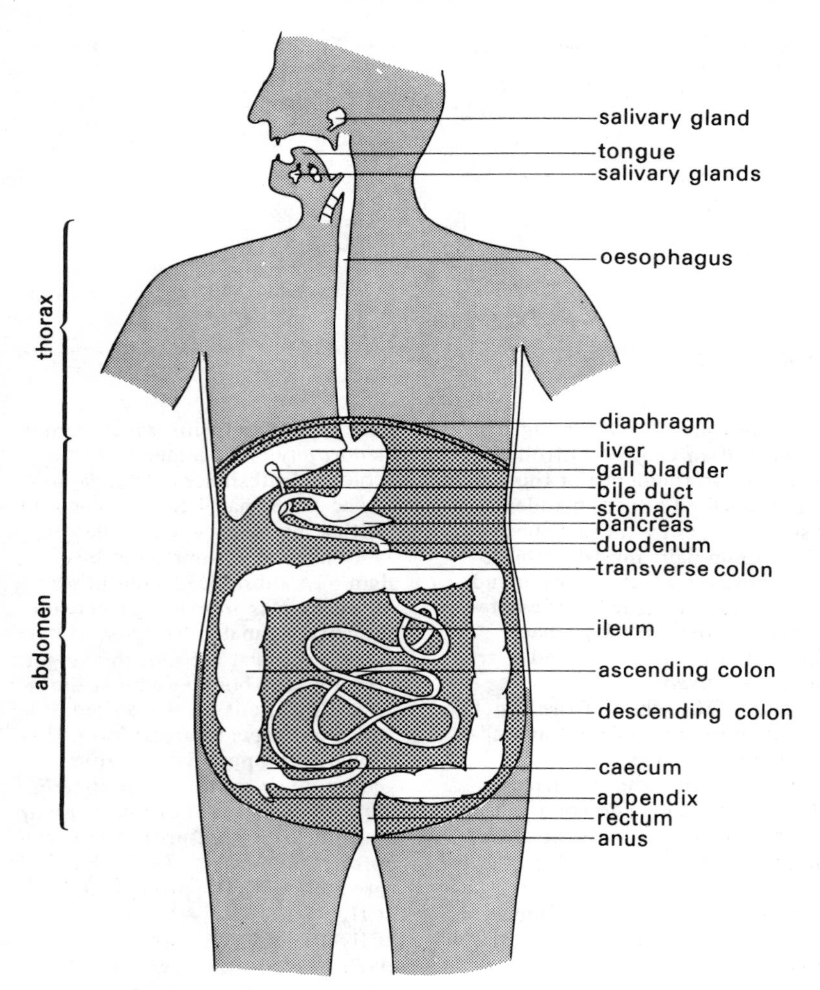

Fig. 5 *The Human Alimentary Canal (simplified)*

by loss of one *electron.* They are all *monovalent.*

alkaline A *solution* with a *pH* greater than 7.

alkaline earth metals The *divalent metals* in Group II of the *periodic table,* beryllium, *magnesium, calcium, barium,* strontium and *radium.* The term is used more particularly for calcium, strontium and barium whose *hydroxides* are soluble and form *alkaline* solutions.

alkanes *Saturated hydrocarbons* whose *molecules* consist of *straight* or *branched chains* of CH_2 groups and having the general *formula* C_nH_{2n+2}. They form a *homologous series* of chemically *stable,* relatively *inert* and *flammable* compounds. The lower members, methane, CH_4; ethane, C_2H_6; propane, C_3H_8; butane. C_4H_{10}, are gases, the next twelve are liquids (present in petrol and mineral *oils*) and the remainder are solids (waxes). Also called paraffins.

alkenes *Unsaturated hydrocarbons*

in which at least two carbon *atoms* are joined by a *double bond,* e.g. *ethene* (ethylene), C_2H_4 or $H_2C{=}CH_2$. The general *formula* is C_nH_{2n}. The lower members are gases and all are fairly *reactive* and form *alkanes* (paraffins) i.e. *saturated* hydrocarbons, by addition (*addition compounds*) of hydrogen. Also called olefins.

alkynes *Unsaturated hydrocarbons* in which two of the carbon *atoms* are joined by a *triple bond,* e.g. *ethyne* (acetylene), C_2H_2 or $HC{\equiv}CH$. The general *formula* is C_nH_{2n-2} and the *compounds* are *unstable* and reactive. Also called acetylenes.

allantois One of the *membranes* around an *embryo.* It is used in birds and *mammals* for *respiration* and *excretion.*

allele One of a pair of *genes.* The paired alleles may be the same or differ in detail, e.g. the height of a plant may depend on which of the two alleles is present.

allelomorph One of the forms of a *gene.* A pair of *homologous chromosomes* will bear a pair of alleles. These will decide a phenotypic (*phenotype*) character in the *organism.*

allotrope Some *elements* can exist in forms which are physically different but chemically the same; these forms are called allotropes and give rise to the same *chemical compounds,* e.g. *sulphur* may be *rhombic* or *monoclinic* in its *crystals,* but if equal *masses* of the two are burnt they will each form the same amount of sulphur dioxide. Also called a polymorph (*polymorphism*).

allotropy The *property* of an *element* whereby it can exist in two or more physically different forms while retaining the same chemical properties. The different forms are usually in the same state (solid, liquid or gas). Examples are *carbon* as diamond or graphite; oxygen as oxygen, O_2, or ozone, O_3. Allotropy is a special case of *polymorphism* where the substance is an element.

alloy A mixture of two or more *elements* which has metallic properties and is generally more useful than the individual elements, e.g. brass contains copper and zinc, steel contains iron, carbon and a little silicon plus traces of other elements to give it special uses.

alnico An *alloy* (of aluminium, nickel, cobalt, copper and iron) used for making strong permanent magnets.

alpha decay When a *radioactive* substance gives out an *alpha particle* it changes to a new element. The *atomic number* falls by two and the *mass number* by four. E.g. $^{226}_{88}Ra \rightarrow {}^{222}_{86}Rn + {}^{4}_{2}He$ (alpha particle). See also *decay.*

alpha particle Certain radioactive materials (like americium-241) give out alpha particles. These are the same as the *nuclei* of helium atoms, being two *neutrons* and two *protons.* Alpha particles have a positive electric charge, change direction when passing through a magnetic field and are stopped by a sheet of paper or a few centimetres of air.

alternating current A type of electric current whose strength varies regularly with time, as shown in figure 6.

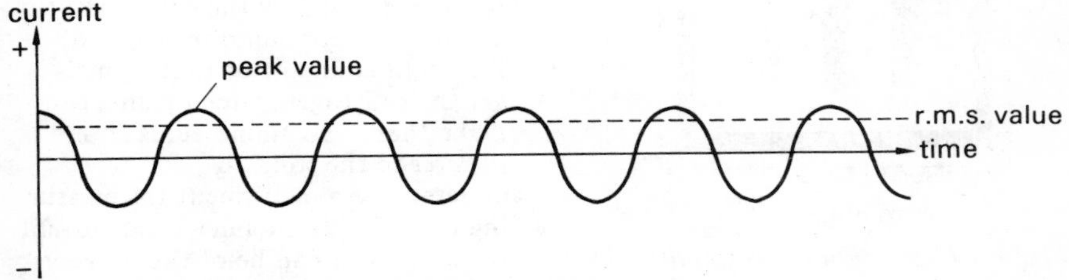

Fig. 6 The Pattern of an Alternating Current

The number of times the current goes from zero to its greatest value in one direction, through zero to its greatest value in the other direction and back to zero in one second is called the *frequency* of the alternating current (a.c.) Typical commercial values of a.c. frequency are 50 or 60 *hertz.* See also *peak value, root mean square value.*

alternation of generations The name given to the kind of *life history* of a plant such as a fern where during one stage of its life it produces only *spores* and during another stage of its life it produces *gametes. Asexual reproduction* and *sexual reproduction* are carried out at different times.

alternator A type of *generator* which produces *alternating current.* The simplest form is a coil of wire turning in a steady magnetic field and connected by *slip rings* to the rest of the circuit.

altimeter A type of *aneroid barometer* calibrated to measure height in, for example, an aeroplane. It works because the *atmospheric pressure* falls as the aeroplane rises higher.

aluminium Al At. No. 13. R.A.M. 27. A light white *metal* which is a very good *conductor* of electricity. The metal and its *alloys* are used for cooking utensils such as saucepans and for aircraft construction where lightness is an advantage.

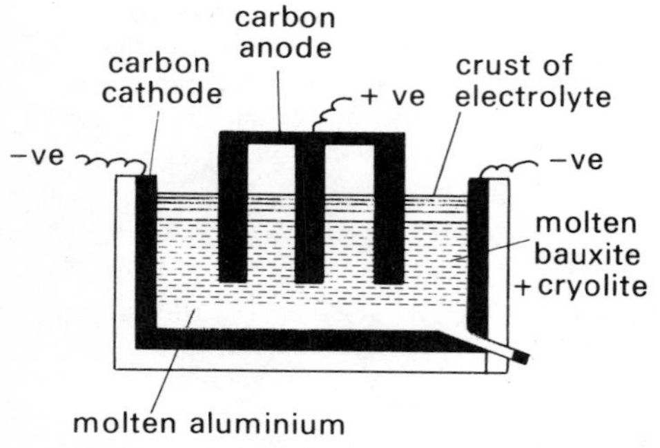

Fig. 7 Extraction of Aluminium by Electrolysis

alums A general name for a class of *salts* consisting of the *sulphates* of two different *metals* and having a general *formula* of $M_2SO_4.R_2(SO_4)_3.24H_2O$, where M is a *monovalent* metal (or the *ammonium ion*) and R is a *trivalent* metal, e.g. iron alum, $K_2SO_4.Fe_2(SO_4)_3.24H_2O$, ammonium alum, $(NH_4)_2SO_4.Al_2(SO_4)_3.24H_2O$, potash alum (common alum), $K_2SO_4.Al_2(SO_4)_3.24H_2O$. All alums are isomorphous, i.e. they all *crystallize* in the same *octahedral* form.

alveolus (*pl.* Alveoli) The *air sacs* which make up most of the *lungs* (Fig. 109ii) of *mammals. Oxygen* passes from the air into the blood and *carbon dioxide* passes from the blood into the air during breathing.

amalgam An *alloy* containing *mercury.* It is usually a *mixture,* but sometimes a *compound* is formed. Amalgams containing cadmium and zinc are used as dental cements for filling teeth.

amber *Resin* which has become a *fossil.* Used for ornaments.

amino acids *Organic fatty acids* in which one or more *hydrogen atoms* in the radical have been replaced by amino ($-NH_2$) groups, e.g. ethanoic (acetic) acid is CH_3COOH and CH_2NH_2COOH is amino-ethanoic acid or glycine. Because they contain both a basic radical ($-NH_2$) and an acid radical ($-COOH$) they combine with each other to form complex *polymers* called *proteins.* When proteins, e.g. lean meats, are eaten, they are broken down into amino acids which are taken into the blood and later used to build up the proteins needed by the eater. All amino acids contain nitrogen and it is from amino acids that animals get their nitrogen. Green plants can make their own amino acids from *nitrates* in the soil.

ammeter An instrument for measuring the strength of an electric current in *amperes.* It can be of the *moving coil* or *moving iron* type. An ammeter is always connected so that the current to be measured flows through it, although moving coil ammeters may contain a *shunt* (a low value resistor) which lets most of the current miss the coil itself. Changing the shunt will change the range of the meter.

ammonia NH_3 A colourless, strong-smelling gas made by combining *nitrogen* (1 volume) and *hydrogen* (3 volumes) in the presence of a *catalyst* (*Haber process*). It is very soluble in water to form an *alkaline* solution containing a small amount of *ammonium hydroxide* (NH_4OH). It is produced naturally when *organic* matter such as farmyard manure and straw decays. It is the only common alkaline gas.

ammoniacal Containing or smelling of *ammonia*. Ammoniacal liquor is produced during the manufacture of gas from *coal* and it contains dissolved ammonium *compounds*.

ammonia-soda process A process for the manufacture of anhydrous *sodium carbonate* (Fig. 150) from *ammonia, brine* and *carbon dioxide*. *Ammonium chloride* is obtained in solution and *sodium hydrogencarbonate* is precipitated (*precipitation*) and *filtered* off. This is then roasted to produce anhydrous sodium carbonate. Also called the Solvay process.

ammonium chloride NH_4Cl White, soluble *crystalline salt* which readily *sublimes* on heating. It is used in dry batteries (*dry cell*).

ammonium hydroxide NH_4OH When *ammonia*, NH_3, is passed into water it forms in the main a physical *solution*, but a small number of ammonium *ions* (NH_4^+) are formed and the solution is *alkaline*. It behaves in a similar way to one containing *hydroxyl* ions (OH^-). Such a solution is often referred to as ammonium hydroxide, but a better term is *aqueous* ammonia. The solution readily decomposes into ammonia and water when it is heated.

ammonium ion NH_4^+ A *monovalent radical* or *ion* which only exists as part of a compound (e.g. ammonium chloride) where it behaves like an *alkali metal* ion. When *ammonia* dissolves in water some ammonium ions are formed.

ammonium salts Taken by plants from the soil and used to build up their bodies. Mammals remove *ammonia* from their bodies in the *urine*. The ammonia is changed to *urea* before it goes to the *kidney*.

ammonium sulphate $(NH_4)_2SO_4$ A *salt* produced industrially by the *neutralization* of *sulphuric acid* with *ammonia* and then *evaporating* and crystallizing. It is a very important *nitrogenous fertilizer*.

amnion The inner of the two *membranes* which surround the *embryo* of a bird or *mammal*.

amorphous A solid substance which is not *crystalline* and has no definite shape, e.g. wax, glass, rubber. It has no definite *melting point*, but softens over a range of *temperature*.

ampere The *SI unit* of electric current. It is a basic SI unit and the symbol is A. If this amount of current flows in each of two very long, parallel wires kept one metre apart in a vacuum, then there will be a force of 2×10^{-7} *newton* (two ten-millionths of a newton) on each metre length of wire.

ampere-hour A *unit* used for measuring the *electric charge* or *quantity of electricity* stored, for example, in an *accumulator*. If the accumulator can give a current of two *amperes* for twenty hours, then it has stored two times twenty or forty ampere-hours. The ampere-hour is not part of the *SI* system. It is equal to 3 600 *coulombs*.

amphibia The group name given to the frogs and toads which spend the first part of their life in water as *tadpoles* and later change to the land living form. See *metamorphosis*.

amphibious Any animal (or machine) which is able to move on

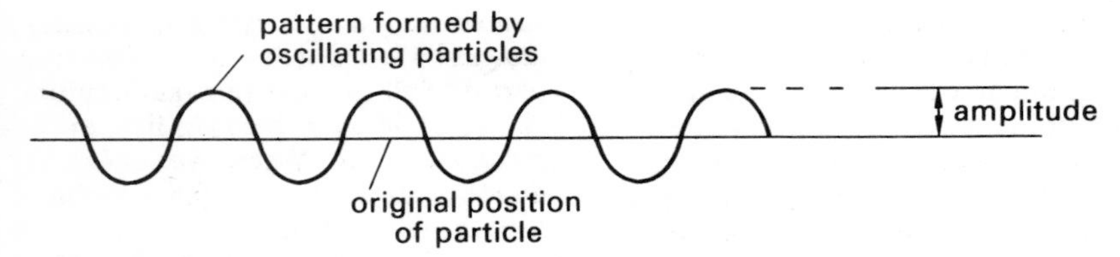

Fig. 8 The Amplitude of Oscillating Particles

land and in water, e.g. crocodile, duck, most humans.

amphoteric A term used for those *compounds* (*oxides* and *hydroxides* of certain *metals*) which can behave as both *acid* and *base*, e.g. zinc hydroxide dissolves in *sulphuric acid* to form zinc sulphate, $ZnSO_4$, and also in excess *sodium hydroxide* to form another salt, sodium zincate, Na_2ZnO_2.

amplifier An instrument, usually containing *valves* or *transistors,* which makes an electric current or potential difference larger.

amplitude In any oscillating (*oscillation*) system, particles move to and fro about their original position. The greatest distance moved from the original position is called the amplitude. (Fig. 8.)

ampulla The most sensitive part of the inside of the *ear* (Fig. 59). The movements of the head are changed to *nerve impulses* in the ampulla by the movement of the watery liquid inside. This helps to keep the balance of the body.

amylase An *enzyme* made in the *salivary glands* and in the *pancreas.* It helps to change *starch* to *sugar* in *digestion.*

anabolism A word used to include all the building up processes in a living body, e.g. the making of sugar and starch from simple *carbon dioxide* and water, or the building up of *proteins* from *amino acids.* The opposite, the breaking down processes, is *catabolism.*

anaemia A condition where *blood* is poor, often due to lack of iron.

anaerobic respiration The release of *energy* in the *cell* without the use of *oxygen.*

anaesthetic A *drug* which produces insensibility to touch, pain or heat. It may act to produce unconsciousness (e.g. ether or trichloromethane) or its effect may be limited to a small area. It acts by making certain *cells* incapable of responding to *stimuli.Chloroform* or *ether* is used in operations to put the patient to sleep. *Cocaine* is used by dentists.

analysis Analysis is the procedure used in determining the composition of a compound or mixture. It may be required to find out not only which elements or groups are present, but also the actual percentage of each. See also *qualitative* and *quantitative analysis.*

anatomy The study of the structure of a plant or animal body.

androecium The botanical (*botany*) name for the *stamens* in a *flower.*

aneroid The word means 'without liquid'. It is usually applied to a type of *barometer* which does not contain mercury. An aneroid

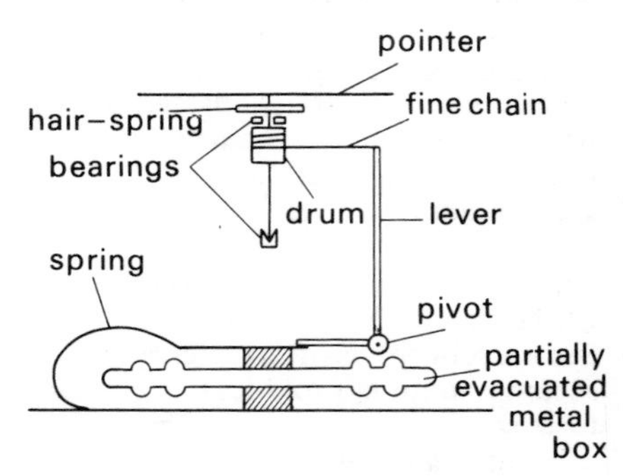

Fig. 9 The Aneroid Barometer

barometer contains a metal drum in which there is a partial vacuum and which changes shape slightly when *atmospheric pressure* changes. This small change is made greater by levers and moves a pointer over a scale.

Angiospermae The group name of the flowering plants.

angstrom unit A *unit* of length which is not part of the *SI* system. It is equal to 10^{-10} metre (one ten-thousand-millionth of a metre) and was used for measuring, for example, the *wavelength* of light and atomic and molecular sizes. The symbol is Å. It is being replaced by the *nano*metre.

angular velocity The angle turned through by a rotating object in one second. The *SI unit* is *radians* per second and the symbol is ω.

anhydride A *compound* which can be obtained from certain *acids* by removing the elements of water or which will dissolve in water to form an acid, e.g. SO_3 (sulphur trioxide) is the anhydride of H_2SO_4 (sulphuric acid), CO_2 (carbon dioxide) is the anhydride of H_2CO_3 (carbonic acid) and $(CH_3CO)_2O$ (acetic anhydride) is obtained from (ethanoic) acetic acid by the loss of one *molecule* of water from two molecules of the acid.

anhydrite $CaSO_4$ *Anhydrous calcium sulphate* obtained from naturally-occurring deposits by mining. It is used in the manufacture of *plaster of Paris* and can be used if required for the production of *sulphur dioxide* for the *contact process* of making *sulphuric acid.*

anhydrous 'Without water'. An *oxide* or a *salt* which does not contain *water of crystallization,* e.g. sodium carbonate or copper sulphate *crystals* on heating lose all their water of crystallization and become anhydrous.

animal A living *organism* which is not a plant. Nearly all animals are able to move in order to get food which they cannot make for themselves as plants can. Most animals have digestive (*digestion*) *organs* and have better *sense organs* than plants.

animal charcoal Material obtained by heating blood, flesh, bones etc., until charring (blackening) occurs. It is mainly *calcium phosphate* with about 10% of *carbon* and is used as a decolorizing agent.

animal fats *Saturated esters* found in such *fats* as butter, lard, suet, tallow, etc.

anion The *negatively-charged ion* or *radical* present in *ionic compounds,* i.e. in *salts* or *acids,* e.g. chloride, Cl^-; sulphate, SO_4^{2-}; hydroxyl, OH^-. Anions are attracted to the positive electrode (*anode*) during *electrolysis.*

Annelida The group name of the worms which are divided into *segments,* e.g. earthworms.

annuals Plants and animals whose life is completed within one year, e.g. mosquito, bean.

annual rings The circular markings in the trunk of a tree, marking the amount of new wood formed during one year.

annulus 1. The ring of torn material on the stem of a *fungus.* 2. A group of *cells* which help to scatter the *spores* of mosses and ferns.

anode 1. The *positive* (+) pole or plate of a *battery.* 2. The *electrode* of a voltameter or thermionic valve, etc., connected to the above pole. 3. The electrode at which *negatively* (−) charged ions are discharged during electrolysis.

antagonistic muscles *Muscles* which work together to cause a movement. When one of a pair tightens, the other loosens, e.g. when the elbow bends the *biceps* muscle tightens and the *triceps* muscle loosens.

antenna (plural antennae) The sensitive feelers on the head of insects. They are sensitive to touch and smell. In insects which live in the dark, e.g. cockroaches, the antennae are very long and are the main *sense organs* of the body.

anterior The front end of an animal. E. g. the head is at the anterior end of a snake.

anther The part of a *stamen* which produces the *pollen.* See *flower* (Fig. 76iii).

antheridium The name given to the male *organs* of some plants such as mosses and ferns. The antheridium produces the male *cells.*

anthracite A hard, black, shiny form of *coal* containing about 85–90% of *carbon* and very little *volatile* matter. It gives a lot of *heat* with very little *flame,* is slow-burning and leaves little ash. It is used in central heating installations.

anthrax A disease which may be caught by handling the hair of animals, e.g. sheep or horses. The disease is due to *bacteria* which are very difficult to kill. The signs in humans are black spots on the skin.

antibiotic A substance made by a *fungus* which stops the growth of some harmful *bacteria,* e.g. *penicillin,* Penicillin is one of the most useful chemicals used by doctors in fighting disease.

antibody A substance produced by blood in a living body to help destroy harmful *bacteria* which might get in through cuts and scratches. See *antigen.*

antichlor A substance, generally *sodium sulphite* or *thiosulphate,* used to remove excess *chlorine* after *bleaching* has been carried out.

antifreeze An *aqueous* solution, usually of *glycol* (ethane.1.2. diol, CH_2OH-CH_2OH), used in cooling *internal combustion engines* and radiators in wintry conditions because its *freezing point* is lower than that of water alone. Because of this, it remains liquid at much lower temperatures.

antigen A substance produced by *bacteria* which is usually poisonous. When this substance is in the blood, *antibodies* and *antitoxins* are produced by the blood.

antinode A point where the *amplitude* of a *stationary wave* reaches its greatest value.

antiseptic A chemical which is poisonous to *bacteria.* It is usually spread over a wound to prevent bacteria getting in, e.g. *iodine* solution.

antitoxin Like an *antibody.* Produced by the blood to make harmless the poisons of *bacteria.*

anus The end of the *alimentary canal* (Fig. 5). through which the waste matter passes to the outside. The opening is controlled by muscles.

anvil One of the three small bones in the *middle ear* which help to pass the movements of the *tympanum* to the *inner ear.* These movements are caused by *sound waves.* Also called the *incus.*

aorta The main *artery* of the body. It leaves the left *ventricle* of the *heart* (Fig. 91ii) and then branches to all parts of the body. The *dorsal* aorta runs backwards until it forks into the two main leg arteries. It carries blood rich in *oxygen.*

aperture The size of the reflecting (*reflection*) or refracting (*refraction*) surface, of a spherical mirror or lens, which is actually used. In a *camera* an aperture of f/8 means that the diameter of the part of the lens actually in use is $\frac{1}{8}$ of its *focal length.*

apparent depth When an object is seen through a substance other than air it appears to be nearer than it actually is, due to the *refraction* of the light. See also *real and apparent depth.*

appendicular The name given to the parts of the *skeleton* other than the *skull* and *vertebral column.*

appendix Part of the *alimentary canal* (Fig. 5). It is a tube with a closed end growing out of the *caecum.* It is a *vestigial structure* in the human. A diseased appendix causes appendicitis.

aquadag A *suspension* in water of *graphite* ground very small. It is used as a *lubricant.* If the suspension is in *oil* it is known as Oildag. The process was developed by the Acheson Company. Also called Dag or Deffloculated Acheson Graphite.

aquatic Having to do with water, e.g. a fish is an aquatic animal.

aqueous Connected with water, e.g. an aqueous solution is one made up in water. Aqueous copper sulphate or copper sulphate (aq) is a water solution of copper sulphate.

aqueous humour The watery liquid found in the front of the *eye* (Fig. 68) It keeps the *cornea* pushed out so that it does not touch the *lens.*

arable Used to describe farming land which is ploughed up and on which crops are grown. It is opposite to *pasture.*

Arachnida The group word for animals of the spider type. Scorpions and mites are arachnids. All have four pairs of legs.

arboreal Having to do with trees.

arc light When a very high *voltage* is produced between two *carbon* rods the air, *gas* or *vapour* between them is *ionized.* A *current* of electricity flows across the gap causing a very bright light to be produced. The *electrodes* vaporize (*vapour*) in the intense heat.

archegonium The name given to the femal *organs* of some plants such as mosses and ferns. The archegonium produces the female *cells.*

Archimedes' principle An object which is all or partly below the surface of a *fluid* experiences an *upthrust* equal to the weight of fluid pushed aside to make room for it.

area A measure of surface. The symbol is *A* and the *SI unit* is the square metre.

arene A *hydrocarbon* with the general formula, C_nH_{2n-6}, also known as aromatic hydrocarbons. The first three members of this homologous series are *benzene,* C_6H_6; *toluene* (methyl benzene), C_7H_8; *xylene* (dimethyl benzene, three *isomers*), C_8H_{10}.

argon Ar At. No. 18 R.A.M. 40 A *noble gas* present in the *atmosphere* to about 0.8% (8 parts in 1000). Used in gas-filled electric lamps and *fluorescent* tubes.

argonons The *noble gases* present in the atmosphere, *helium, neon, argon, krypton* and *xenon* and also *radon.* These *elements* make up Group O of the *periodic table.*

armature The coil system of an *electric motor* or a *generator.* It is usually the moving part.

arteriole A small branch of an *artery.*

artery One of the main *vessels* carrying *blood* away from the *heart.* The blood is *oxygenated* except in the *pulmonary* arteries, where the blood is *deoxygenated.*

Arthropoda One of the main groups of *invertebrate* animals. All arthropods have a body divided into *segments* and have jointed legs. The main groups of arthropods are *insects* (3 pairs of legs), *arachnids* (4 pairs), *crustaceans* (5 pairs of walking legs) and *myriapods* (many pairs of legs).

articulation The fitting together of bones to form a movable *joint,* e.g. the *femur* articulates with the *pelvis* to form the hip joint.

asbestos A *mineral fibre* which will not burn and which has *crystals* so extremely thin as to be elastic (*elasticity*). It is mainly calcium magnesium silicate and its *properties* make it very suitable for use as a heat insulating material and for fire-proof fabrics.

ascorbic acid The chemical name of *Vitamin C.*

aseptic Where *germs* are absent from the air or material or a wound.

asexual Without *sex.* Used in types of *reproduction* where there is no *fertilization,* e.g. *spore* production in a *fungus.* See *binary fission* (Fig. 17).

aspirator A large storage jar or bottle with a tap at the bottom and filled with a *solution* of some kind which is fairly frequently used. For another purpose, it may be filled

Fig. 10 Aspirator

information is then used to cause other activities, e.g. the smell of meat causes *saliva* to be produced.

astigmatism A fault in a *lens* (including the lens of the eye) which causes an *image* to appear to be correct in one direction but blurred in a direction at right angles to this first direction. It is caused by the lens not being shaped in the same way all over its surface. If the eye lens has this fault spectacles with cylindrical lenses have to be worn.

astronomical telescope A *telescope* made with two *converging lenses.* The *objective* lens (nearer the object) has a longer *focal length* than the *eyepiece* (nearer the eye). As the *image* produced is upside down the telescope is only suitable for looking at objects like the stars where this does not matter.

atlas The first bone of the *vertebral column.* It supports the *skull* and allows for the nodding of the head.

with water and as this runs out a current of air is drawn through a suitable piece of apparatus connected to the top of the jar.

assemblage A regular arrangement of *ions* in a *crystal lattice* or of *atoms* in a giant *molecule.*

assimilation The process of building up. Carbon assimilation is another name for *photosynthesis.* Food is assimilated when it is used to build up the body.

association 1. The joining together of the *molecules* of a *compound* to form larger groups under certain conditions. If nitrogen dioxide gas (NO_2) is cooled down the molecules associate in pairs to form liquid dinitrogen tetroxide (N_2O_4); this change is *reversible.* Ethanoic (acetic) acid exists as single molecules CH_3COOH when dissolved in water, but as double molecules $(CH_3COOH)_2$ when dissolved in benzene. 2. One of the *functions* of the *central nervous system.* Information from *sense organs* is collected and sorted out. That

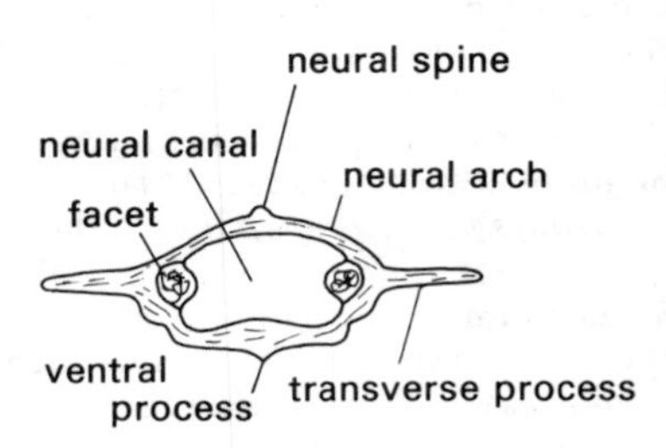

Fig. 11 The Atlas Vertebra

atmosphere 1. The gases which surround the earth. 2. A unit of *pressure* which is not part of the *SI* system. It is about the pressure caused by the weight of the gases which surround the earth, but it is agreed that it is equal to the pressure at the bottom of a column of mercury 760mm high under certain conditions. This equals 101 325 *pascals.* See also *atmospheric pressure, barometer.*

atmospheric pressure The actual pressure caused by the weight of the gases which surround the earth. It should be given in *pascals* (*newtons* per square metre) but is often stated in millimetres of mercury, a *unit* of pressure which is not part of the *SI* system. See also *atmosphere, barometer.*

atom The smallest part of an *element* which can exist and still keep all the *properties* of that element. It is the smallest unit of matter able to take part in a *chemical change.* Atoms have a diameter of about 10^{-10} m (one ten-thousand-millionth of a metre). They have a central *nucleus,* made of *protons* and, in every case except that of hydrogen, *neutrons.* The nucleus therefore has a positive *electric charge,* and is surrounded by negatively charged *electrons* in orbit such that the atom, as a whole, has no total electric charge. As electrons have little mass compared to a proton or a neutron, most of the mass of an atom is in the nucleus.Both the nucleus and the electrons are very small compared to the atom, most of the atom is empty space.

atomic Related to the *atom*, or to some particular *property* of it.

atomic bond An imaginary *valency bond* representing the attraction between atoms (*co-valent bond*) or ions (*ionic bond*) in a compound. The link is formed by the interaction of the valency electrons.

atomic energy The *energy* obtained when a heavy *atom* splits into two lighter ones of lower total mass. The *mass* lost is converted into energy according to the equation

Energy = Mass x (Velocity of light)^2

atomic mass unit A *unit* used for stating the mass of any particular type of atom. It is equal to one twelfth of the mass of an atom of carbon-12 and equal to 1.660×10^{-27} kg. It is not part of the *SI* system.

atomic number The number of *protons* in the *nucleus* of an *atom.* It is also equal to the number of electrons in an atom which has no total *electric charge.* In the *periodic table* elements are arranged in order of their atomic numbers. If it is necessary to show a *nuclide* with its atomic number, it is written at the bottom of the element symbol, e.g. $_1H$ or $_6C$ or $_8O$. See also *mass number.*

atomic pile The old (and still popular) term for a *nuclear* chain *reactor* in which *energy* is continually released in a controlled manner.

atomic weight See *relative atomic mass.*

A T P See *adenosine triphosphate.*

atrium Another name for *auricle.*

atto- A prefix which means one million million millionth or 10^{-18}. It can be joined in front of any *SI unit,* e.g. 1 attogram = 10^{-18} gram. The symbol is a.

audible limits The *frequency* of the lowest and highest notes which can be heard by the ear.

auditory Anything to do with the sense of hearing, e.g. the auditory *nerve.*

auricle A part of the *heart* (Fig. 91ii). It receives *blood* from the *veins.* The left auricle receives blood from the lungs and the right auricle receives blood from the rest of the body.

autonomic The part of the *nervous system* which is not under the control of the brain, e.g. heart beat control.

auxin A plant *hormone.* There are different kinds of auxins. They are all concerned with controlling the rate of growth of plant *cells.*

Avogadro's constant Also called Avogadro's number. The number of atoms, etc. in one *mole* of a substance, equal to 6.022×10^{23}. The symbol is L or N_A.

Avogadro's hypothesis The statement put forward by Avogadro that "Equal volumes of all gases under the same conditions of *temperature* and *pressure* contain the same number of *molecules.*" The *molar volume* of any gas, corrected to 0 °C and a pressure of 76 cm of mercury, is 22.4 litres and contains the *Avogadro constant* of molecules, 6×10^{23}.

axial skeleton The *skull* and *vertebral column.*

axil The angle between a leaf stalk and a stem. There is usually a bud (axillary) at this place.

axis 1. An imaginary line through an object about which it turns or can be imagined to turn. 2. The name given to the second neck *vertebra.* It fits into the *atlas* and allows for the turning of the head.

axon Part of a *neuron. Nerve impulses* pass through an axon towards or away from the cell body. See *synapse* (Fig. 155).

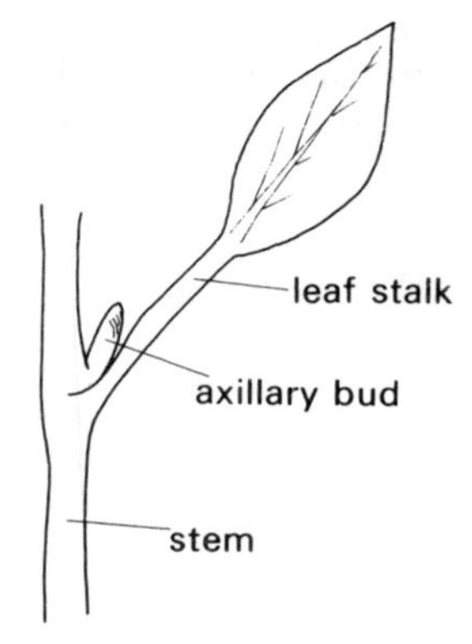

Fig. 12 *Axil*

back e.m.f. 1. An *electromotive force* set up in a polarized (*polarization*) *cell* which acts in the opposite direction to the original e.m.f.
2. When an *electric motor* is turning it acts like a *dynamo* as well as a motor and sets up a back e.m.f. opposing the flow of the electric current through the motor.

backbone The ordinary name for the *vertebral column.*

bacteria One-celled fungi (*fungus*). They are often the cause of disease, but some kinds are very useful in the soil (see *nitrogen bacteria*). They are all very small and occur in great numbers. The study of bacteria is bacteriology.

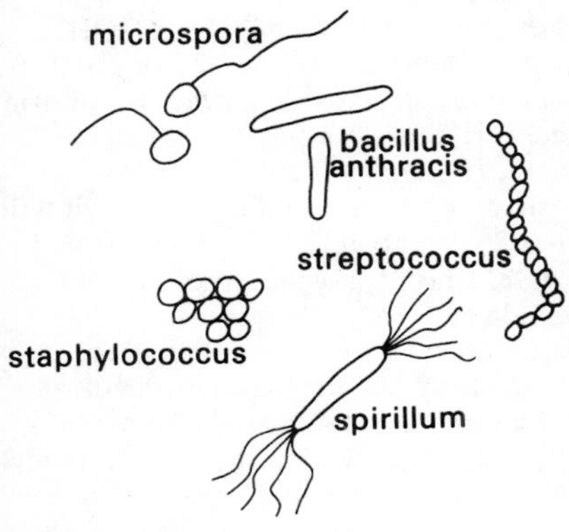

Fig. 13 Bacteria

bacteriophage A *virus*, even smaller than *bacteria,* which feed on and destroy bacteria.

bakelite The trade name of one of the early forms of *thermosetting plastics.* It is made from *phenol* C_6H_5OH and *formaldehyde,* HCHO.

baking-powder A substance used in making cakes. It contains *sodium hydrogencarbonate,* some sugar and a solid acid such as *tartaric acid* (or cream of tartar). When it is heated *carbon dioxide* and *steam* are given off and these cause the mixture to rise and become less dense (*density*).

baking soda See *sodium hydrogencarbonate.*

balance An instrument for finding the *weight* (and so the *mass*) of an object. Many types are in use, e.g. a *chemical balance* compares the weight of the object on one pan to that of masses placed on another pan. The most modern type is a direct reading balance where the object is placed on a pan and the weight is shown on a lighted scale, sometimes after certain knobs have been moved. See also *spring balance.*

balanced reaction A *reversible chemical reaction* at its point of *equilibrium,* e.g. in the reaction $A + B \rightleftharpoons C + D$ there will be a point reached at which A & B are reacting to form C & D at exactly the same rate as C & D are re-combining to form A & B: at this point the reaction is balanced.

balancers The very small second pair of wings of a fly. They help it to keep its balance when landing.

ball and socket A *joint* (Fig. 97ii) such as the shoulder or hip. It allows very loose movements in all directions.

bar A *unit* of *pressure* which is not part of the *SI* system. It is chiefly used in describing *atmospheric pressure* on weather maps. 1 bar = 10^5 *pascals* (*newtons* per square metre).

barb One of the fine hair-like parts of a *feather.*

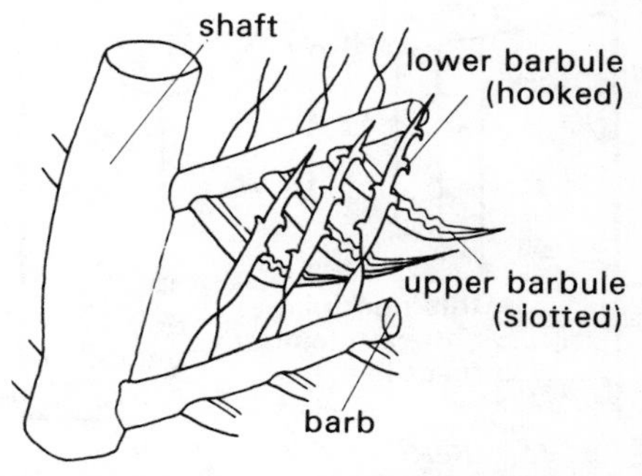

Fig. 14 Section of Feather (magnified)

barbule The hooks on a barb which act like a zip fastener to keep a *feather* firm.

barium Ba At. No. 56 R.A.M. 137.5. A silvery-white *metal* forming *compounds* similar to those of *calcium*. Barium nitrate is used in some fireworks to give a green colour to the flame.

bark The corky outside of a tree or bush.

barometer An instrument for measuring *atmospheric pressure*. The simplest form is made from a long glass tube by closing one end, filling it with mercury and then putting the open end in a container of mercury. The mercury falls to a level which depends on atmospheric pressure, making a *Torricellian vacuum* above it. Atmospheric pressure is measured by the vertical height of this level above that in the container. See also *fortin barometer, aneroid.*

Barton's pendula A set of pendula (*pendulum*) hanging from the same horizontal string. One has a heavy *bob*, the others have lighter bobs but are of different lengths. When the one with the heavy bob is made to swing, it makes the others swing with the same *frequency*. Also the pendulum with a light bob, which has the same length as the pendulum with the heavy bob, swings most, showing *resonance*.

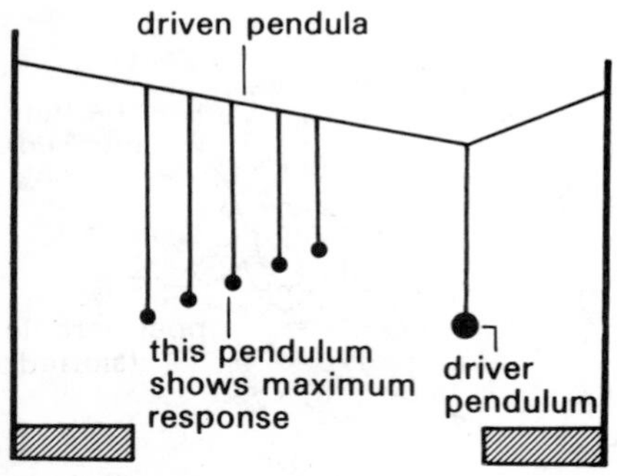

Fig. 15 Barton's Pendula

base A *compound* which reacts with *acids* to give a *salt* and water only. *Oxides* and *hydroxides* of *metals*, e.g. copper oxide, sodium hydroxide. A base which is soluble in water gives an *alkaline* solution containing *hydroxyl*, OH^-, *ions*. Sodium oxide does this and forms the *alkali*, sodium hydroxide. Ammonium hydroxide, NH_4OH, is also a base (and an alkali).

base-exchange See *ion-exchange*.

base metal A *metal* which *corrodes*, *tarnishes* or *oxidizes* when exposed to air, water or heat. A metal such as copper, zinc, tin or sodium.

basic Having the properties of a *base*, e.g. copper oxide is a *basic oxide* and sodium hydroxide is a basic hydroxide.

basicity A term used to denote the number of *hydrogen atoms* in the *formula* of an *acid* which can be replaced by *metal* atoms, e.g. hydrochloric acid, HCl, is monobasic as is ethanoic (acetic) acid, CH_3COOH, where only the last hydrogen atom can be replaced in this way. Sulphuric acid, H_2SO_4, is dibasic, phosphoric acid, H_3PO_4, tribasic.

basic oxide A *metal oxide* which will react with an *acid* to form a *salt* and water only, e.g. calcium oxide, zinc oxide.

basic salt A *salt* which has been formed by the partial *neutralization* of a *base*, i.e. it still contains some of the base from which it has been made. Malachite is basic copper carbonate, $CuCO_3.2Cu(OH)_2$ and basic lead carbonate is $Pb(OH)_2.2PbCO_3$.

bast The *phloem*.

Battersea crucible A large *crucible* made of rough unglazed *earthenware*.

battery A group of *cells* connected together. If the cells are connected one after the other (series) the total e.m.f. (*electromotive force*) is the sum of the separate e.m.f.s. If similar cells are connected side by side (parallel) the total e.m.f. is equal to the e.m.f. of any one but a larger current can be supplied. The words battery and cell are often confused in non-scientific use. See also *accumulator* (for car battery).

bauxite Hydrated aluminium oxide, $Al_2O_3.xH_2O$; the most important *ore* of *aluminium*.

beaker A glass or plastic cylindrical vessel used for holding liquids. It may be *calibrated,* but is not intended for great accuracy.

beam 1. The horizontal bar of a *chemical balance* which is supported centrally on a vertical shaft and from the ends of which the pans are suspended. 2. A wide band of *radiation,* e.g. light going in a certain direction.

beats When two similar *waves* of slightly different *frequencies* happen together, then they interfere (*interference*) and the total *amplitude* rises and falls with a frequency equal to the difference between the frequencies of the original two waves.

beehive shelf A device made of glass or earthenware and used for supporting gas jars when collecting gases over water.

beet sugar See *sucrose.*

Benedict's solution A *solution* containing copper sulphate and sodium citrate (two *salts*) and used in testing for reducing (*reduction*) *sugars;* these convert the copper sulphate into a reddish-yellow precipitate (*precipitation*) of copper(I) oxide, Cu_2O.

benzene C_6H_6 ⬡ or ⌬ An organic compound with a closed-chain ring structure. It is the first member of a *homologous series* of *hydrocarbons.* It forms *compounds* known as aromatics, because of certain properties. Benzene is a colourless, flammable, poisonous liquid with a characteristic smell.

beri beri A disease of the *nervous system* caused by not having enough *Vitamin B.*

berry A fleshy *fruit* with many seeds, e.g. tomato or orange.

Bernoulli's Law When a *fluid* is flowing smoothly an increase in speed means a decrease in *pressure.*

Bessemer process A method of converting *iron* to *steel* by pouring it, when molten, into a pear-shaped vessel (a Bessemer converter) through which hot air or oxygen is blown to *oxidize* most of the impurities which are present. See *L–D process* (Fig. 100).

beta decay When a *radioactive* substance gives out a *beta particle* there has been a change in the *nucleus.* A *neutron* has become a *proton* plus the *electron* (which is the beta particle). This means that the *atomic number* rises by one while the *mass number* does not change. E.g. ${}^{24}_{11}Na \rightarrow {}^{24}_{12}Mg + {}^{0}_{-1}e$ (beta particle). See also *decay.*

beta particle Certain *radioactive* materials, like Strontium-90, give out beta particles. These are very high speed *electrons* coming from the *nucleus* of the *atom* as a *neutron* changes to a *proton* plus the electron. Beta particles have a negative *electric charge,* change direction when passing through a magnetic field and are stopped by a few millimetres of aluminium.

bi- A prefix indicating the *acid salt* of a *dibasic acid,* e.g. sodium bisulphate, $NaHSO_4$; now replaced by the term '*hydrogen*', e.g. sodium hydrogensulphate.

bicarbonate ion See *hydrogencarbonate ion.*

biceps The *muscle* which is on the upper part of the arm between the shoulder and elbow. When it tightens it causes the arm to bend. See *joint* (Fig. 97i).

biconcave A lens in which both surfaces are curved inwards.

biconvex A lens in which both surfaces are curved outwards.

bicuspid A *valve* in the *heart.* (Fig. 91ii). It prevents blood from the *left ventricle* passing back into the *left auricle.*

biennial A plant which makes growth in the first year, flowers in the second year and then dies, e.g. cabbage.

bile The liquid made in the *liver* which helps to digest fat.

bile duct The tube through which *bile* passes from the *liver* to the *duodenum.* See *alimentary canal* (Fig. 5).

bile salts The chemicals in *bile* which split up the fatty foods.

billion In American use a thousand million, 10^9. In British use a million million, 10^{12}.

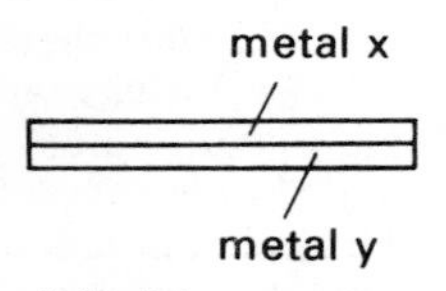

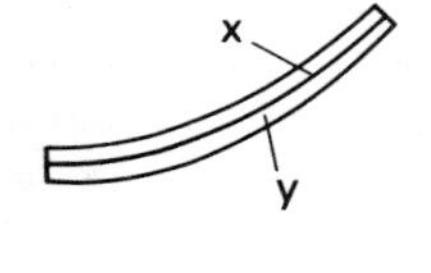

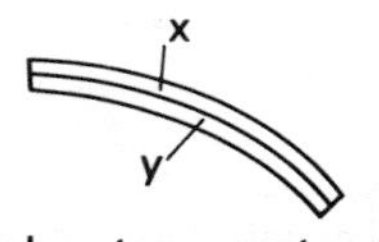

Fig. 16 Effects of Temperature on Bimetallic Strip

bimetallic strip Two thin strips of different metals joined together. When the temperature changes the strip becomes curved. If a bimetallic strip is made of two metals X and Y where metal X has the higher *expansivity,* then they behave as shown. Bimetallic strips are used in thermostats and fire alarms etc.

binary A *compound* containing only two *elements.*

binary fission Dividing into two. A form of *asexual reproduction,* e.g. bacteria, Amoeba.

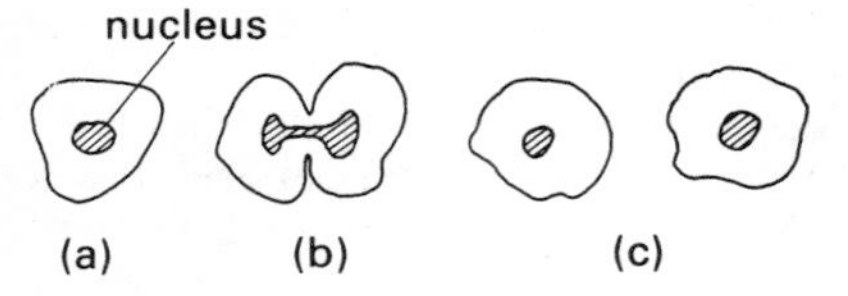

Fig. 17 Reproduction through Binary Fission

binocular vision Seeing with both eyes. It allows a person to judge shape and distance.

binoculars An instrument for looking at distant objects, which makes them appear closer. It is a pair of *astronomical telescopes* with two reflecting (*reflection*) *prisms* in each. These prisms make the last *image* the right way up and also reduce the length of the instrument.

biological control Using one living thing to control the numbers of other harmful plants or animals, e.g. moth caterpillars control the prickly pear in Australia; ladybird beetles control greenfly and scale insects.

biology The study of living things.

biomass The word used to measure the whole amount of life—plants and animals—in any one place. E.g. all the trees, plants, large and small animals, bacteria in a forest added together.

biotic factor The action of one kind of living thing on another, e.g. A tree affects other plants close by, a fly can carry disease to a human. The tree and the fly are biotic factors in the lives of the plants or the human.

bismuth Bi At. No. 83 R.A.M. 209 White *brittle metal;* poor *conductor* of electricity. Used in *alloys* where a low *melting point* is desirable.

bisulphate ion See *hydrogensulphate ion.*

Biuret test A means of showing the presence of a *protein* in a substance. An *aqueous* solution of the substance is mixed with an equal volume of *sodium hydroxide* solution and a few drops of *copper sulphate* solution are added. A violet colour forms if a protein is present.

bivalve A two shelled *mollusc,* e.g. oyster, mussel.

black iron oxide Fe_3O_4 *iron(II) iron(III) oxide*; used as a black *pigment.*

black lead Powdered *graphite;* it is used for making vessels such as *crucibles* which have to resist high *temperatures.* Mixed with *clay* in varying proportions it forms the 'lead' of 'lead' pencils and by itself it is a good *lubricant.*

bladder A container in the body, e.g. *gall bladder* which stores the *bile,* or the *urinary bladder* which stores the *urine.*

bladderworm The early stage of a tapeworm, found sometimes in the lean meat of a pig.

blast furnace A vertical *retort* or furnace used for extracting iron or some other metal from its *ore*. The iron ore, with *limestone* and *coke*, is put in at the top and air or oxygen—already heated to a high temperature—is blown in from the bottom. The iron and *slag* are tapped off separately.

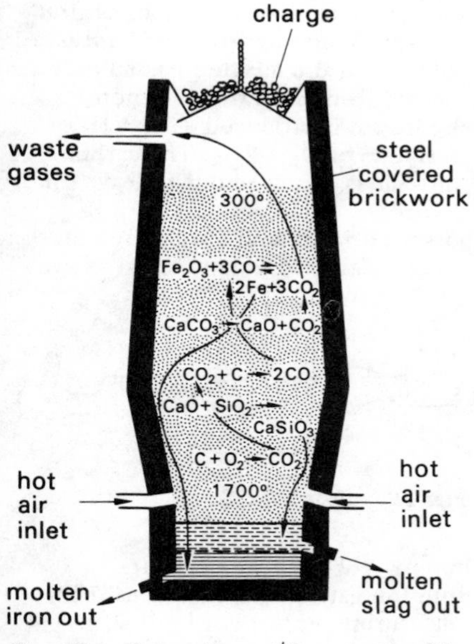

Fig. 18 Extraction of Iron in the Blast Furnace

bleaching Bleaching is a process whereby a coloured compound loses its colour. This occurs when a new, colourless compound is formed by a *chemical change* and involves either *oxidation* (as with chlorine) or *reduction* (as with sulphur dioxide).

bleaching powder When *calcium hydroxide* (slaked lime, $Ca(OH)_2$) reacts with *chlorine* a white powder is obtained from which the chlorine can be released when required by adding some dilute *acid:* this is bleaching powder. It is sometimes called chloride of lime and represented by the formula, $CaOCl_2$, but it is much more complicated than this suggests. It contains about 35–38% by mass of available chlorine (i.e. can be released by acid) when freshly made, but this amount falls on exposure to moist air.

blend A term applied to a mixture of *petrols* of different *octane numbers.*

blight A general name for plant disease caused by some kinds of *fungus.*

blind spot A position on the inside of the *eye* (Fig. 68) where the *optic nerve* passes out.

block and tackle A set of *pulleys* and ropes used for raising heavy loads.

blood A liquid *tissue* carrying *cells.* The liquid is *plasma.* The cells are (*a*) *erythrocytes,* the red cells containing *haemoglobin,* (*b*) *leucocytes,* the white cells and (*c*) the platelets, small parts of cells. Red cells have the *function* of carrying *oxygen,* white cells are mainly concerned with defending the body against *disease* and platelets play a part in healing wounds.

blood group The type of blood in the body. There are four main groups, A, B, AB, and O. The blood group is important in blood transfusion, which is when blood has to be given to a person in the case of too much bleeding. The wrong blood group can cause death.

blood sugar *Blood* always contains several kinds of *sugar,* the main one being *glucose.*

blow pipe 1. (Mouth) A narrow metal tube placed in the mouth so that a *flame* can be directed to where it is needed—usually for a reduction on charcoal experiment. 2. (Torch) An arrangement of tubes whereby gas and air or oxygen can be mixed and burned to produce a very high temperature.

blubber The thick oily fat of whales, found under the skin. It is used in making soap and margarine and for burning in lamps.

blue vitriol *Crystalline copper sulphate,* $CuSO_4.5H_2O$.

blue water gas A gaseous mixture obtained by passing *steam* over *coke* at a temperature of about 1000 °C. It consists mainly of *hydrogen* and *carbon monoxide* and has a *calorific value* of about 11–12 mega*joules* per cubic metre or 300–320 *British thermal units* per cubic foot.

Board of Trade Unit This is often called just a 'unit'. It is a measure of electrical energy equal to one *kilowatt-hour.* It is a name used in law and commerce rather than in science, and is not part of the *SI* system.

boat A small vessel made of *metal, fireclay* or *porcelain.* Substances which have to be heated in a stream of gas can be placed in it and the boat put in a tube (combustion tube) which is heated while the gas is passed through.

bob The heavy object which is tied to a piece of thread, etc., to make a simple *pendulum.*

body fluid The name given to the watery liquid which keeps the inside of the body moist. It contains many different chemicals and foods.

boiling When a liquid is heated its *temperature* rises until the bubbles of *vapour* formed inside the liquid break through the surface. The temperature now remains constant and the liquid is said to be boiling.

boiling point The temperature at which a liquid boils under a definite *pressure,* i.e. when its *saturation vapour pressure* equals the outside pressure. It is usually stated for a pressure of 760 millimetres of mercury, (one *atmosphere*).

bole The trunk of a tree.

bolus The soft rounded food as it is swallowed.

bond See *chemical bond.*

bone The hard material which makes the *skeleton* (Fig. 148) of most *vertebrates.* Bone is a living *tissue*. It consists of *cells* surrounded by a *matrix* containing *calcium carbonate* and *calcium phosphate.* Bones give support and protection and enable an animal to move with the help of muscles.

bone ash The ash left after bones have been heated strongly in air: it consists mainly of calcium phosphate.

bone black Fine black *pigment* obtained by heating bones strongly in the absence of air. It contains about 10% of *carbon* and is also used in *sugar* refining.

borax Sodium tetraborate, $Na_2B_2O_7.10H_2O$. White soluble *crystalline salt* occurring naturally as tincal. Used in *glass* making and as a weak *antiseptic.*

Bosch process An industrial process for making *hydrogen* from *blue water gas* which consists mainly of *carbon monoxide* and hydrogen. *Steam* is added and the mixture passed over a heated iron *catalyst* when more hydrogen is produced. $(CO + H_2) + H_2O \longrightarrow CO_2 + 2H_2$. The carbon dioxide is dissolved out to leave the hydrogen.

bosshead A metal device for fastening a clamp to a stand.

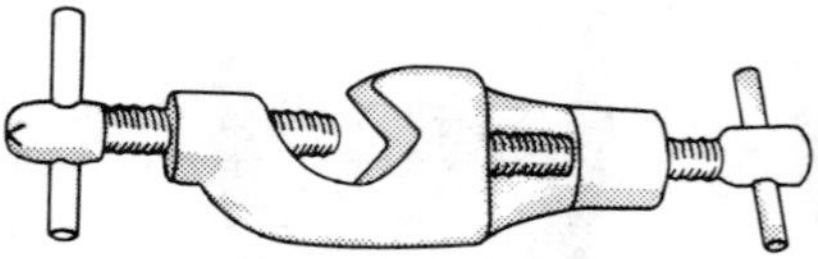

Fig. 19 Bosshead

botany The study of plants.

Bourdon gauge An instrument for measuring *pressure.* A hollow, coiled tube uncoils slightly when the gas pressure inside it increases. This small change is made greater by levers, etc., and moves a pointer over a scale.

Bowman's capsules The small cup-like structures inside a *kidney*. They *filter* out the *urine* substances from the blood.

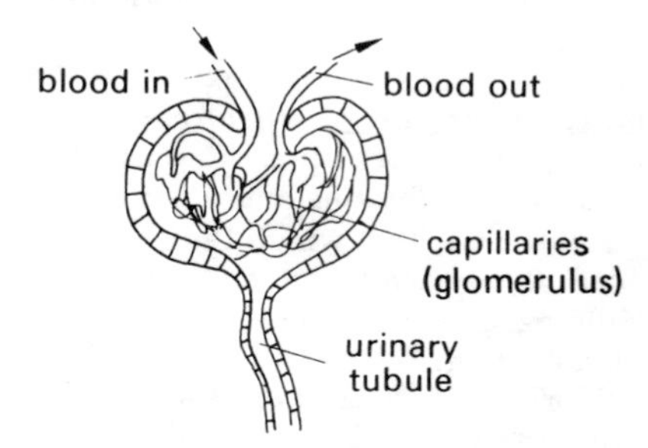

Fig. 20 Bowman's Capsule in cross-section

Boyle's Law The *volume* of a fixed *mass* of gas is *proportional* to one over the pressure if the *temperature* remains *constant.* In equation form this can be written: $\text{pressure}_1 \times \text{volume}_1 = \text{pressure}_2 \times \text{volume}_2$. See also *gas laws, ideal gas equation, perfect gas.*

brachialis muscle The muscle which helps in the arm movements. It works with the *biceps.*

bract A small leafy structure at the base of a flower stalk.

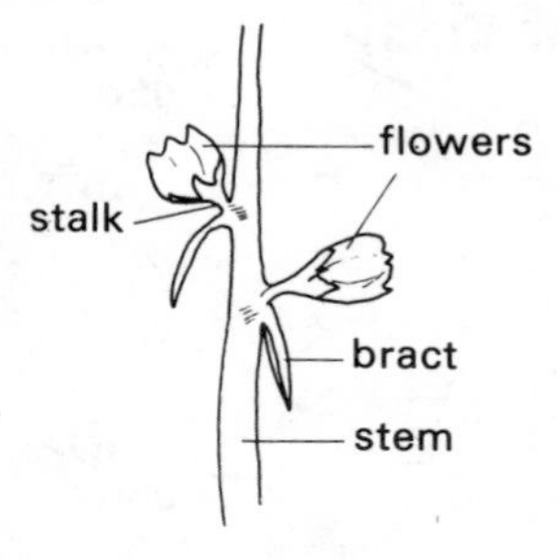

Fig. 21 Bract

brain The main part of the *central nervous system* (Fig. 34i) inside the *skull.* It receives messages from the sense organs and sends out orders to the muscles and *glands.* It contains many millions of *cells* which are joined together to form pathways. In man the brain is used for thinking and remembering. Some very small animals have no brain and some others, e.g. earthworm, have many small 'brains' instead of one main one.

branched chain *Aliphatic compounds* such as the *alkanes* may contain a number of $-CH_2-$ groups joining the first and last $-CH_3$ groups. These may be linked in a *straight chain* such as $CH_3-CH_2-CH_2-CH_3$ or in a branched chain such as $CH_3-CH_3-CH_2-CH_3$.

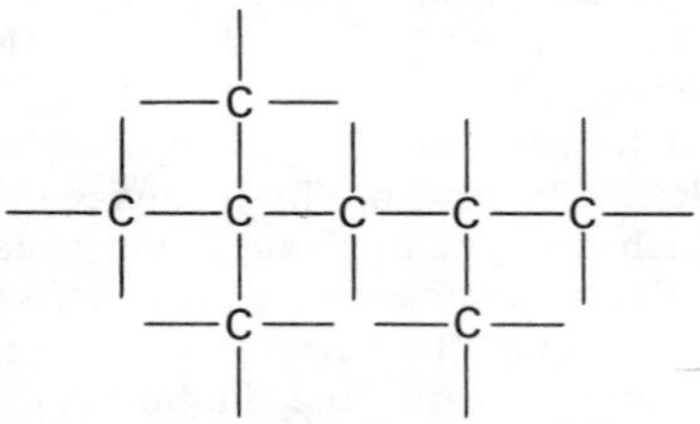

Fig. 22 Branched Carbon Chain

branched circuit At any point in an electric *circuit* where several wires join, the total electric current coming to the point equals the total leaving it.

brass A large class of *alloys* consisting principally of *copper* and *zinc*, the main one containing about 60% of copper and 40% of zinc.

breastbone The *sternum.* The flat bone in the front of the chest.

breathing The movements which cause air (or water) to be passed in and out of the body. *Oxygen* is absorbed and *carbon dioxide* is released.

brimstone *Sulphur fused* into blocks or rolls.

brine A *solution* consisting mainly of water and *sodium chloride.*

British thermal unit Btu The amount of *heat* needed to raise the *temperature* of one pound of water by 1 °F. The *calorific value* of a *fuel* is expressed (in British industry) in Btu/lb for solid or liquid fuels and in Btu/ft^3 for gaseous fuels. 1 Btu = 1055J and 100 000 Btu = 1 *therm.* Gradually this unit of heat is being replaced by the kilo*joule,* as it is not part of the *SI* system.

brittle Easily broken, as opposed to ductile.

bromine Br At. No. 35 R.A.M. 80 A non-metallic *element* of the *halogen* family in Group VII of the *periodic table.* Discovered in 1825 by Balard. A dark-red, poisonous and fuming (*fumes*) liquid with a strong *corrosive* action on the skin. Its *vapour* is irritating and choking.

bronchiole One of the small air tubes inside the *lung* (Fig. 109ii).

bronchus One of the two main air tubes in the *lung* (Fig. 109i).

bronze Correctly, an *alloy* made from *copper* and *tin,* but the term is also used for alloys not containing tin, e.g. aluminium-bronze (copper and aluminium) and phosphor-bronze.

brood chamber Where the queen bee lays her eggs and where the young *larvae* grow.

Brownian movement The continuous movement of very small particles in rapidly changing directions as they are hit by the *molecules* of the gas or liquid around them. For example the motion of smoke particles in air or particles of gamboge (a *pigment*) or of Indian ink on the surface of water. The motion is usually followed through a microscope.

brown ring test A means of identifying the presence of a *nitrate* (NO_3^-) ion (group) in a *compound*. The compound is dissolved in water and placed in a test tube along with some iron(II) sulphate solution and a little *dilute sulphuric acid*. Some *concentrated* sulphuric acid is run gently down the inside of the tube without shaking and a brown ring is formed where the layers meet if a nitrate is present. With a *nitrite* the brown colour appears when the dilute acid is added.

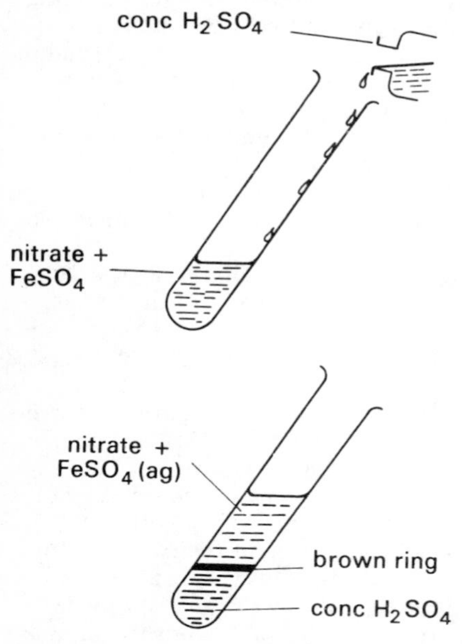

Fig. 23 Brown Ring Test

Bryophyta The group name given to *mosses* and *liverworts*.

buccal cavity The space inside the mouth.

Buchner funnel A *porcelain funnel* with a horizontal base perforated with small holes and used for *filtering* by suction. A *filter paper* is placed over the base ensuring a large area of filtration.

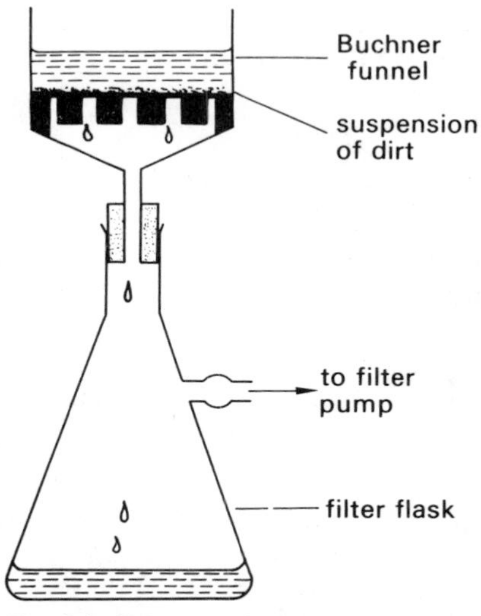

Fig. 24 Filtering through aBuchner Funnel

bud In plants a young undeveloped shoot or flower. In animals, e.g. coral, a young animal which grows on its mother and later breaks off.

bulb A large underground *bud* which contains much food and can remain in a resting condition until ready to grow, e.g. lily.

Bunsen burner A gas burner in which gas is led through a small jet at the bottom of the barrel and burned at the top. The gas is mixed with air, which comes through an adjustable inlet at the bottom, by means of which the type of *flame* is controlled. A 'flame-retention' ring may be fixed at the top of the barrel to prevent the flame 'lifting-off' and the burner going out.

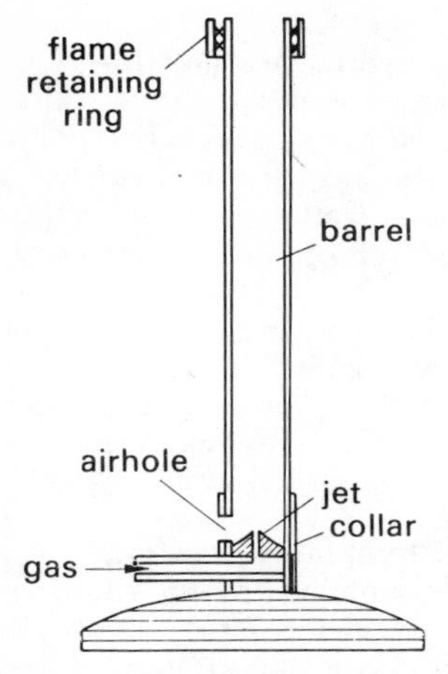

Fig. 25 *Bunsen Burner*

Bunsen valve A one-way *valve* which allows the passage of a gas under *pressure,* but prevents water or air being sucked back if the pressure falls. In its simplest form it consists of a small piece of rubber tubing, closed at one end, attached to the tube through which the gas is issuing. A slit about 6 mm long is made through the rubber. Examples of its use are (*a*) in the preparation of a solution of a very soluble gas; (*b*) in the preparation of a solution of a salt such as iron(II) sulphate from iron and dilute sulphuric acid in a flask. The hydrogen evolved can escape, but air is not drawn in on cooling and so *oxidation* of the salt is prevented.

buoyancy The upward push of a liquid or gas on an object that is placed in it. See also *Archimedes' principle.*

burette A long tube, fitted with a tap or *pinch-cock* at the bottom, used for delivering a definite volume of liquid or for measuring the volume of liquid run out in *volumetric analysis.* Burettes usually have a capacity up to 50 cm^3 and are accurately *graduated* to read to 0.1 cm^3.

burning The chemical *combination* of a substance with *oxygen* or some other gas such as *chlorine,* during which heat, light and *flame* are produced by the speed of the reaction and new products are formed.

butane C_4H_{10} A member of the *alkane* series of *hydrocarbons;* present in crude *petroleum.* It is sold in steel cylinders (under pressure) as calor gas for use in heating and lighting.

by-product A substance produced in a *chemical reaction* which is not the main product.

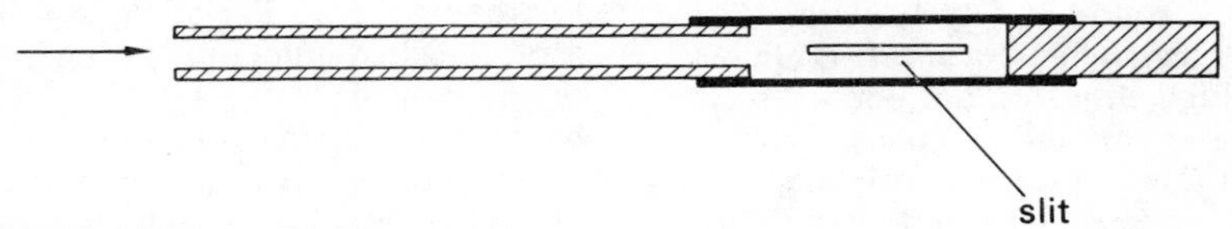

Fig. 26 *Bunsen Valve*

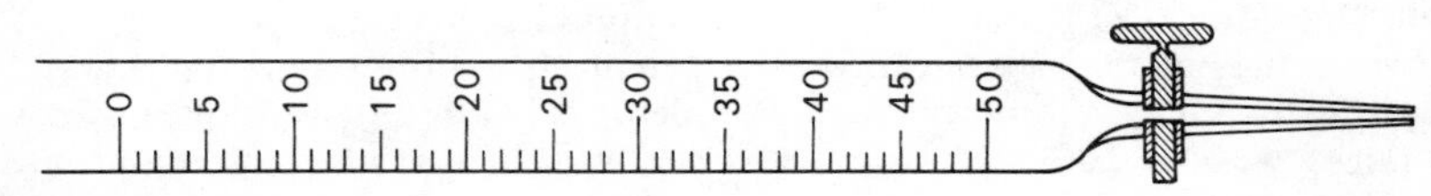

Fig. 27 *Burette*

caecum Part of the intestine which joins the *ileum* to the *colon.* In some animals, e.g. rabbit, it is large and is important in digestion. In others, it acts as a *valve* between the two parts of the intestine. The *appendix* is part of the caecum. See *alimentary canal* (Fig. 5).

calamine $ZnCO_3$ Zinc carbonate. A paste of calamine made with water is used for relieving the pain from sunburn.

calcite $CaCO_3$ A naturally-occurring *crystalline* form of *calcium carbonate.*

calcium Ca Atomic Number 20 Relative atomic mass 40 A dull, grey, brittle *metal* discovered by Davy in 1808. It belongs to the *alkaline earth* series of metals in Group II of the *periodic table.* It is very reactive with water and *tarnishes* rapidly in air. It is placed high in the *electrochemical series.*

calcium bicarbonate See *calcium hydrogencarbonate.*

calcium carbide CaC_2 A greyish solid (colourless when pure) made by heating *lime* with *coke* or hard coal in an electric *furnace* at 3000 °C. On addition of water, *acetylene* gas (*ethyne*), C_2H_2, is given off.

calcium carbonate $CaCO_3$ The most commonly-occurring *calcium compound,* found mainly as *marble, limestone, calcite* and *chalk.* It is almost insoluble in water, but does dissolve in water containing dissolved *carbon dioxide* when it forms *calcium hydrogencarbonate* and gives rise to *temporary hardness* in water. It has many uses including the production of *lime* and *cement.*

calcium chloride $CaCl_2.6H_2O$ A *by-product* in the *ammonia-soda* (Solvay) process for making *sodium carbonate.* When heated above 200 °C it becomes *anhydrous* or *fused* and as such is used as a *drying agent* for gases in laboratories and in industry. Spread on dirt road surfaces it absorbs (*absorption*) moisture from the atmosphere and prevents dust formation. It is also used to slow down the 'setting' of cement.

calcium hydrogencarbonate $Ca(HCO_3)_2$ Known only in *solution* in which it is prepared by passing *carbon dioxide* through a *suspension* of *calcium carbonate* in water: the latter gradually dissolves. Rain water contains carbon dioxide dissolved out from the air and takes in much more from the soil through which it flows. It is then capable of dissolving *chalk* and *limestone* in the rocks it meets and this gives rise to *temporary hardness* in the water.

calcium hydroxide $Ca(OH)_2$ A white powder obtained by the addition of water to *calcium oxide;* an *exothermic* reaction. Addition of more water gives a thick *suspension* known as milk of lime. A clear, filtered *solution* of the hydroxide is called limewater and this is used for identifying *carbon dioxide* which turns it milky, *calcium carbonate* being formed. Used by builders as one of the *components* of *mortar.* It is often called slaked lime.

calcium oxide CaO White solid produced by heating limestone (calcium carbonate) in a *limekiln* to above 825 °C. It reacts violently with water to produce *calcium hydroxide* (slaked lime) with much heat. It is used in making *mortar* and *cement* and for improving the quality of the soil. It is often called quicklime or lime.

calcium phosphate $Ca_3(PO_4)_2$ A naturally-occurring *mineral* from which *phosphorus* is obtained. It is present in bones, bone-ash and bone-meal. When treated with sulphuric or phosphoric acids it forms phosphatic *fertilizers.*

calcium salts One of the main chemicals required by living things. In plants calcium salts form part of the *cell wall.* In animals calcium salts are important in the blood and in the *skeleton.*

calcium stearate $Ca(C_{18}H_{35}O_2)_2$ or $CaSt_2$ When *soap* (sodium stearate) is placed in *hard water* (i.e. water containing dissolved *calcium salts*) a white *suspension* of calcium stearate particles is formed. When mixed with dirt it forms a '*scum*' on the water.

calcium sulphate $CaSO_4$ *A compound* of *calcium* occurring naturally as *anhydrite,* $CaSO_4$, and *gypsum,* $CaSO_4.2H_2O$. The former is used for making *sulphuric acid* and the latter is the basis of *plaster of Paris.* It is slightly soluble in water and causes *permanent hardness* in water.

calibrate 1. To mark (*graduate*) an instrument in divisions for accurate measurement, e.g. a thermometer or a galvanometer. 2. To determine the correct value of graduations already made on an instrument.

caliche Impure deposits of *sodium nitrate* (Chile saltpetre) found in the deserts of northern Chile. It was formerly used in immense quantities as a *fertilizer,* but the demand is less than it was because of the many artificial fertilizers now available.

callipers An instrument which helps one to measure the inside or outside diameter of an object like a tube.

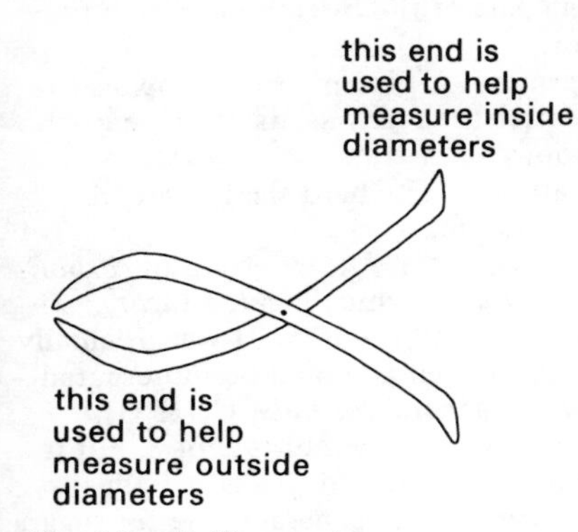

Fig. 28 Callipers

calor gas *Butane* stored in steel cylinders under pressure for use in heating, lighting and cooking.

calorie A *unit* of heat energy which is not part of the *SI* system. It is the amount of heat energy needed to raise the temperature of one gram of water through one *kelvin*. The *SI* unit of heat energy is the *joule:* 1 calorie = 4.18 joules. The word calorie (sometimes large calorie or kilocalorie), especially when used in connection with food, can also mean the kilogram calorie, equal to one thousand of the calories just defined.

calorific value The amount of heat energy set free when unit mass of a liquid or solid *fuel* or unit volume of a gaseous fuel is all burnt. In *SI* units it will be given in *joules* per *kilogram* or joules per cubic metre.

calorimeter A vessel in which experiments on heat energy are carried out. In its most simple form it is a metal can, usually made of copper or aluminium.

calyx The *sepals* of a flower. They protect the flower in bud. Sometimes they form part of the fruit.

cambium The ring of tissue in plants where growth takes place by means of the formation of new cells. Primary cambium in stems and roots forms *xylem* and *phloem.* Secondary cambium in woody stems forms *bark.*

camera An instrument which is used to record optical *images.* It is based on a box which keeps out light and has a lens at one end and a light sensitive film at the other. When the part, the shutter, covering the lens is removed (often for only a very short time) light from the object is focused (*focus*) on to the film. The film needs to receive a definite amount of light to be correctly exposed and this is controlled, in better cameras, by the time for which the shutter is open and by the size of the *aperture* or hole through which the light enters the camera. The lens can also be moved in and out so that light from objects at different distances can be focused on the film. Cameras and films are also made which are sensitive to *radiations* other than visible light.

camouflage Outside colouring or shape which matches the background, as in many forest and desert animals. Insects are often shaped to look like leaves or bark.

camphor $C_{10}H_{16}O$ A white, *crystal-*

line solid with a *characteristic* smell. Used in *solution* in *alcohol* as a *solvent* for some *plastics.*

canal A hollow tube, e.g. *auditory* canal—carrying sound waves into the ear; *conjugation* canal—carrying *gametes; eustachian* canal—a tube joining the cavity of the ear with the mouth; *haversian* canal—carrying blood in bones; *semicircular* canal—carrying liquid which has to do with balance, in the ear.

candella The *SI unit* of *luminous intensity.* It is a basic SI unit. The symbol is cd. If a completely black surface (i.e. one which absorbs (takes in) all radiation falling on to it) is heated to the temperature of freezing platinum under one *atmosphere* pressure, then the luminous intensity is one candella at right angles to each $\frac{1}{60}$ cm^2 of the surface.

candlepower The *luminous intensity* of a source of light, now given in *candella.* It may be measured in one direction or averaged over all directions.

cane sugar The sort of *sugar* which is stored in plants, e.g. sugar cane, sugar beet. It is obtained from these plants for man's use. See also *sucrose.*

canine Relating to the dog tribe. Canine teeth are the long pointed teeth next to the *incisors.* See *tooth* (Fig. 159ii).

capacitance If a *capacitor* had a capacitance of 1 unit (the *farad*) then the *potential difference* between its plates would rise by one *volt* if the capacitor gained an *electric charge* of one *coulomb.* The larger the area of the plates of the capacitor and the closer they are together, the greater will be the capacitance. The material between the plates is also very important. The symbol is *C.*

capacitor An electrical *component* containing, for example, two metal plates of large area which are close together but separated by an *insulator.* It stores *electric charge* when there is a *potential difference* between the plates. It used to be called an electrical *condenser.*

capillarity If a narrow tube is placed with one end in a liquid, then the liquid will rise (or sometimes fall) inside the tube. A similar effect occurs when one end of a piece of blotting paper is put into water or a wick into paraffin. The liquid rises in the narrow spaces between the fibres. Capillarity is important in young plants as a way of supplying water from the roots to the leaves. See also *surface tension.*

capillary A very narrow tube. Blood is sent to the cells of the body from the *arteries* through the capillaries and collected from them into the *veins.*

capillary tube A glass tube which has a very small inside diameter (a small bore). It can be bought ready made or a piece of ordinary glass tubing may be softened in a flame and then pulled out so that it becomes longer but thinner.

capitulum 1. The head of a daisy type flower made up of small *florets.* 2. The rounded head of a bone, e.g. the part of the thigh bone which fits into the hip joint.

capsule A type of fruit. It opens to scatter the many seeds inside, e.g. poppy.

carapace The hard shell of a crab-like animal.

carbohydrates 1. A group of *carbon compounds* which include *sugars, starches* and *cellulose.* It was originally thought that all could be represented by the general *formula,* $C_x(H_2O)_y$, where $x = 6$ or a multiple of 6, but it is now known that this is not always the case. 2. A general name for sugars of which there are different types such as (*a*) simple sugars like *glucose* and *fructose,* $C_6H_{12}O_6$ and (*b*) double sugars such as *sucrose* and lactose, $C_{12}H_{22}O_{11}$. They exist as *isomers.* Many carbohydrates are formed by plants as a result of *photosynthesis* and are used in making cell walls in young plants. Carbohydrates are also used by the body as a food for the supply of *energy.*

carbon C At. No. 6 R.A.M. 12 An

element, often called 'the backbone of life' as, along with hydrogen, it is present in all living matter. As an element it occurs naturally as *diamond* and *graphite* which *crystallize* in different forms. It is also obtainable in many other varieties such as *charcoal, lamp black, gas carbon, soot,* etc., some of which are *amorphous.* Its *valency* is always four and it melts at a temperature higher than 3500 °C. Combined with other elements, carbon is widely distributed in nature as a *constituent* of animal and vegetable matter. In the air it occurs as *carbon dioxide* and it is also found in rocks in combination as calcium carbonate, (*limestone, chalk* and dolomite), and in many water supplies as *calcium hydrogencarbonate*. People and animals obtain their *energy* by the *oxidation* of carbon compounds eaten as food. *Organic* chemistry is concerned with the study of the compounds of carbon other than its *oxides* and the metal *carbonates.*

carbon assimilation Another name for *photosynthesis.*

carbon cycle 1. A continuous process in which *carbon dioxide* from the air is taken in by plants to build up their *tissue* by *photosynthesis* and is later returned to the air as a *decomposition* product of the plants or of the animals which have eaten them.
2. A *nuclear reaction* which is thought to be the main source of *energy* in the larger stars.

carbon dioxide CO_2 The colourless, odourless gas formed when *carbon* or *carbon monoxide* is burnt completely in air or oxygen. It is also one of the products when any *fuel* (other than hydrogen) or *organic compound* is burnt. In the laboratory it is usually prepared by the action of a dilute *acid* on *calcium carbonate.* It is present in air to the extent of 0.03–0.04% and is used by green plants in building-up their structure (see *photosynthesis*). Air breathed out contains about 100 times as much of the gas as the air breathed in. The gas is fairly unreactive, only supporting the *burning* of such substances as magnesium which do so at a sufficiently high temperature to break down the gas into its *elements.* It is fairly *soluble* in water

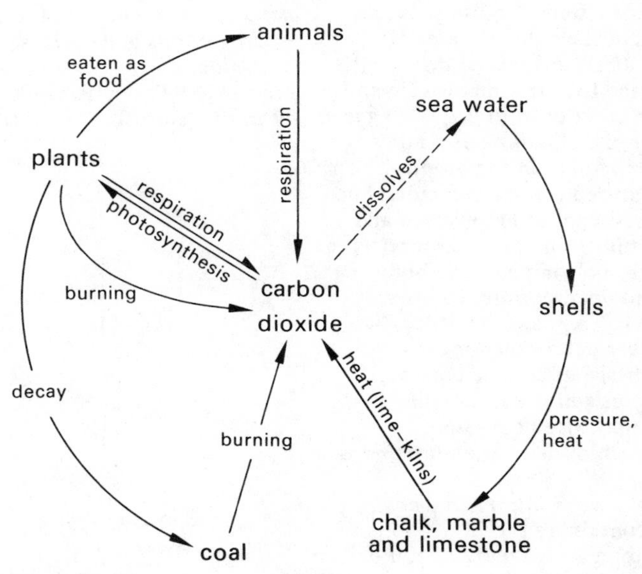

Fig. 29 *Carbon Cycle*

to form *carbonic acid,* but much more so in a solution of sodium or potassium hydroxide, a *caustic alkali,* with which it reacts to form *carbonates.* It is produced and used in *fire extinguishers.* Liquid carbon dioxide is contained under pressure in small steel bulbs ('sparklet' bulbs) and used for making *soda-water,* while solid carbon dioxide is used for preserving foods at low temperatures. It is the main source of the carbon essential to living things.

carbon disulphide CS_2 An evil-smelling, *flammable* liquid, colourless when pure, but usually yellow because of impurities. It is made by passing *sulphur vapour* through red-hot *coke.* It is used as a *solvent* for sulphur, phosphorus and rubber, in the extraction of perfumes, as an *insecticide* and in the manufacture of *rayon.*

carbon monoxide CO A colourless, odourless gas produced when *carbon* or carbon *compounds* are burnt in a limited amount of air. It is usually prepared in the laboratory by passing *carbon dioxide* through red-hot *charcoal* or by the action of concentrated sulphuric acid on *formic* or *oxalic acids.* It is extremely poisonous and is present in the exhaust gases of *petrol* or *diesel* engines. It burns with a blue flame to form the dioxide and is a *constituent* of both *producer gas* and *water gas.* It combines readily with *haemoglobin* in the *blood* to form bright-red *carboxyhaemoglobin.* This process cannot be reversed and as the haemoglobin can no longer take up *oxygen,* poisoning of the body occurs and death usually follows.

carbon steel A *steel,* the properties of which are determined mainly by the percentage of *carbon* present; small amounts of other metallic elements may also be present.

carbon tetrachloride See *tetrachloromethane.*

carbonates *Salts* of *carbonic acid,* H_2CO_3, containing the CO_3^{2-} ion (group) along with a *metallic ion* or *organic radical.* The only common ones soluble in water are those of sodium, potassium and ammonium; these can be prepared by passing *carbon dioxide* into a solution of the metal *hydroxide* or of *ammonia.* Other metal carbonates are usually prepared by *double decomposition.*

carbonic acid H_2CO_3 When *carbon dioxide* is passed into water some of it dissolves to form just a *solution* of the gas and some combines with the water to form a solution of carbonic acid. If the gas is forced in under *pressure* more dissolves to form *soda water.* Carbonic acid is a very *weak acid* and is only known in solution; the solution is *unstable* and easily breaks down on warming to produce carbon dioxide again.

carbonization The heating of *organic* substances in the absence of air to form *carbon*, e.g. wood on such heating forms *charcoal* and coal leaves *coke* and gives off *coal gas.* See also *destructive distillation.*

carboxyhaemoglobin See *carbon monoxide.*

carburetted water gas *Blue water gas* is passed into a heated chamber in which *gas oil* is sprayed. The oil *vaporizes* and enriches the gas with *hydrocarbons* and increases its *calorific value.*

carnassial The scissors-like side teeth of *canine* animals. Useful for cutting

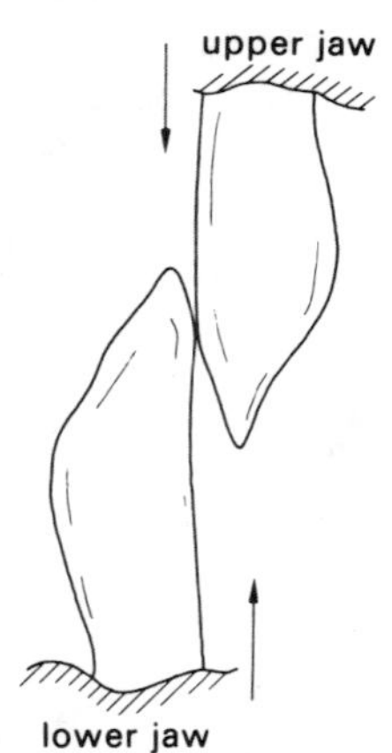

Fig. 30 Carnassial Teeth

meat foods into pieces small enough to be swallowed.

carnivore An animal which feeds on the flesh of other animals.

carotene The orange-coloured substance found in carrots and tomatoes. It is also found with *chlorophyll* in the green parts of plants. It helps to make *Vitamin A*.

carpel A part of the *flower* (Fig. 76ii). One or more carpels make the *ovary* in which are the *ovules*. Later the carpels form the fruit.

carpus The bones in the wrist. They make a free moving joint. See *skeleton* (Fig. 148).

cartesian diver This is a small, hollow, glass figure with a hole in it. It is made so that it just floats in water. If it is placed in water in a bottle with a well fitting cork and the cork is pressed in the *pressure* in the bottle rises. This forces water into the diver. Once the weight of the diver, the air and water it contains is greater than the weight of the water it pushes aside, it sinks. If the pressure from the cork is made less the diver rises again. See also *Archimedes' principle*.

cartilage The *gristle* found on the ends of bones. It acts as a cushion to soften movements. Cartilage forms the whole skeleton in fish like the shark and is found in other parts of the human body, e.g. the *pinnae* of the ears. It is strong but will bend. See *joint* (Fig. 97).

casein The *protein* of milk.

cast-iron An impure *brittle* form of *iron* which usually contains up to 4.5% of *carbon* and some manganese, phosphorus, sulphur and silicon. It is produced in the *blast furnace* where molten iron is run from the bottom into moulds called pigs and there cools down. It is usually converted into *steel* or *wrought iron* before use. It is also called pig-iron.

Castner-Kellner cell A process for the electrolytic (*electrolysis*) production of *sodium hydroxide*. Two *cells* are used, the first containing aqueous sodium chloride with a carbon *anode* and a flowing mercury *cathode*. *Sodium* is discharged to form an *amalgam* with the *mercury* and this is carried to the second cell where it reacts with water to form a solution of sodium hydroxide.

catabolism The breaking down of chemical substances in the body. *Energy* is always set free. *Respiration* is the main catabolic happening.

catalyst A catalyst or catalytic agent is a chemical which helps a *chemical reaction* to take place more easily, but is found chemically unchanged at the end; it may be an *element* or a *compound*. The process is called catalysis. The catalyst may make it possible for a process to be carried out at a lower temperature than would otherwise be the case or it may increase (or decrease, sometimes) the speed of the reaction. It cannot alter the position of *equilibrium* in a chemical reaction which is *reversible* nor the *mass* of the product finally obtained. Common catalysts are manganese(IV) oxide (dioxide) in the production of oxygen from potassium chlorate (the temperature needed is lowered) or from hydrogen peroxide (the rate is increased). Platinum or vanadium(V) oxide (pentoxide) helps in the manufacture of sulphuric acid by the *contact process*. *Enzymes* are biological catalysts.

catapult field When a wire carrying an electric current is placed in a uniform *magnetic field* the total magnetic field is stronger on one side of the wire than the other. If the *lines of force* (field lines) are imagined to be like elastic threads, then they look very much like the elastic in a catapult, trying to shoot the wire out of the field. (Fig. 31.)

caterpillar The larval stage of *butterflies* and *moths*.

cathode 1. The *negative* (–) pole or plate of a *battery*. 2. The electrode of a voltameter or thermionic valve, etc., connected to the above pole. 3. The electrode at which positively (+) charged ions are discharged during

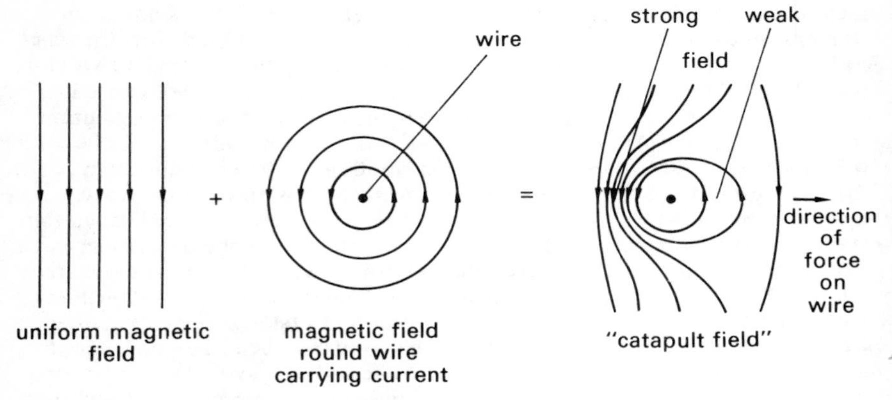

Fig. 31 Action of a Catapult Field

electrolysis.

cathode-ray oscilloscope (abbr. C.R.O.) An instrument built round a *cathode-ray tube* and used to show the pattern of rapidly changing electric *potential differences,* and therefore of anything else which is changing, if it can be made to give an electric potential difference. The tube, the power supplies and a variety of other electronic equipment are built into one box. The commonest electronic components are *amplifiers* (to increase small signals coming to the instrument) and a time base. The time base causes the spot to travel steadily across the screen, then very quickly back to the start and then repeat the pattern again and again. The speed at which the spot moves can usually be changed over a very wide range. Cathode-ray oscilloscopes are widely used in teaching, research, industry and medicine for many different purposes and are closely related to the television set.

cathode rays High speed *electrons* sent out by the *cathode* in a high *vacuum* tube when a large *potential difference* is set up between the cathode and the *anode.* The number of electrons is greatly increased if the cathode is heated. See *thermionic emission.*

cathode-ray tube A tube made to study or use the properties of *cathode rays.* In a high *vacuum* tube cathode rays are sent out from the heated *cathode.* They are made to move quickly by the accelerating *anodes* and travel down the tube, striking the fluorescent screen (one which gives out light when struck by cathode rays). The focusing *electrodes* cause the cathode rays to come together on the screen to give a small spot of light.

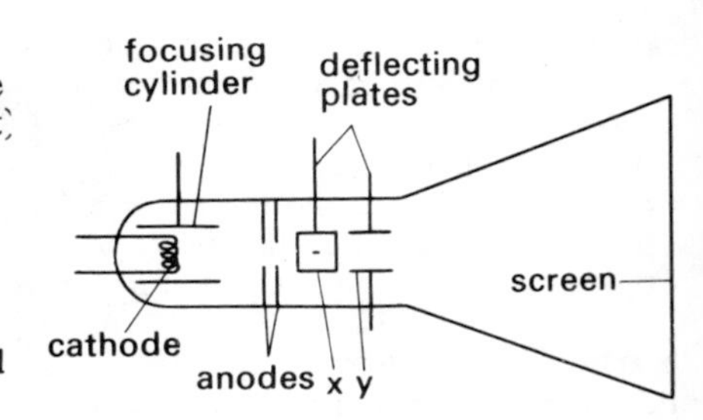

Fig. 32 A Cathode-ray Tube

If *potential differences* are set up between the X-plates, the (negative) cathode rays will move sideways towards the X-plate which is positive. In the same way the Y-plates can cause an up and down movement. In this way the spot can be moved to any point on the screen. The cathode ray tube is the basic part of the *cathode-ray oscilloscope* and the television set. Sometimes coils, outside the tube, are used instead of the plates to move the cathode-ray beam.

cathodic protection A way of protecting *iron* and *steel* structures again against *rusting.* Rusting occurs by loss of *electrons* (oxidation). The iron is made the *cathode* in an electrolytic (*electrolysis*) *cell* and a small electric current (d.c.) is passed between it and a suitable anode in the surrounding ground or water. It continually receives electrons and rusting is thus prevented.

cation A positively-charged *ion* (e.g. H^+ or Cu^{2+}) which moves to the *cathode* during the *electrolysis* of a *solution* or melt.

caudal Having to do with the tail, e.g. the caudal *fins* (Fig. 71) of fish.

caustic A term applied to substances other than acids which are *corrosive* towards *organic* matter, i.e. they cause blisters on the skin if in contact. Examples are sodium hydroxide (caustic soda). potassium hydroxide (caustic potash) and phenol.

caustic alkalis Solutions of sodium and potassium hydroxides which are *corrosive* and cause 'burning' of the skin. 880 ammonia is sometimes included in this class because it behaves in a similar manner.

caustic curve When a wide beam of light falls on to a *concave mirror* or a *converging lens* it is not brought to a true point *focus.* Instead a bright curve (a cusp) is formed with its centre at the true focus. This is called a caustic curve. Such a curve is often seen on the surface of a cup full of liquid.

caustic potash See *potassium hydroxide.*

caustic soda See *sodium hydroxide.*

cell 1. An electrical *component* which sets up a *potential difference* by chemical action. In a complete *circuit* it is a source of electric current. It often contains two different plates or rods dipping into a liquid. See also *accumulator, Daniell cell, dry cell, Leclanché cell, simple cell, standard cell, voltaic cell.* 2. The unit of living matter. Very varied in appearance according to its work. Some plants and animals are unicellular, that is, have only one cell, e.g. amoeba. Most plants and animals are multicellular– with many cells–e.g. human. Each single cell has a *nucleus* and *cytoplasm.* Always some food particles are present. Plant cells have a wall of *cellulose* and liquid filled spaces inside called *vacuoles.* Animal cells have no walls and are softer. Some cells can move around, e.g. *white blood cells* and *sperms.* (Fig. 33.)

cell body The part of a *neuron* which contains the *nucleus.*

cell division A series of happenings, starting in the *nucleus,* which results in one cell dividing into two. Essential for growth. See *mitosis.*

cell sap The liquid, mostly water, found in the *vacuoles* of a cell.

celluloid A *thermoplastic* material made from *cellulose* nitrate and camphor. It can be produced in very thin sheets which are quite clear and is used for one type of photographic film. One trade name is Xylonite.

cellulose $(C_6H_{10}O_5)_n$ A naturally-occurring *polymer* of high *molecular mass* which consists of a long *chain* of *glucose molecules* joined together. It forms the *cell* walls of many plants and is obtained chiefly from cotton fibre or wood pulp and is used in the manufacture of paper, rayon, explosives and plastics. It may change to wood in older cells.

Celsius temperature A scale of *temperature* originally based on the *melting point* of pure ice (0 °C) and the *boiling point* of pure water (100 °C) both under one *atmosphere pressure.* It is now defined from the *absolute temperature:* Celsius temperature = absolute temperature – 273(.15). See also *centigrade temperature.*

cement 1. A grey powder made by heating *limestone* or *chalk* with *clay* which, after mixing with water, sets to a hard mass and is used in building construction work. Cement powder, coarse sand, washed small stones and water are used as concrete, a building material of considerable strength.

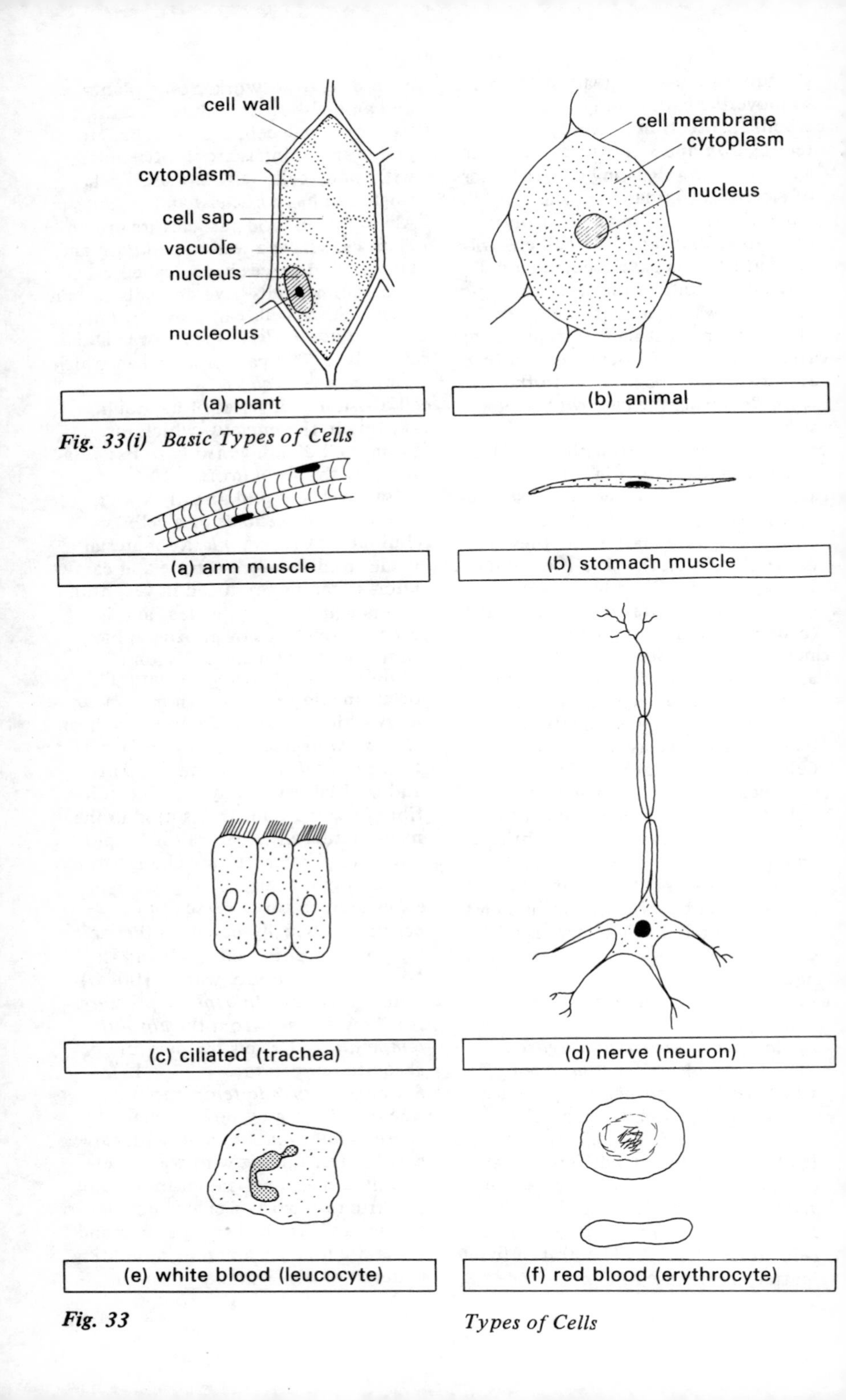

Fig. 33(i) Basic Types of Cells

Fig. 33 *Types of Cells*

2. The hard substance which is on the outside of the root of a *tooth* (Fig. 159i). It fixes the tooth firmly into its socket.

centi- A prefix which means one hundreth or 10^{-2}. It can be joined in front of any *SI unit,* e.g. 1 centimetre = 10^{-2} metre. The symbol is c.

centigrade temperature The old name for *Celsius temperature.* Although the change of name was officially made in 1948 the older name is still in very common use.

central nervous system (Abbr. CNS) The main part of the nervous system. It is made up of the *brain* and the *spinal cord. Nerves* from other parts of the body come to and go from the CNS, which acts as a centre to receive, understand and reply to messages coming from the *sense organs.*

centre of curvature The centre of the sphere of which a curved surface is part. The term is very often used in connection with *lens* and *spherical mirror* surfaces. See also *principal axis, radius of curvature.*

centre of gravity The point in, or near, a body through which its total *weight* always appears to act whatever position the body is in.

centrifugal force See *centripetal force.*

centrifuge A machine in which balanced tubes containing a *suspension* can be rotated at very high

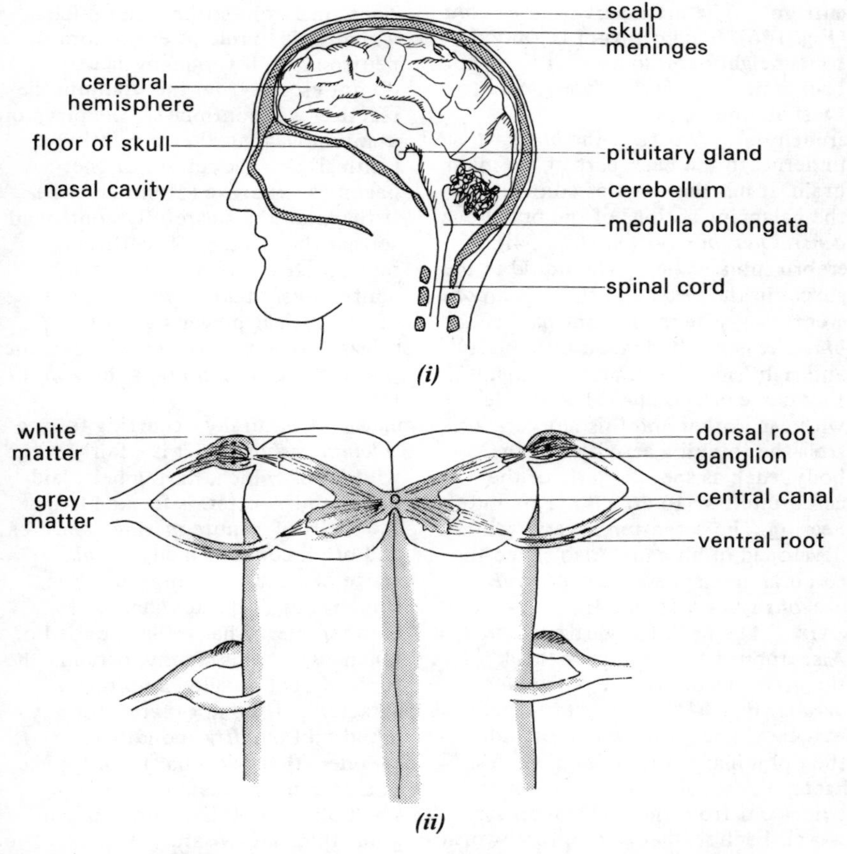

Fig. 34 *(i) Section of Human Head* *(ii) Spinal Cord and Nerves*

speeds, so causing the solid to be packed at the bottom of the tube. The liquid (centrifugate) can then be removed by a *teat pipette.* In *analysis* centrifuging often replaces filtration.

centripetal force The *force* which has to act on an object to keep it moving in a circle. For example, if a stone is swung round on a piece of string, the string pulls the stone into a circular path. This pull is the centripetal force. If the string breaks the stone continues in a straight line because the inward force has gone. The pull on the persons' hand, equal and opposite (see *Newton's* third *Law of Motion*) to the pull on the stone, is an example of *centrifugal force* (this last phrase is often wrongly used).

centrum The main part of a *vertebra* (Fig. 166). Each centrum is jointed to its neighbours, so making the whole backbone (*vertebral column*) able to twist and bend.

cerebellum A part of the *brain.* It is underneath the back part of the main brain. It has the work of controlling the balancing actions of the body. See *central nervous system* (Fig. 34i).

cerebro-spinal fluid The liquid in the spaces in the *brain* and the *spinal cord.*

cerebrum The main front part of the *brain.* It is partly divided into a left and right side. There are very many tiny nerve cells (*neuron*) connected with each other and this network controls the thinking activities of the body, such as speech. Parts of the cerebrum have to do with sight and hearing. The cerebrum is very well developed in humans. Also called the cerebral hemispheres. See *central nervous system* (Fig. 34i).

cervix The neck (cervical *vertebrae*). Also applied to the narrow 'neck' of the *uterus* (womb). See *reproductive organs* (Fig. 141i).

Cetacea The group which includes the dolphins, porpoises and whales.

chaeta A small pointed rod which comes out from the body of an earthworm. Each *segment* of an earthworm has 8 chaetae. They grip the earth and allow the body muscles to pull the earthworm along. Specially large chaetae are used when two earthworms *mate.* Also known as seta.

chain A series of carbon-hydrogen groups joined together in an *organic compound,* e.g. $CH_3-CH_2-CH_2-CH_2-CH_2-CHO$. Chains may be long or short, *straight* (as above), *branched* or closed as in the *benzene* ring.

chain-dial balance A *chemical balance* in which *masses* of less than one gram are added by means of a hanging chain controlled by a graduated dial.

chain reaction When the *nucleus* of a uranium-235 atom is struck by a *neutron* the nucleus splits into two main parts but also sets free two or three more neutrons and energy. These neutrons strike other nuclei which split to release even more neutrons. So if too many neutrons do not escape from the uranium the reaction will continue. If the piece of uranium is larger than a certain (critical) size the release of energy becomes explosive (atomic bomb). If the reaction is carefully controlled so that the number of neutrons remains steady then energy can be continuously taken away (the basis of the nuclear power station).

chalaza The part of the white of the *egg* (Fig. 61) which holds the *yolk* in the middle.

chalk A naturally-occurring form of *calcium carbonate.* It is a fairly soft, white rock which, having been laid down under water, is formed from the shells of minute marine creatures and often contains many *fossils.*

change of state When one of the three *states* of matter changes to another energy has to be supplied or taken away. This energy, often in the form of heat is called *latent heat.*

character Used in *genetics* for any inherited (*heredity*) condition.

characteristic Peculiar to or typical of a particular substance, e.g. the smell of ethanol, the colour of sulphur, the taste of salt.

charcoal A *microcrystalline* impure

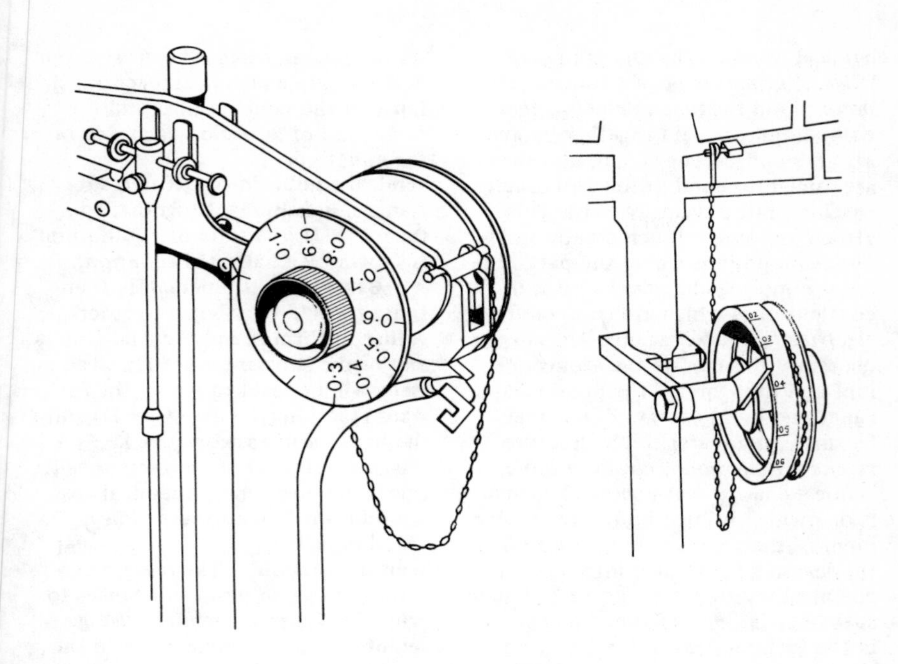

Fig. 35 *The Dial for Fractional Weights on a Chain-dial Balance*

form of *graphite,* often referred to as '*amorphous*' *carbon.* It can be made by heating many animal or vegetable substances (e.g. wood, sugar) strongly in the absence of air, or by adding concentrated *sulphuric acid* to them. Charcoal has a very large surface area compared to its mass and because of this and its porous nature it is much used for adsorbing (*adsorption*) gases or *decolorizing* impure liquids.

charging by induction A way of giving an *electric charge* to an insulated (*insulator*) *conductor*. Another charged object is brought near to the conductor and the conductor is connected to the earth, usually by touching it, for a moment. Then the charged object is taken away leaving the conductor with a charge opposite in sign to that of the first charge.

Charles' Law The *volume* of a fixed mass of gas at *constant pressure* is directly *proportional* to the *absolute temperature.* The volume therefore increases by $\frac{1}{273}$ of the volume at 273 K (O °C) for every *kelvin* rise in temperature. In equation form:

$$\frac{\text{volume}_1}{\text{absolute temperature}_1} = \frac{\text{volume}_2}{\text{absolute temperature}_2}$$

chemical balance An instrument designed for finding the *mass* of an object. In a two-pan balance the object is placed on the left-hand pan and accurate masses (weights) are added to the right-hand pan until the beam of the balance is horizontal. Many balances have different methods of adding these masses (*chain-dial balance, triple-beam balance*) and some only have one pan, the counterweights being inside the balance case. A type which gives a direct reading of the mass of the object is the *top pan balance.*

chemical bond The *atoms* in *molecules* of *elements* or of *covalent compounds* and the *ions* in *electrovalent* compounds are held together by an attractive force, suggesting that they are joined by bond or link, although this does not physically exist. This attractive force is much greater in *ionic* compounds where the particles carry opposing charges than it is in covalent ones which are composed of *electrically neutral* atoms. The *energy* necessary to separate the atoms or ions and thus 'break' the bond is called the bond-energy. Bonds may be single, double or treble. See also *valency, ionic bond, covalent bond.*

chemical change A chemical change is one which brings about a re-arrangement of the *atoms* or *molecules* of the reacting substances into new and chemically different structures. A new substance is always formed and *energy* in the form of *heat* or *light* may be given out; in some cases heat is taken in, but quite often no heat change is noticed.

chemical compound A substance composed of two or more *elements* chemically united in fixed proportions by *mass.* A compound has a fixed composition and cannot be broken down into its elements by simple *physical changes.* Simple compounds are water, sulphuric acid, sodium chloride (salt). Each can be represented by a definite *formula,* e.g. H_2O, H_2SO_4, NaCl.

chemical equation A method of indicating by means of *formulae* the *chemical changes* which occur during a chemical reaction, e.g. $Zn + 2HCl \rightarrow ZnCl_2 + H_2$ indicates that 1 *mole* of zinc atoms reacts with 2 moles of hydrochloric acid to form 1 mole of zinc chloride and 1 mole of hydrogen molecules. If it is written $Zn(s) + 2HCl(aq) \rightarrow ZnCl_2(aq) + H_2(g)$ it also shows that the zinc is a solid, the hydrogen is *gaseous* and the hydrochloric acid and zinc chloride are in water *solution.* Other details may also be included by means of suitable signs. As matter can neither be created nor destroyed in a chemical reaction, there must be the same number and the same kind of *atoms* on each side of the equation.

chemical equilibrium If two substances, A & B, react together to form C & D, the state of equilibrium is shown by a balanced *equation:* $A + B \rightleftharpoons C + D$. Immediately C and D are formed they begin to react to gether to form A and B again. Unless any of the substances are removed a stage will be reached where the forward reaction ($A + B \rightarrow C + D$) and the backward reaction ($C + D \rightarrow A + B$) are going on at the same speed; this is the position of chemical equilibrium. See also *reversible reaction.*

chemical reaction The *reacting* together of two or more substances to bring about some *chemical change.*

chemotaxis The name given to the movement of animals in *response* to a chemical *stimulus,* e.g. a shark swims towards blood.

chemotropism The growth of part of a plant towards or away from a chemical substance, e.g. a *pollen tube* grows away from *oxygen* and towards *sugar.*

chinagraph pencil A special type of pencil which will write on *glass* or *porcelain* (or other) surfaces. Often used for marking *crucibles* when several of them are being put into a *furnace* as the mark is still visible afterwards.

chitin The tough waterproof substance which makes the skins (*exoskeletons*) of insects.

chloride A salt of *hydrochloric acid,* HCl. Chlorides may be formed by (*a*) direct *combination* of an *element* with *chlorine;* (*b*) reacting a *metal,* a *basic oxide* or *hydroxide* or a metal *carbonate* with hydrochloric acid or (*c*) *double decomposition* if the chloride is insoluble in water. The chloride *ion,* Cl^-, present in metal chlorides, carries a single *negative* charge.

chloride of lime See *bleaching powder.*

chlorine Cl At. No. 17 R.A.M. 35.5 A greenish-yellow poisonous *gas* with a choking, irritating, smell; obtained industrially by the *electrolysis* of *brine* or in the laboratory by the *oxidation* of *hydrochloric* acid. It occurs combined in many *compounds*, principally *sodium chloride,* common salt. As a member of the *halogen* family it occurs in Group VII of the *periodic table.* There are two *isotopes* of *relative atomic masses* 35 and 37 respectively. It is used in the manufacture of *bleaching powder, PVC, disinfectants, insecticides* and hydrochloric acid. It is often recognized by its bleaching action on moist *litmus* paper.

chloroform See *trichloromethane.*

chlorophyll The green substance found in most plants and a few animals. It is able to absorb *light energy.* In a living cell this energy is used to make *glucose.* Chlorophyll contains *magnesium.* It is similar to *haemoglobin* of blood which instead contains iron. Ordinary chlorophyll is made up of green, yellow and red *pigments.* It can be dissolved out of leaves with warm *methylated spirits.*

chloroplast One of the small structures in a cell which contains the *chlorophyll.*

choke An electrical *component* which is a coil of wire, often wound round an iron centre part (core). Such a component opposes the flow of a rapidly changing electric current as an e.m.f. is set up within it by *electromagnetic induction* and acts against the current. However it does not oppose the flow of a steady (direct) current and will therefore help to separate *alternating current* from *direct current.*

chordae tendinae The strong *tendons* which are attached to *heart* (Fig. 91ii) *valves* to prevent the valves turning inside out.

choroid 1. Part of the *eye* (Fig. 68). The middle of the three layers in the eyeball skin. It contains the blood supply to the other parts. 2. Choroid plexus. The skin, containing blood, of parts of the brain.

chromatogram A column of *clay* or aluminium oxide or a strip of *filter paper* on which coloured *dyes* present in a liquid have been adsorbed (*adsorption*) in different places.

chromatography A means of identifying different coloured substances present in a solution. A drop of the liquid is placed near the end of a strip of *filter paper* which is hung vertically with the end just touching a suitable *solvent.* As the solvent rising up the paper passes the spot it carries the different colours with it. Since these travel at different speeds they become separated and may be identified in different places on the paper. Alternatively, the liquid mixture may trickle down a column of adsorbing (*adsorption*) material such as *chalk* or *clay* and the coloured *components* of the liquid appear in different positions in the column.

chromium Cr At. No. 24 R.A.M. 52 A blue-white *metallic element* which forms brightly-coloured *compounds.* It is used in stainless *steel alloys* and for chromium *plating* of other metals.

chromosome One of the rod-like structures which show up when the *nucleus* of a cell is going to divide. The chromosomes carry the *genes* which pass on the characters of the parent cell to the young ones. See *mitosis* and *meiosis.*

chrysalis The *pupa* of a butterfly or moth.

chyle A milk-like liquid made of *lymph* and oil drops. It is taken in through the wall of the *intestine* and later into the blood.

chyme The partly *digested* food passed from the *stomach* to the *intestine.*

cilia Very fine hair-like outgrowths from the surface of some living cells. Cilia can move and set up a current in water. On the surface of very small animals (*protozoans*) they cause swimming movements. In the *trachea*

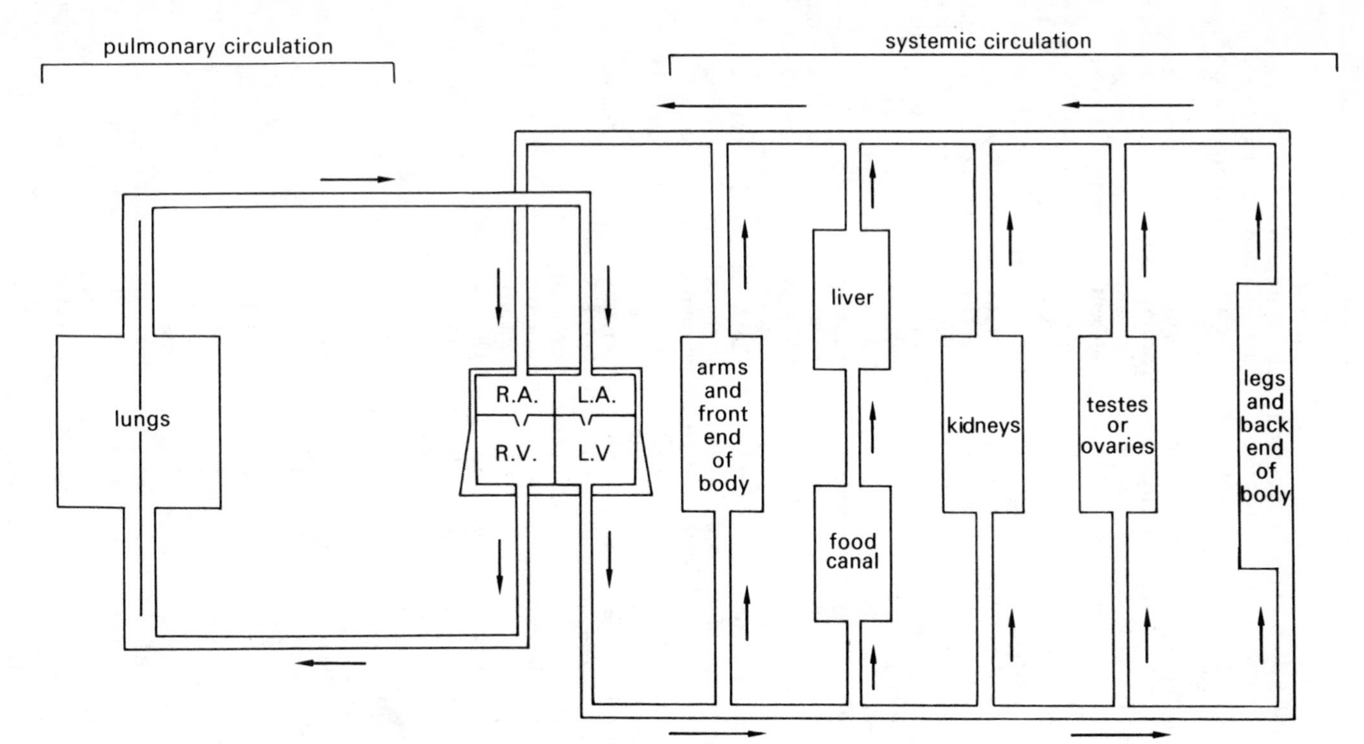

Fig. 36 *Circulation of Blood* A = *auricle* V = *ventricle*

ANIMALS

Invertebrates	**Vertebrates**
Protozoa eg. amoeba	**Chordata**
Coelenterata eg. jellyfish	Craniata
Platyhelminthes eg. liverfluke	(i) Pisces eg. shark
Nematoda eg. hookworm	(ii) Amphibia eg. frog
Annelida eg. earthworm	(iii) Reptilia eg. snake
Echinodermata eg. starfish	(iv) Aves eg. vulture
Arthropoda	(v) Mammalia eg. dog
(i) Crustacea eg. crab	Homo sapiens –man
(ii) Insecta eg. bee	
(iii) Arachnida eg. spider	
(iv) Myriapoda eg. centipede	
Mollusca eg. snail	

Viruses

PLANTS

Thallophyta

- (i) Algae eg. seaweed
- (ii) Fungi eg. mould
- (iii) Bacteria
- (iv) Lichens

Bryophyta

- (i) Liverworts
- (ii) Mosses

Pteridophyta

- (i) Ferns
- (ii) Horsetails

Spermatophyta

- (i) Conifers eg. monkey puzzle
- (ii) Flowering plants
  - (a) Monocotyledons eg. orchid
  - (b) Dicotyledons eg. bean

Fig. 37 Classification of Main Groups

the cilia keep the *mucus* moving.

ciliary body Part of the *eye* (Fig. 68). It is a circular *muscle* which controls the shape of the *lens*. It is joined to the lens by the *suspensory ligament*.

circuit A number of electrical *components* joined together to make a complete path round which an electric current can flow.

circular motion If an object is to move round a circular path then it must be continually pulled in by a *force* acting to the centre of the circle. See also *centripetal force*.

circulation of blood The way by which the blood is kept flowing around the body. The *heart* is the pump. The blood flows from the heart through the *arteries*, passes around the *tissues* in the *capillaries* and returns to the heart through the *veins*. Where there are *gills* as in a fish, the blood passes through the heart once every circulation (single circulation), but where there are lungs, the blood passes through the heart twice (double circulation). (Fig. 36.)

citric acid $C_6H_8O_7$ A white *crystalline organic acid* used in the preparation of *effervescent* drinks. It causes the sour taste in many fruits in which it occurs, e.g. in lemons. It is also present in strawberries, raspberries, tomatoes and oranges along with *sugar*.

claret colour The colour of claret wine, a red with a tinge of purple. *Carbon dioxide* turns *neutral litmus* solution claret colour.

Clark process A means of softening (*soft water*) *temporary hardness* in water by the addition of *calcium hydroxide*. *Calcium carbonate* is precipitated (*precipitation*). $Ca(OH)_2 + Ca(HCO_3)_2 \longrightarrow 2CaCO_3 + 2H_2O$.

class One of the main divisions of the animal kingdom. E.g. the frog belongs to the class *Amphibia*.

classification The way by which plants and animals are put into groups to show likenesses and relationships. The largest group is a *phylum* and the smallest is a *species*. In between are *classes*, *orders*, families and genera (*genus*). (Fig. 37.)

clavicle Part of the *skeleton* (Fig. 148). The collar bone, joining the shoulder to the top of the ribs.

clay 1. Part of the *soil*. It helps to hold water in the soil because the particles are very small. When wet, clay soil is sticky and when dry, very hard. 2. A fine-grained 'earth' consisting mainly of aluminium silicates. It is used in the manufacture of pottery and of bricks.

cleavage The splitting of a *crystal* along certain definite *planes* when it is subject to stress or strain.

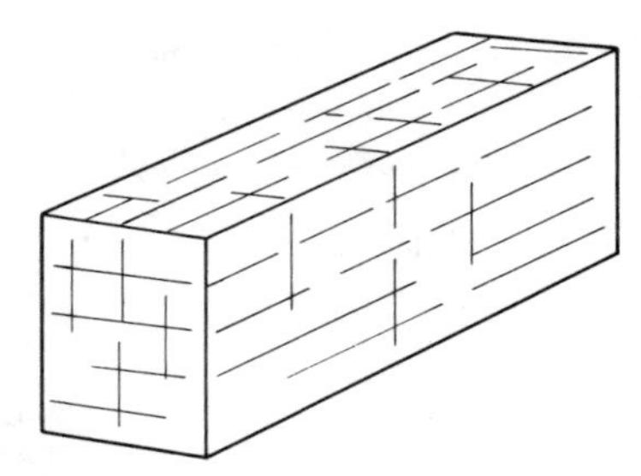

Fig. 38 Crystal of Iceland Spar showing Planes of Cleavage

clinical thermometer A special *thermometer* used for measuring the temperature of the human body. It has a very narrow hole (bore) through the glass, so that the scale is well spread out but limited in range; and an even narrower place (constriction) near the bulb so that the mercury cannot flow back when the thermometer is taken from the body. The thermometer is prepared for use again by shaking the mercury back past this constriction.

clinker When *coke* is burnt in a grate or furnace, the impurities present usually *fuse* together to form a hard mass of unburnable material known as clinker. It is used in road-making.

clinostat An instrument used to show that the direction of growth of shoots and roots is under the control of the force of *gravity*. (Fig. 39.)

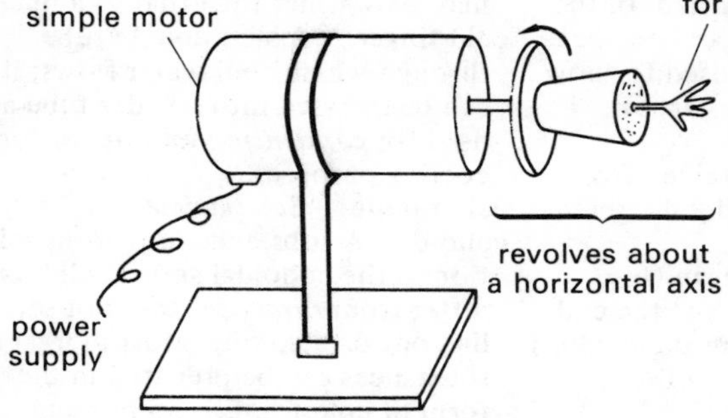

Fig. 39 *Clinostat*

clitellum The swelling on an earthworm when it is ready to mate. It later becomes a container for eggs.

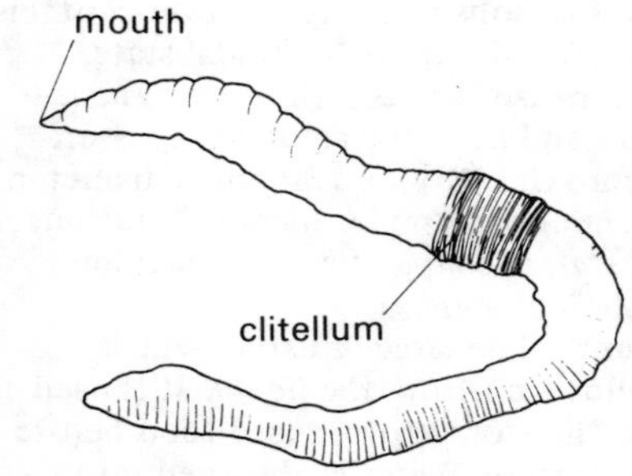

Fig. 40 *Clitellum*

cloaca A cavity into which the *reproductive* and *excretory* products are passed before being sent outside. Found in fishes and frogs.

clone A set of plants grown from one parent by *vegetative reproduction,* e.g. the plants grown from a set of *cuttings* from one plant.

cloud A mass of water droplets suspended in the atmosphere. If the cloud is caused to rise, these are generally precipitated as rain.

cloud chamber An instrument for showing the paths taken by, for example, *alpha particles.* As the alpha particle passes through the cloud chamber it produces *ions* and the *vapour* in the cloud chamber condenses on these ions to form drops of liquid which are large enough to be seen. The many liquid drops show the paths of the particles. The correct conditions for condensation may be caused by expanding the gas in the cloud chamber or by keeping part of it very cold.

coagulate To cause particles in *suspension* to join together in larger groups either by heating or by *precipitation.*

coal A black or brown organic rock formed from the fossil remains of trees and plants which grew many millions of years ago. The energy of these plants has been kept in the coal and is set free when the coal is burned. It is mainly *carbon* and *decomposition* of the original material is thought to have occurred in this order: *peat, lignite,* ordinary coal, *anthracite.*

coal gas The gas obtained by heating *coal* strongly in the absence of air, usually in closed iron *retorts.* It is a very good *fuel,* consisting mainly of *hydrogen, methane* and *carbon monoxide.*

coal tar When *coal* is heated in *retorts* the *components* are driven off and a sticky, black, oily liquid–coal tar–can be collected. *Distillation* and purification yield many compounds such as *benzene, phenol, naphthalene* and *pitch.*

cobalt chloride paper A strip of *filter paper* used in the detection of moisture. It is dipped in a *solution* of cobalt chloride (pink) and this becomes pale blue when dried. In the presence of water the paper becomes pink again, e.g. it can be used to show that leaves give out water *vapour* (*transpiration*).

cocaine A substance obtained from the coca plant. It is used by doctors to prevent pain.

coccyx Part of the *skeleton* (Fig. 148). The few small bones at the end of the backbone which are the remains in man of the tail bones of other animals.

cochlea Part of the *ear* (Fig. 59). It is a twisted tube which contains a liquid, *endolymph.* Sound *waves* cause the liquid to move against nerves, which then carry messages to the brain.

cocoon An egg case or a *pupal* case.

Coelenterata A group of animals of the jellyfish type.

coelom The space inside the *abdomen.* It is filled with coelomic liquid, which prevents the parts rubbing against each other.

cohesion The force of attraction between like molecules.

coition A joining together as in the *mating* process of many animals.

coke A hard, *porous, brittle* greyish solid formed by heating *coal* in the absence of air so that all the *volatile* products are driven off. It is left behind in the *retorts* after the manufacture of *coal gas,* but is also produced in *coke ovens.* It contains about 80% of *carbon.* It is used as an industrial and household *fuel* and for the *reduction* of *metallic oxides,* e.g. in the *blast furnace* production of iron. *Water gas* and *producer gas* are both made from coke.

coke ovens A *retort* specially made for heating *coal* at a lower *temperature* than is used for the manufacture of *coal gas;* as a result *coke* is the main product.

cold blooded Used to describe the blood heat of those animals whose temperature changes as the outside temperature changes. A wrong term, since in hot sun a crocodile's blood may be warmer than that of a man.

cold finger A slim double tube through which cold water flows; it can be inserted into a wider tube and used for *condensing vapours* or for cooling purposes.

collar bone See *clavicle.*

colloid A substance present in solution in the colloidal state. Colloids differ from *suspensions* in not settling out on standing. Most inorganic substances can be prepared in the form of colloids. E.g. *metals* and their *compounds* and, as such, the particles are electrically charged. When heated some colloids, e.g. the white of an egg, become solid. Protoplasm is colloidal. See *colloidal solution.*

colloidal solution A *solution* containing particles in the colloidal state, i.e. in size between those of a *suspension* and of a true solution; they are approximately 1–100 nm in diameter. Common examples include solutions of *starch, soap,* glues, *albumen* and colloidal *metals.*

colon The large *intestine* which follows on from the *ileum.* It is used for the storage of unused food before *defecation.* Water is absorbed into the *blood* through the wall of the colon. See *alimentary canal* (Fig. 5).

colonization The growth of young plants from *seeds* or *spores* on bare soil. *Lichens* can colonize bare rock.

colony A group of plants or animals which live close together.

colour The colour of light is fixed by its *frequency* and so by the *response* of the human eye and brain to that frequency. As different people respond to the same frequency in different ways, these people may not agree on the description of the 'same' colour. The colour of an object is fixed by the colour of the light it reflects to the eye of the person looking at it.

colour addition When lights of different *colours* are shone on to the same area then the resulting colour is formed by addition. Typical *responses* from the human eye and brain to simple coloured light additions are: red + blue = magenta (purple), red + green = (compound) yellow, blue + green = turquoise (cyan or peacock-blue), red + blue + green = white.

colour blindness A condition where a person cannot tell the difference between two or more colours, e.g. red-green colour blindness. It usually affects men and boys only, but can be passed on by the mother. It is an example of *sex-linkage.*

colour filter A thin sheet of material which only allows certain *frequencies* or *colours* of light to pass through it.

colour subtraction A *colour filter* removes some *frequencies* of the light passing through it. A second filter would remove more frequencies so when both are used only the frequencies (or *colours*) passed by both are seen. Mixed paints work in the same way, only colours which are reflected by both reach the eye.

combination The joining together of *elements* to form *compounds* or of small *molecules* to form larger ones, e.g. water and *calcium oxide* combining to form *calcium hydroxide.*

combustion See *burning.*

combustion value See *calorific value.*

commutator A revolving switch which allows the current in the coil of an *electric motor* to change direction at the correct moment. This keeps the turning forces in one direction and so allows the motor to keep revolving. It may also be used with a *generator* to produce a current which always flows in one direction. Basically it consists of a metal ring split into two (or more) parts with two spring connections (or brushes) pressing on to them.

companion cell A cell that lies at the side of a *sieve tube* (Fig. 146). It probably controls the liquid carrying powers of the sieve tubes.

compass An instrument containing a magnetic needle which is carefully pivoted so that it always points to magnetic north. Often all the other directions are marked, either on the case, or on a card fastened to the magnet. Better compasses are also filled with a liquid to steady the movement of the magnet.

complementary colours Two colours of light which, when added together, make white light, e.g. (compound) yellow and blue, magenta and green, and turquoise (cyan) and red.

complex Complicated, not simple. Used of *molecules* which contain many atoms or groups.

component 1. (Electrical) The parts from which an electric *circuit* is built up, e.g. switches, bulbs, resistors, capacitors, ammeters, voltmeters, etc. 2. (Vector) A *vector* quantity may be split up (resolved) into two or more components which together have the same effect as the original vector. 3. (Chemical) One of the parts which make up the whole, e.g. the *elements* present in a *compound* or the separate substances which make up a mixture.

Compositae The large group of flowering plants where each *flower* has a number of small *florets,* e.g. sunflower, daisy.

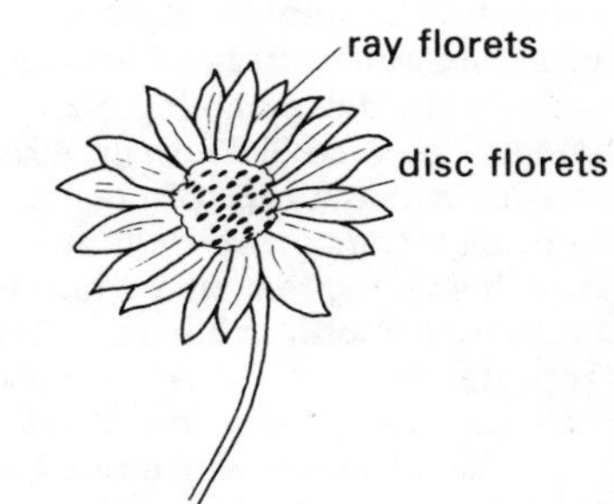

Fig. 41 *Compositae*

composition The way in which a *compound* is made up, i.e. the nature and proportions of the *elements* present.

compound See *chemical compound.*

compound eye The kind of eye found in the *Arthropoda.* Each eye is made up of a large number of *lenses.* (Fig. 42.)

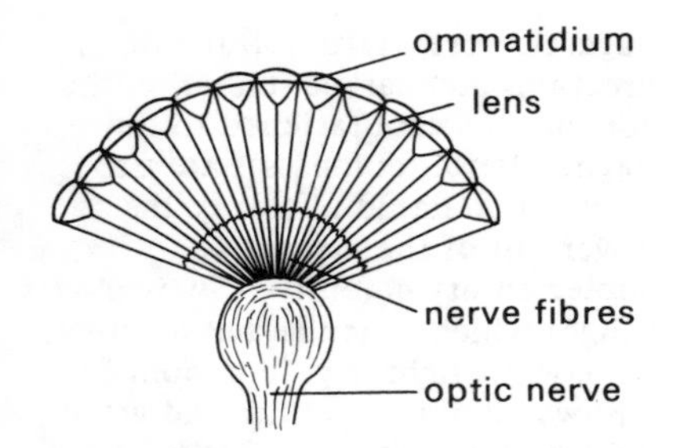

Fig. 42 Compound Eye in section

concave A curved surface which is shaped like the inside of a ball.

concentrate To increase the proportion of *solute* in a *solution* by *evaporation* of some of the *solvent,* e.g. water from *brine.*

concentrated A *solution* in which the proportion of *solute* to *solvent* is high; if the solution is an *aqueous* one then it contains only a small amount of water: the opposite of *dilute.*

concentration The amount of *solute* in a definite *volume* or *mass* (weight) of a *solvent,* usually expressed as *grams* or *moles* per *litre.*

concentric Concentric tubes, such as those used in the *Frasch process* for the extraction of sulphur, are ones having a common centre and arranged as shown in the diagram. (Fig. 82.)

conchoidal Shaped like a shell; term used to describe the type of breakage which often occurs in *amorphous* solids such as glass, the curved face of the breakage sometimes having a shell-like appearance.

concrete A mixture of *cement, sand,* water and small stones which sets hard on standing and is used as a building material.

condensation 1. The *liquefying* of a *gas* or *vapour* on cooling or on being subject to increased *pressure* which forces the *molecules* closer together. 2. A *chemical change* in which two molecules react together to form a new and more complex substance along with a small simple molecule such as water or ammonia. 3. The joining together of simple molecules to form a chain, i.e. *polymerization.*

condense To cause a *gas* or *vapour* to *liquefy* by cooling or by applying an increased *pressure.*

condenser 1. (Electrical) The old name for an electrical *capacitor.*
2. (Optical) A *lens* system which directs light on to the film or slide, etc., in an optical instrument (e.g. a *microscope* or a projector).
3. (Chemistry) A device for cooling a *gas* or *vapour* during *distillation* so that it liquefies. Water or air are usually used for cooling. See also *Liebig condenser, cold finger.*

conduction The passing of heat or electricity through a substance, without that substance having to move.

conductivity A measure of the ability of a substance (solid, liquid, gas or in solution) to conduct an *electric current.*

conductor A substance which allows heat (thermal conductivity) or electricity (electrical conductivity) to flow through it. Metals are good conductors and therefore offer a low resistance to the flow of heat or electricity, but non-metals are usually bad conductors or *insulators.*

cone 1. The 'flower' of fir trees. When ripe it is made of small hard brown leaves between which are the *seeds.* 2. A colour sensitive cell in the *retina* of the *eye.*

conjugate foci Two points near a *lens* or *spherical mirror* such that light from one of them is brought to a focus at the other. Thus an object placed at one point will give a *real image* at the other.

conjugation A method of *sexual reproduction.* It is not possible to tell male from female. Found in simple plants, e.g. spirogyra.

conjunctiva Part of the *eye* (Fig. 68). The *transparent* skin over the front of the eye. May become diseased (conjunctivitis).

consciousness Knowing what is happening. A conscious action is one where the brain gives the orders, e.g. kicking a football.

conservation of mass and energy In any system, considered as a whole, the total amount of *mass* and *energy* remains *constant.* The various forms of energy may change from one to another or may be changed into mass, or mass changed into energy, but there can be no total loss or gain. It used to be thought that there were two separate conservation laws, one for mass and another for energy, but from *Einstein's equation,* these are seen to be connected.

conservation of momentum When two (or more) objects hit each other, the total *momentum* does not change.

constant A constant quantity is one which does not change in value.

Constant Composition, Law of All pure *compounds,* no matter how made, always contain the same *elements* joined together in the same fixed proportions by *mass* (weight). This is also called the Law of Definite Composition, Definite Proportions or

Constant Proportions, Law of See *Constant Composition, Law of.*

constant volume air thermometer In this instrument a dry gas is kept in a large bulb connected by a *capillary tube* and a rubber tube (in the shape of a letter U) to a final, open, glass tube. Mercury fills the final part of the capillary tube, all the rubber tube and part of the open tube. The mercury level in the capillary tube is kept fixed by moving the open glass tube up or down. The difference in height between the two ends of the mercury allows the gas *pressure* to be found, if *atmospheric pressure* is known. The apparatus can be used to watch the behaviour of the gas or as a *thermometer* if it is *calibrated* at the *fixed points.*

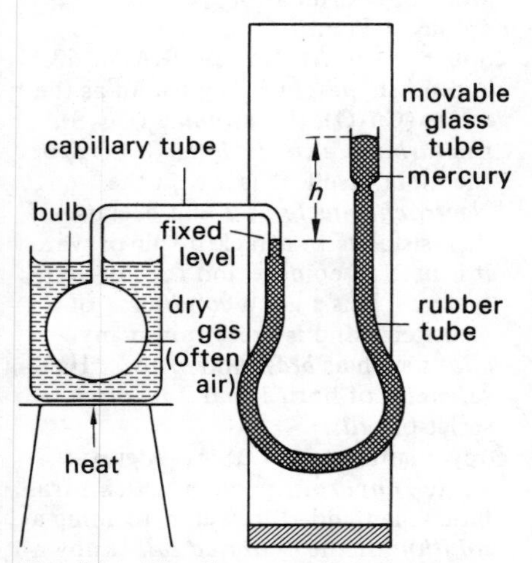

Fig. 43 The Constant Volume Air Thermometer

constantan The trade name of a copper-nickel alloy (approximately 40% nickel). It is often made into a wire and used for making *resistors* as its *resistance* does not change very much as its temperature changes. The same alloy is also called contra or eureka.

constituent See *component.*

contact process An industrial process for the preparation of *sulphuric acid.* Dry *sulphur dioxide* and excess dry air are mixed and passed at *atmospheric pressure* and a *temperature* of about 450 °C over a *catalyst* of vanadium(V) oxide. The *sulphur trioxide* formed is absorbed in sulphuric acid already produced to form oleum ($H_2S_2O_7$) and this then reacts with water to give more sulphuric acid.

contact stimulus A change in the surroundings of the skin, which produces a *response,* e.g. if something hot touches the skin of the hand, the hand is pulled away.

continuous spectrum The *spectrum* formed from *white light.* It has bands of the six (or seven) spectral colours, red, orange, yellow, green, blue, (indigo) and violet, gradually changing from one to the next without any gaps.

contractile root Roots on *bulbs* or *corms* which shorten after a time, so pulling the bulb or corm down into the ground. (Fig. 44.)

contractile vacuole A small space inside some one-*celled* animals. It fills and empties all the time and gets rid of water which has been sucked in. (Fig. 45.)

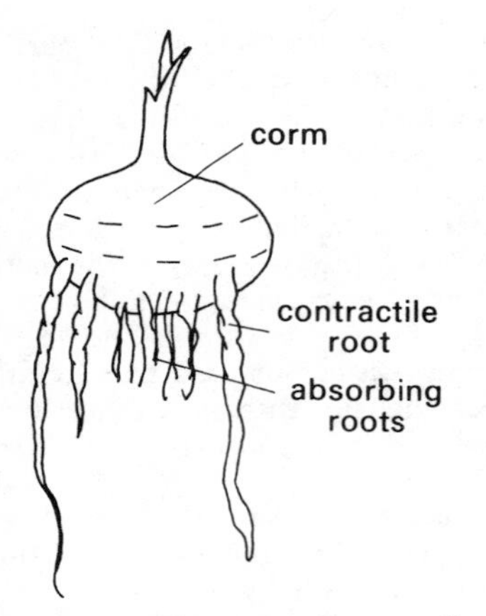

Fig. 44 Corm showing Contractile Root

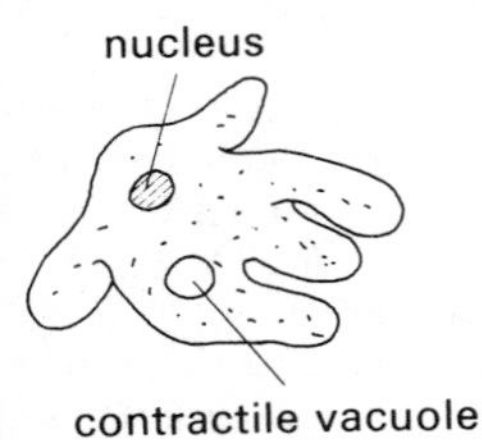

Fig. 45 Amoeba showing Contractile Vacuole

control experiment An experiment used to compare ordinary conditions with different conditions, e.g. growing plants in normal light (control) and comparing them with plants grown in darkness.
convection If a *fluid* (liquid or gas) is warmed it will expand and become less dense (*density*). Then, if it can, it will rise and cooler fluid will take its place. This sets up a flow of fluid called a convection current. This carrying of heat energy by a moving fluid is called convection.
converging lens A *lens* which brings parallel rays of light to a focus after they have passed through it. Such lenses used to be called convex lenses as one or both surfaces are *convex*.
convex A curved surface which is shaped like the outside of a ball.
coolant 1. A *fluid,* usually water or air, which, by passing through or over a *condenser,* causes a *vapour* to *liquefy* or prevents a *liquid* from vaporizing. 2. A fluid used to keep motor car engines (internal combustion engines) cool, e.g. air, water, ethylene glycol (*antifreeze*).
cooling by evaporation *Molecules* need energy to escape from a liquid and if this is not supplied from outside the energy is taken from the liquid itself, making the liquid cooler. Using the *kinetic theory* this can be explained in another way. A liquid evaporates (*evaporation*) when some of its faster moving molecules escape through the liquid surface to become part of the vapour. The loss of the faster moving molecules reduces the average energy of those which remain, so they must be cooler.
cooling curve A *graph* of *temperature* against time for a substance which is allowed to cool freely. If the graph has a horizontal part this suggests that the substance was a liquid which is turning to a solid.
coordination Being able to make different parts of the body work together to produce a movement, such as swimming. The *brain* has the work of coordination.
copious Plentiful.
copper Cu At. No. 29 R.A.M. 63.5 A reddish *metal* usually found as the *oxide* (Cu_2O), the *sulphide* (Cu_2S), the *carbonate* (*malachite,* $CuCO_3$) or the metal itself. It is low in the *electrochemical series* and because of its resistance to attack by air or water it is used in coinage and for hot-water systems. It is a good *conductor* of electricity and is present in many *alloys* such as *brass* and *bronze*. Having *valencies* of both 1 and 2 it forms two series of *salts*.
copper sulphate $CuSO_4$ A grey-white *anhydrous* powder which turns blue when added to water, forming a *solution* of the *hydrated salt* (known also as blue vitriol), $CuSO_4.5H_2O$.
coracoid Part of the *skeleton*. One

of the bones of the *pectoral girdle* in fish and frogs.

coral A rock-like substance, mainly *calcium carbonate,* which is made by a *colony* of tiny sea animals whose *skeletons* join together. When the animals die, the hard skeleton remains.

cork A light, soft, elastic, waterproof substance made by cork *cambium* on the outside of plants. On the trunk of a tree it forms the *bark.* Most cork used by man comes from the bark of the cork oak tree.

cork cambium The plant *cells* which produce *cork.*

corm An underground *stem* which stores food for *perennation.* See *contractile root* (Fig. 44).

cornea Part of the *eye* (Fig. 68). The *transparent* layer in front of the *lens.*

corolla The *petals* of a *flower.*

coronary artery The *blood vessel* which takes blood to the *muscles* of the *heart.* If this artery is shut off, a heart attack follows.

corpuscle A small *cell,* e.g. a red corpuscle of the *blood.*

corpus luteum The tissue left on the *ovary* of a *mammal* after the *egg* has ripened. It produces the *hormone progesterone* if *pregnancy* follows.

corrode To become subject to *corrosion.*

corrosion A chemical action which causes the surfaces of *metals* to be eaten away. Usually caused by air and moisture together or by certain chemicals. The action is mainly *electrolytic,* involving *oxidation.* Metals high in the *electrochemical series* corrode most rapidly although some, such as *aluminium,* soon get coated with a layer of the *oxide* which stops further action. See also *rusting.*

corrosive Capable of causing *corrosion* or destroying a substance; e.g. *concentrated acids* and *alkalis* or air containing *acidic fumes* such as are present in industrial districts.

cortex 1. (Brain) The outside *tissues* which contain many *nerve cells* (neurons). 2. (Kidney) The outside tissues which contain the *filter* cups (*Bowman's capsules*). 3. (Plant stems and roots) The group of *cells* which lie between the outside skin and the *vascular bundles.* Used for storing water and *starch.*

cosmic radiation Particles (mainly *protons*) of high energy which reach the earth's atmosphere from outer space.

cost of electricity The cost of electrical energy is found by multiplying the *power* in kilowatts (*kilo-, watt*) of whatever is being used, by the time of use (in hours) and by the price per unit or *kilowatt-hour,* i.e. cost = power x time x price per unit.

cotyledon Part of a *seed. Monocotyledons,* e.g. corn, have one, *dicotyledons,* e.g. bean, have two. Fir trees have several. Cotyledons are used to store food or to make food for the young *seedling.* If they come up above the ground, they turn green.

coulomb The *SI unit* of *electric charge* or of *quantity of electricity.* The symbol is C. It is equal to the amount of electricity passing a point in a circuit in one second when a steady current of one *ampere* is flowing. Thus one coulomb is one ampere-second.

counterpoise 1. To balance by means of certain masses (weights). 2. A mass (weight) used to balance another one.

couple A couple is made up of two equal *forces* acting on an object in opposite directions but not along the same line. A couple will cause the object to turn unless another equal couple acts in the opposite sense.

covalency The joining together of *atoms* by sharing of *electrons,* each atom contributing one electron to each *covalent bond* formed.

covalent Not *ionic* or *electrovalent;* not forming *ions,* but sharing *electrons.*

covalent bond A *bond* between *atoms* formed by the sharing of *electrons,* each atom contributing one electron to the bond, e.g. in water

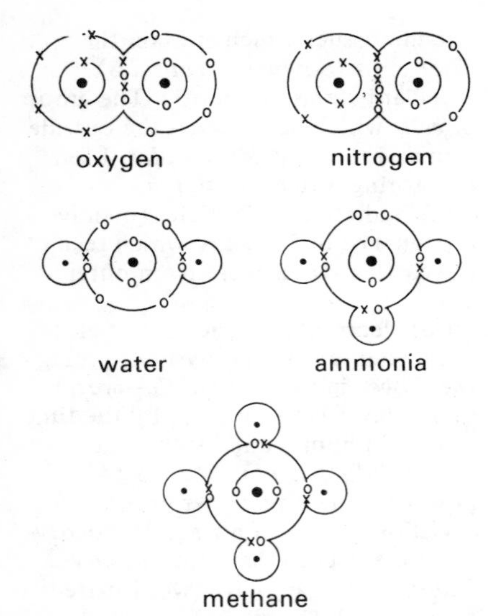

Fig. 46 Covalent Molecules (the electrons are represented by x and o for convenience only in showing to which atoms they originally belonged; they are, of course, all the same)

there are two covalent bonds formed as in the diagram by the sharing of the two electrons from the two *hydrogen* atoms and two from the outer *shell* of the *oxygen* atom. If one atom contributes both electrons the bond is variously called a co-ionic or dative bond or a co-ordinate link.

covalent compound One having the *properties* associated with a *covalent bond,* i.e. it is a *gas,* a *liquid* with a low *boiling point* or a *solid* which melts easily; it does not *conduct* electricity when dissolved in water.

cow pox A disease of cows, affecting the making of milk. It is caused by a *virus*, similar to the *smallpox* virus in man. It can be used as a *vaccine.*

coxa The section of the leg of an *insect* nearest the body.

cracking The breaking down by *heat* of large *molecules* into smaller ones, e.g. if *ethanol vapour* is passed over heated *porous pot* it is cracked or broken down into the smaller units of *ethene* and water. Cracking is of particular importance in producing *petrol* from the higher boiling fractions obtained in *oil refining.* Cracking is often helped by the use of a *catalyst* and is then called catalytic cracking.

cramp *Muscle* tightening caused by cold or over use.

cranium Part of the *skeleton* (Fig. 148). The bony box in which the *brain* is found. It is made up of many bones joined together.

cretinism A condition caused by lack of a properly grown *thyroid gland* in babies. The body and the *brain* become under developed.

critical angle The angle of *incidence* of the *critical ray.*

critical ray When a ray of light is passing from a material like glass or perspex into air it is part reflected

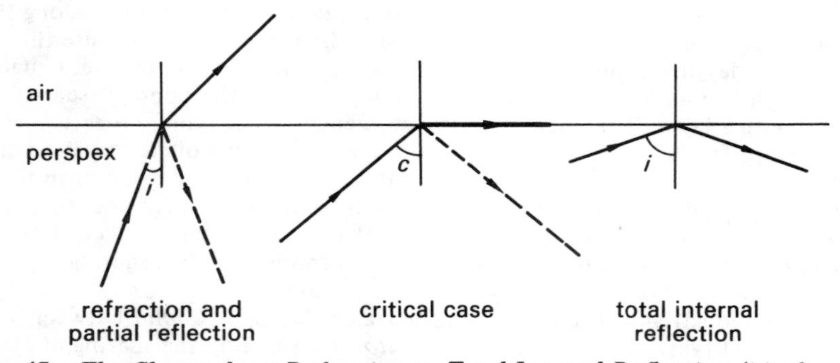

Fig. 47 The Change from Refraction to Total Internal Reflection (i is the angle of incidence; it has the value c in the critical case)

(*reflection*) and part refracted (*refraction*) if the angle of incidence is small, but it is totally internally reflected (*total internal reflection*) when the angle of incidence is large. The ray of light which comes at the boundary between these two possibilities is called the critical ray, after refraction it travels along the surface of the glass etc.

critical temperature That temperature above which it is impossible to *liquefy* a gas simply by increasing the *pressure.* For carbon dioxide this is 31 °C, for nitrogen it is −147 °C and for hydrogen −240 °C.

crop rotation A system where different crops are grown in turn on the same piece of land. In this way, the soil is best used.

crossing over The joining up of parts of *homologous chromosomes* during *meiosis.* An exchange of *alleles* takes place.

cross pollination When the *pollen* used to *fertilize* an *ovule* comes from the *stamens* of another *flower* of the same sort.

crucible A small vessel made from some heat-resisting material such as *fireclay, porcelain, graphite* or a metal such as nickel or platinum. It is used for heating small amounts of substances to a high *temperature.*

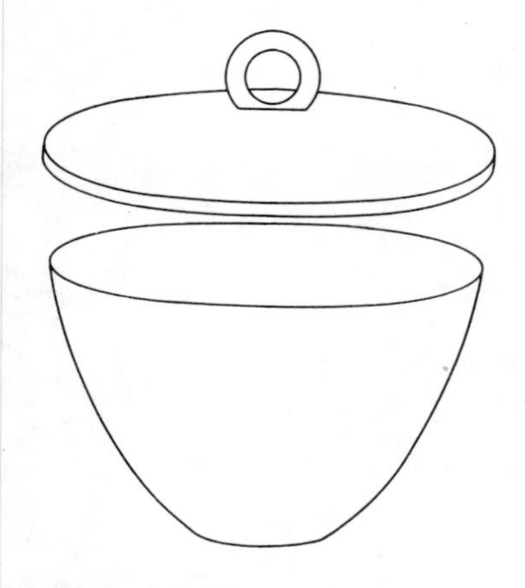

Fig. 48 *Crucible*

Crustacea The group of animals which includes the crab and shrimp.

cryolite Na_3AlF_6 A *compound* of sodium, aluminium and fluorine which is mixed with bauxite (aluminium oxide) in order to lower the *melting point* of the latter. The molten mixture is then *electrolyzed* and aluminium is obtained.

crystal A pure substance which has solidified into a definite geometrical shape, e.g. a cube. Crystals have *plane* surfaces and a regular arrangement of *atoms,* (e.g. diamond) or *ions,* (e.g. sodium chloride).

crystal lattice The regular arrangement of *positive* and *negative ions* in a *crystalline compound* as in, e.g. sodium chloride (see diagram). Also called an ionic lattice.

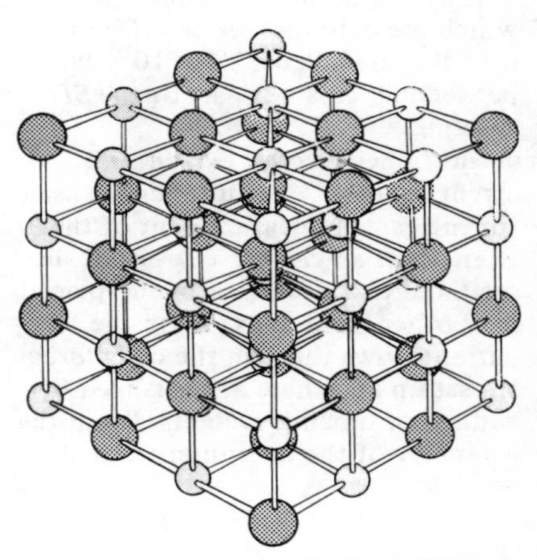

Fig. 49 *Model of a Sodium Chloride Crystal-lattice (The larger darker spheres represent chloride ions and the smaller lighter spheres represent sodium ions)*

crystalline Clear and *transparent* and resembling a *crystal* in shape. However, the term may also be used for substances cut to shape of a crystal and not formed naturally in this way.

crystallize To form *crystals* on cooling or *evaporating* a *saturated solution* of a substance.

crystalloid A substance which has *dissolved* completely in water, e.g. salt.

cubic A *crystal* system in which sodium chloride (common salt) occurs.

culture A growth of *bacteria* or any other kind of living thing, produced under *experimental* conditions.

culture solution A food *solution* used to grow plants in an *experiment.*

cupric A term formerly applied to *copper compounds* in which the *metal* has a *valency* of 2, e.g. cupric oxide, CuO, now called copper(II) oxide.

cuprous A term formerly applied to *copper compounds* in which the *metal* has a *valency* of 1, e.g. cuprous oxide, Cu_2O, now called copper(I) oxide.

curie A *unit* which measures the amount of *radioactive* material present by the number of nuclei (*nucleus*) which break up per second. One curie of material will give 3.7×10^{10} events per second. It is not part of the *SI* system.

current See *electric current.*

current balance An instrument used for the careful measurement of the strength of an *electric current.* Coils are fixed to a balance arm and placed near other fixed coils. When the current flows through the coils *forces* are set up and these are balanced by adding, or moving, weights. From the geometry of the instrument and a knowledge of the forces involved, the current can be found. It is an instrument used for standardizing other *ammeters* rather than for common measurements.

cuticle 1. The waxy covering on the *epidermis* of a leaf. 2. In man, the thin skin at the base of the finger nail.

cutting A part of a stem or root cut off a plant and rooted to form a new plant.

cyanamide $CaCN_2$ Correctly called calcium cyanamide. A *synthetic fertilizer* made by heating *calcium carbide,* CaC_2, in nitrogen at 1000 °C. It is converted into *ammonia* in the soil by water present and the nitrogen can then be used by plants.

cycle A series of events, in a regular pattern, which leaves the process at the same point as it started. See *frequency, hertz.*

cylindrical lens A *lens* which is curved in one direction only as if it had been cut from a cylinder of glass.

cyme The name given to a kind of grouping of *flowers.* The oldest flower is at the top.

cyst (*vb.* Encyst) 1. A hard shell which contains an *egg.* It is found in very small animals. 2. A small growth under the skin of an animal. It contains liquid or fat.

cytoplasm The living part of a *cell* other than the *nucleus.*

damped vibration When an object is made to move rapidly to and fro and then left, the *amplitude* of these vibrations is damped, i.e. it gets gradually smaller and smaller due to a loss of energy.

damping off A *disease* in plants caused by a *fungus.*

Daniell cell A *cell* containing a zinc rod dipping into zinc sulphate solution or sulphuric acid, in a porous pot. This is placed in copper sulphate solution in a copper can. The zinc rod is the negative *electrode* and the copper can the positive one. The *electromotive force* is 1.1 *volts.* The cell cannot be kept when it is made up and so is little used outside the laboratory.

Darwin, Sir Charles (1809–1882) An Englishman who wrote 'The Origin of Species' and many other books. Best known for his Theory of Evolution which suggested that changes in plants and animals take place over a long time, producing new sorts which are better suited to their conditions. See also *natural selection, evolution.*

Davy lamp A type of miner's safety lamp. The flame of an oil lamp is separated from the outside gases (which could be explosive) by a wire gauze. Although the gases may burn inside the lamp the gauze never becomes hot enough to set fire to the gases outside because it is a good *conductor* of heat. Such lamps are now replaced mainly by electric lights.

deamination The removal of the unwanted *amino acids* from food proteins by the liver.

deca- A prefix which means ten times. It can be joined in front of any *SI unit,* e.g. 1 decametre = 10 metres. The symbol is da. It is not used very much in science.

decant To pour a *liquid* from a *suspension* without disturbing the *solid* which has settled down. The process is called decantation.

decarbonize To remove the hard *carbon* deposits which form on the internal surfaces of the *combustion* chamber of internal combustion engines (motor car engines).

decay 1. (Oscillations) If any object is set into vibration and left, the vibration will gradually get less and less, or decay, due to damping. See also *damped vibrations.* 2. (Radioactivity) If any one type of *radioactive* material is separated then it will gradually get less and less, or decay, as time passes. The particular radioactive *nuclide* has decayed into some other nuclide.

deceleration The rate at which the *velocity* of an object decreases with time. See also *acceleration, retardation.*

deci- A prefix which means one tenth or 10^{-1}. It can be joined in front of any *SI unit,* e.g. 1 decimetre = 10^{-1} metre. The symbol is d. It is not used very much in science.

deciduous A tree which drops its *leaves* in the cold season.

decimolar The *molarity* of a *solution* containing one-tenth of a *mole* of *solute* in a *litre* of *aqueous* solution.

declination The angle between the direction of true (geographical) north and of magnetic north at any place.

decolorize To remove the colour from a substance, e.g. *charcoal* will decolorize ink and *reductants* will decolorize *aqueous potassium permaganate* acidified with dilute sulphuric acid.

decomposition 1. The breaking down of a *compound,* usually by the action of *heat* or electricity, e.g. calcium carbonate on heating decomposes to calcium oxide and carbon dioxide and concentrated hydrochloric acid yields hydrogen and chlorine on *electrolysis.* 2. The breaking down of dead plants and animals, usually in the soil.

decrepitation The crackling sound heard when some *crystals* are *heated.* It is often due to the expansion of water within the crystals causing them to break up.

defecation Passing out from the body of the unwanted solid parts of the food.

deficiency disease Trouble caused by a lack of a *vitamin* in the food, e.g. *scurvy* caused by lack of *Vitamin C.*

Definite Composition, Law of See *Constant Composition, Law of.*

Definite Proportions, Law of See *Constant Composition, Law of.*

deflagrating spoon A burning spoon. A small cup or bend at the end of a metal rod in which substances may be placed. They are set alight and the spoon lowered into a jar of *oxygen* or *chlorine* in order to see if the gas will *support combustion* of that substance.

degradation The breaking down of large *complex molecules* into simpler ones, .e.g. converting *glucose* into *ethanol* by *fermentation* or breaking down *proteins* into *amino acids* by the action of dilute acid.

dehydrating agent A substance which is able to remove water or *water of crystallization* from *compounds* and so bring about *dehydration.* Some are able to remove hydrogen and oxygen in the proportions of 2 atoms to 1 *atom* from certain compounds and so produce water, e.g. concentrated sulphuric acid converts glucose ($C_6H_{12}O_6$) into carbon and water and changes formic acid (H_2CO_2) into carbon monoxide and water.

dehydration Removal of *water* or *water of crystallization* from a substance; e.g. *concentrated sulphuric acid* acts as a *dehydrating agent* on compounds such as oxalic and formic acids in which the *oxygen* and *hydrogen molecules* are in the same proportion as in water, 2 : 1. Often heat alone is sufficient to remove water of crystallization and make a crystalline compound *anhydrous.*

deionization The removal of all *cations* and *anions* except *hydrogen* (H^+) and *hydroxyl* (OH^-) from tap water so that it is made completely pure. Metal and other ions are normally present after chemical methods of purification and these can be removed by letting the water run through a column containing certain *resins* which will remove these *ions* and replace them by hydrogen and *hydroxyl* ones.

deliquesce To take up moisture in sufficient quantity when exposed to it as to *dissolve* in the water and form a *solution.* The term is applied to such substances as *anhydrous* (fused) calcium chloride, phosphorus pentoxide and sodium hydroxide.

deliquescence The taking up of water and then dissolving in it to form a *solution.*

deliquescent Having the *property* of becoming a *liquid* on exposure to damp air.

delta The symbol is Δ. A sign used in two ways: (a) to indicate that a *chemical reaction* needs *heat*, e.g. $2HgO \xrightarrow{\Delta} 2Hg + O_2$ means that mercury(II) oxide must be heated to make it *decompose.* (b) to indicate a change in a particular value, e.g. ΔH means the change in the *heat energy* content of a substance when a reaction occurs. In the reaction $C + O_2 \rightarrow CO_2$; $\Delta H = -393$ kJ the change in heat energy of the carbon and oxygen on combining is − 393 kJ for every *mole* (12 g) of carbon burnt. The negative sign indicates that heat is given out.

demagnetization The process of removing magnetism from an object. Sometimes it can be done (or partly done) by heating the object above a certain temperature or by banging it. However it is better to place the object in a *magnetic field* which is always changing direction as it gets smaller and smaller. This occurs in a coil in which a decreasing *alternating current* is flowing.

demineralize To purify commercial *methylated spirit* by *fractional distillation.*

dendrite A branch of a *dendron*. See *synapse* (Fig. 155).

dendron Small branches of a *nerve cell* which join up with other nerve cells.

density The *mass* of unit *volume* of a material. The density of a material is therefore the mass of a given piece divided by its volume. The *SI unit* is kilograms per cubic metre and the symbol is ρ. See also *relative vapour density*.

dentine A substance which makes up the main part of a *tooth* (Fig. 159i). It contains *calcium* salts.

deodorant A substance which removes bad smells or odours.

deoxygenated blood *Blood* from which most of the *oxygen* has been removed for use by the *cells. Veins* usually carry deoxygenated blood.

depolarizer The chemical or chemicals placed in a *cell* to prevent or reduce *polarization*, e.g. maganese dioxide in the *Leclanché* or *dry cell.*

deposit *Solid* precipitated (*precipitation*) from a *suspension* as it settles down.

depth of focus The range of object distances over which a *camera* will produce a clear photograph (without moving the lens).

dermis Part of the *skin.* It contains *blood, nerves* and sensitive *cells.*

desiccant See *drying agent.*

desiccator A glass vessel used for keeping substances dry. It has a tightly-fitting lid and contains *deliquescent* or, preferably, *hygroscopic* material in the base, e.g. *anhydrous calcium chloride* or *silica gel.*

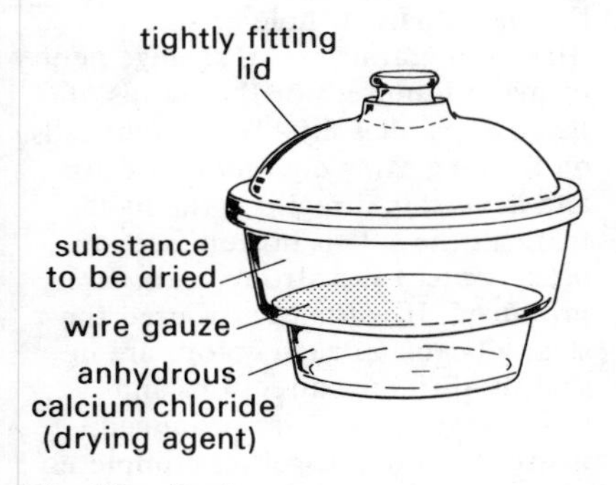

Fig. 50 Desiccator

destructive distillation A term applied to the heating of an *organic* substance such as *coal* or wood in the absence of air. The *gaseous* products *distil* over and some *condense* while others pass on. *Coal gas* is obtained in this way along with *coke* while wood yields wood-tar and ethanoic (acetic) acid.

detect To note the presence of something, e.g. a gas given off in an experiment.

detergent Any cleaning agent for grease and dirt removal such as *soap*, but the term is usually applied to those powders or liquids which form a *lather* and not a *scum* with *hard water.* It may be the *sodium salt* of some *organic acid* with a long-*chain molecule.*

deuterium An *isotope* of hydrogen containing one *neutron* and one *proton* in the *nucleus.* It is also called 'heavy' hydrogen and deuterium oxide is called 'heavy' water. It makes up about 0.016% of 'natural' hydrogen.

deviation The angle between the incident (*incidence*) and emergent rays when light passes through a *prism* or *lens* etc. See also *refraction.*

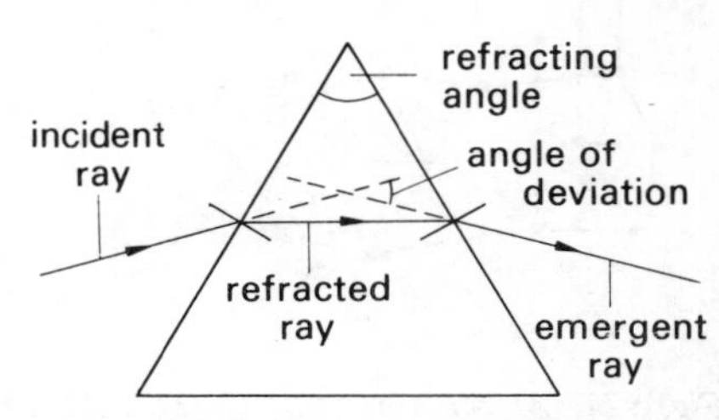

Fig. 51 The Path of Light through a Prism

dew point The *temperature* to which the air must be cooled for dew (drops of water) to start to form.

dextrose $C_6H_{12}O_6$ Also known as grape sugar. See *glucose.*

di- A prefix meaning 2, e.g. divalent = having a valency of 2; carbon disulphide contains 2 *atoms* of *sulphur* in each *molecule,* CS_2.

diabetes The name given to two kinds of *disease.* Diabetes mellitus is due to lack of *insulin.* As a result, sugar *metabolism* is upset and sugar is found in the *urine.* Diabetes insipidus is a nervous disease, leading to too much urine and to a great thirst for water.

diamond A natural *allotropic* form of *carbon* occurring in *octahedral crystals,* found mainly in South Africa. The *atoms* of carbon are joined together *covalently* to give a tetrahedral (*tetrahedron*) arrangement (see diagram). Colourless when pure, diamond is the hardest natural substance known; it is used for cutting and drilling tools and as a gem.

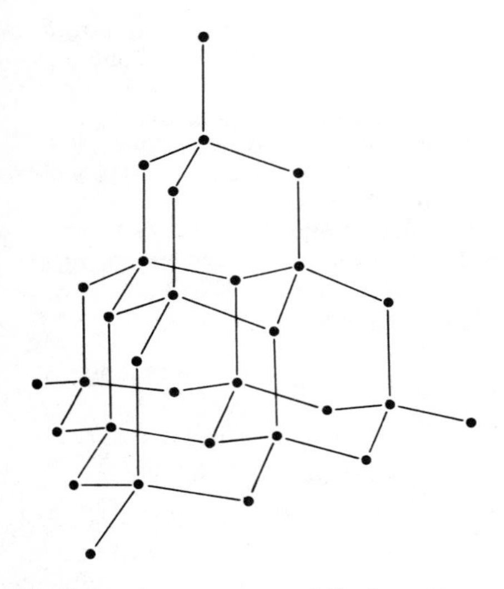

Fig. 52 Arrangement of Carbon Atoms in Diamond

diaphragm The muscular floor of the chest. It is used with the ribs for breathing in and out, by changing the size of the *lungs* (Fig. 109i).

diastole The movement of the *muscles* of the *heart* which makes the spaces inside bigger and so allows *blood* to flow into the heart. See also *systole.*

diatom A very small plant which lives in water. In the sea, the bodies of diatoms fall to the bottom and make an earthlike layer.

diatomic A *molecule* made up of two *atoms* only, e.g. bromine, Br_2 or hydrogen chloride, HCl.

dibasic acid An *acid* which has two *hydrogen atoms* in the *molecule,* each of which may be replaced by a suitable metal atom; e.g. sulphuric acid, H_2SO_4, forms two kinds of *salts* (1) *normal salts* such as sodium sulphate, Na_2SO_4 and (2) *acid salts* such as sodium hydrogensulphate, $NaHSO_4$.

dicotyledon A plant whose *seeds* contain two *cotyledons,* e.g. bean.

dielectric The insulating (*insulator*) material placed between the plates of a *capacitor* to increase its *capacitance.*

diesel oil One of the distillates (*distil*) obtained by the *fractional distillation* of crude *oil* is gas oil (b.p. 200–350 °C). After some refining it is used as a *fuel* in trucks, buses and some railway engines; this is diesel oil. Often known as DERV(diesel-engined road vehicles).

differentiation The changing of *cells* and *tissues* in shape and content so that they can do their work properly.

diffraction The spreading out of a *beam* of light (etc.) when it passes near to the edge of an object or through a narrow hole or slit.

diffraction grating An arrangement of many hundreds or thousands of narrow, parallel slits. When light falls on to the grating *diffraction* occurs. and in certain directions the light from all the slits is in step (see *interference*) and strong beams are produced. If *white light* is used the strong beams of each colour are in slightly different directions and spectra (*spectrum*) are produced.

diffuse To mix together completely, often against the force of *gravity.*

Usually said of *gases* which mix thoroughly when placed together, but if a solid such as potassium permanganate is placed in the bottom of a jar of water, the purple colour gradually diffuses (spreads) throughout the liquid.

diffusion 1. (Solids, liquids and gases) The gradual spread of one material into another. It is a rapid process in gases, slow in liquids and very, very slow in solids. It supports the idea that the *molecules* of which matter is made up are moving. See also *kinetic theory.* 2. (Light) The scattering of light when it passes through frosted (roughened) glass for example or when it is reflected from a surface like that of filter paper.

digestion The changing of food by *enzymes* in the *alimentary canal* so that the food can be used by the body more easily.

digit In *biology*, a finger or a toe.

dilead (II) lead(IV) oxide Pb_3O_4 A 'mixed' oxide of lead containing the metal in *valency* states 2 and 4. It behaves in *chemical reactions* as though it is a mixture of lead(II) oxide, PbO and lead(IV) oxide, PbO_2. With *acids* it forms a mixture of *salts* in which the metal exerts its different valencies. It is also known as tri-lead tetroxide, red lead and minium. It is used as a *pigment* for the initial painting of *steel* structures to prevent *corrosion.*

diluent 1. A *liquid* used to make a *solution* more dilute as when water is added to a concentrated aqueous solution, i.e. the addition of more *solvent.* 2. An *inert gas* used to reduce the *concentration* of another, e.g. if air is used in *chemical reactions* involving *oxygen* the *nitrogen* present acts as a diluent and reduces the effectiveness of the oxygen.

dilute A *solution* in which the proportion of *solvent* to *solute* is large, e.g. a little salt in a lot of water: the opposite of *concentrated.*

dinitrogen oxide N_2O Nitrous oxide; laughing gas. A sweet-smelling colourless gas made by heating ammonium nitrate. It is fairly soluble in water, will relight a glowing splint and so is a good *supporter of combustion.* Mixed with oxygen and carbon dioxide it is used by dentists as an *anaesthetic.*

dinitrogen tetroxide N_2O_4 A yellow-brown liquid at room *temperature* (b.p. 23°C), formed by the *association* of two *molecules* of *nitrogen dioxide,* NO_2, a process which is reversed (*reversible*) on *heating* the liquid.
$2NO_2$ (dark brown) $\rightleftharpoons$ N_2O_4 (yellow-brown)

diode A *valve* containing two *electrodes.* The *filament* (*cathode*) is heated so that it gives out *electrons* which travel to the *anode.* A diode only allows *current* to flow one way through it because if the anode becomes negative it pushes away the electrons and the flow stops. See also *rectification, thermionic emission.*

dioecious A plant where the *pollen-producing flowers* and the *ovule-producing* flowers are on different plants, e.g. stinging nettle.

dioptre The *unit* in which the *power* of a *lens* is given. Power (in dioptres) equals one divided by the *focal length* (in metres).

dip If a magnetized needle is pointing (magnetic) north but is also free to turn vertically it will come to rest at an angle to the horizontal. The angle between the needle and the horizontal is called the angle of dip.

dip circle An instrument for measuring the angle of *dip.* It contains a well balanced magnetic needle which can turn vertically.

direct current An *electric current* flowing steadily in one direction. It is the opposite of *alternating current.*

direct vision spectroscope An instrument which is used to see the *spectrum* formed by a source of light. It may contain a *diffraction grating* or several *prisms.*

disaccharide A sugar which is made from *glucose,* e.g. cane sugar.

discharge tube A tube with two *electrodes* containing gas at low pressure. When a high *potential difference* is applied between the electrodes the gas gives out light. The colour of the light depends on the gas in the tube, or the *spectrum* of the light could be used to find out which gas was present.

discrete Single, individual, distinct, e.g. discrete particles.

disease Anything which goes wrong with the body other than by an accident. It may be due to *germs* or to some part of the body not working properly.

disinfectant A chemical substance used to destroy *germs* or *bacteria,* e.g. sodium hypochlorite, NaOCl, which releases *chlorine.* Disinfectants are never used on the body or taken internally. See also *antiseptic.*

disintegration Breaking-up into smaller particles; term particularly used for the breaking-up of a *radioactive nucleus.*

dispersion The splitting up of *white light* into its separate colours as it passes through a *prism* or *diffraction grating.*

displacement Taking the place of, e.g. zinc will displace hydrogen when placed in a dilute acid, $Zn + 2HCl \rightarrow ZnCl_2 + H_2$ and copper will displace silver from a solution of silver nitrate, the silver being precipitated (*precipitation*),
$Cu + 2AgNO_3 \rightarrow Cu(NO_3)_2 + 2Ag$
In general, any *metal* A will displace a metal B from a solution of one of its *salts* if B is lower than A in the *electrochemical series.* Hydrogen can be included in the list and is displaced from dilute HCl or H_2SO_4 by any metal above it. Displacement reactions are *ionic.*

displacement can See *eureka can.*

dissociate To break down into simpler *molecules* or *ions* as a result of heating or putting into water, the change usually being *reversible.* See also *thermal dissociation, ionization,*

dissociation 1. A *decomposition* brought about by *heat* in which the original substance is re-formed on cooling; e.g. if ammonium chloride is heated in a tube it forms the gases ammonia and hydrogen chloride and these recombine on the cooler parts of the tube: $NH_4Cl\ (s) \rightleftharpoons NH_3(g) + HCl(g)$ 2. The *reversible* breakdown of large *molecules* into smaller ones by heat, e.g. dinitrogen tetroxide molecules change to the simpler ones of nitrogen dioxide when heated: $N_2O_4 \rightleftharpoons 2NO_2$.

dissolve Said of a *solid* which breaks down into infinitely small invisible particles when placed in a suitable *solvent* and so forms a *solution,* e.g. sugar dissolves in water, iodine dissolves in ethanol. Many gases also dissolve in water and some other liquids. The term is often used, though incorrectly, when *metals react* with *acids* and a solution (of a new substance) is formed.

distal Further away from the centre, e.g. the hand is a distal part of the *skeleton.*

distil 1. To heat a *solution* so as to *vaporize* the *solvent* and then to *condense* the *vapour.* The pure solvent can be collected as the distillate, the whole process being one of distillation. The *solute* is left behind. 2. To separate a mixture of *liquids* having different *boiling points* by heating and condensing the vapours at different temperatures. See *fractional distillation, refine* (Fig. 135).

distillate See *distil.*

distilled water Pure water prepared by *distillation* or *deionization* of tap water. It is much used in laboratories for making solutions and in general for 'topping-up' car batteries or *accumulators.*

distribution of charge When a *conductor* is given an *electric charge* this charge spreads over the surface of the conductor. If the surface changes shape the charge is greatest

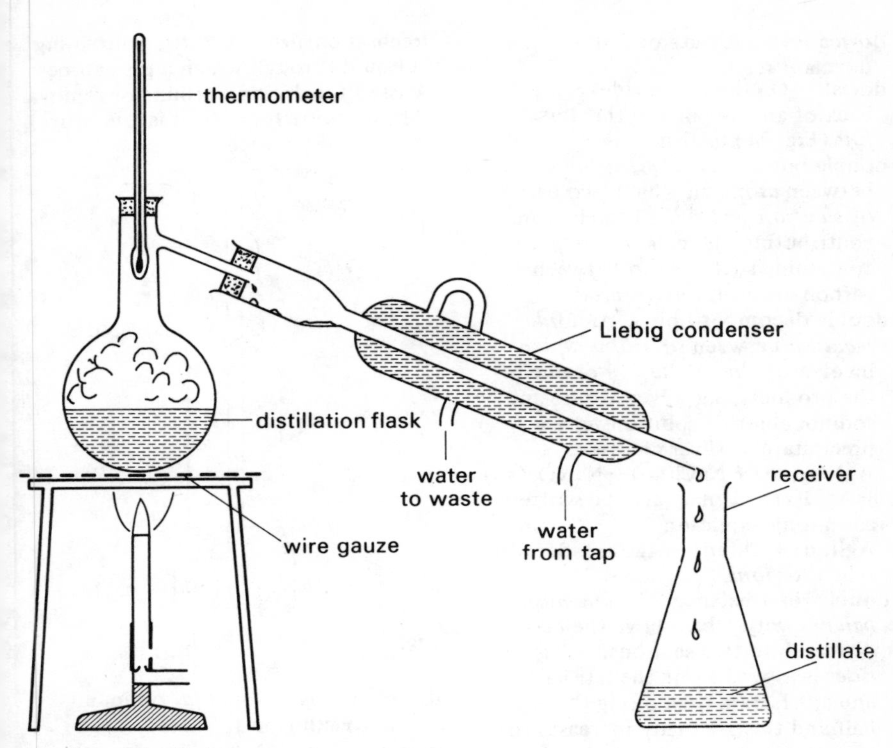

Fig. 53 Distillation

where the surface curves most, e.g. at any points or corners.

divalent Having a *valency* of two.

diverging lens A *lens* which makes parallel rays of light spread out, as though they had come from a point, after they have passed through it. Such lenses used to be called *concave* lenses as one or both surfaces are concave.

DNA The substance in the *nucleus* of a *cell* which is important in *heredity*. When cells divide or when *gametes* are formed, the DNA passes on information to the next set of cells. The letters stand for deoxyribonucleic acid.

dolomite A white or grey mineral rock composed of *calcium* and *magnesium carbonates*, which form much of the mountainous area of central Europe. It is used as a lining for *retorts* and *furnaces*.

domain theory A magnetic material like steel contains many small areas, called domains, which are naturally magnetized. If the steel has not been magnetized the domains point in different directions and cancel out one another's effects. However if a magnetic field is applied some domains grow while others turn or get smaller, and the material becomes magnetized.

dominance When one of a pair of *alleles* in a *heterozygous organism* hides the other. E.g. if the alleles for brown eyes and blue eyes are present in a child, only the brown will show.

Doppler effect The change in *frequency* of the note heard by a listener if the instrument giving out the note or the listener is moving. The same effect happens in light and is seen as a change in colour.

dormancy A state of rest, e.g. a dormant *seed.*

dorsal On the upper side, e.g. the back of an animal, and the dorsal *fins* (Fig. 71) of fish.

double bond A *covalent bond* between *atoms* in which two pairs of *electrons* are shared, each atom contributing one pair. A *compound* containing such a bond between carbon atoms is *unsaturated.*

double decomposition An *ionic reaction* between *solutions* which involves the *precipitation* of one of the products, e.g. silver nitrate and sodium chloride solutions give a precipitate of silver chloride: $AgNO_3(aq) + NaCl(aq) \rightarrow NaNO_3(aq) + AgCl(s)$ This may also be written as an ionic equation $Ag^+(aq) + Cl^-(aq) \rightarrow AgCl(s)$ See also *spectator ions.*

double-rider balance A *chemical balance* with a bar above the *beam* divided into two sections. A 1 gram *rider* is moved along the left half and a 0.1 gram rider along the right half and this gradually increases the effective *mass* of the right-hand (weights) pan.

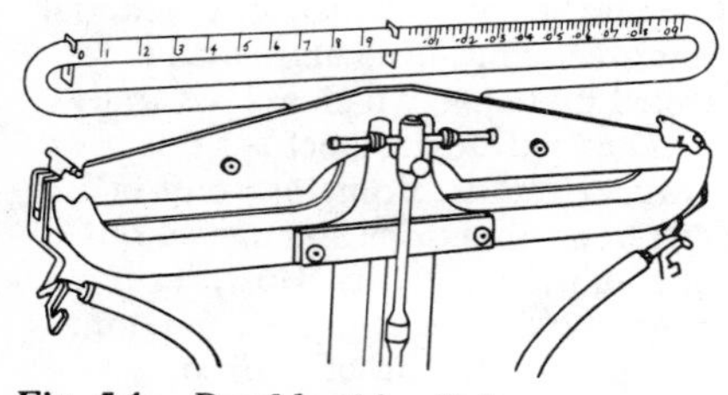

Fig. 54 Double-rider Balance (showing markings for fractional weights)

double sugar A *sugar* or *carbohydrate* with the formula $C_{12}H_{22}O_{11}$, e.g. *sucrose* (cane sugar) which can be converted by heating with a dilute acid to a *simple sugar* such as *glucose,* $C_6H_{12}O_6$.

dough A mixture of flour and water used in making bread. It is made to rise by adding *yeast* which causes the production of carbon dioxide.

Drechsel bottle A bottle containing a liquid through which a gas can be passed for drying, washing or removal of one *constituent* (if it is a mixture).

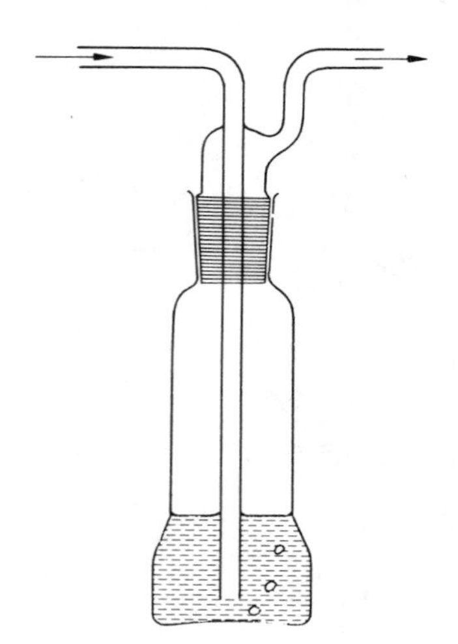

Fig. 55 Drechsel Bottle

dripping An *animal fat* obtained from roasting meat.

drone The *male* bee.

dropping bottle A small bottle fitted with a *teat pipette* which is used for dispensing small amounts of a liquid.

dropping funnel A *funnel* with a tap used for (a) adding a *liquid* drop

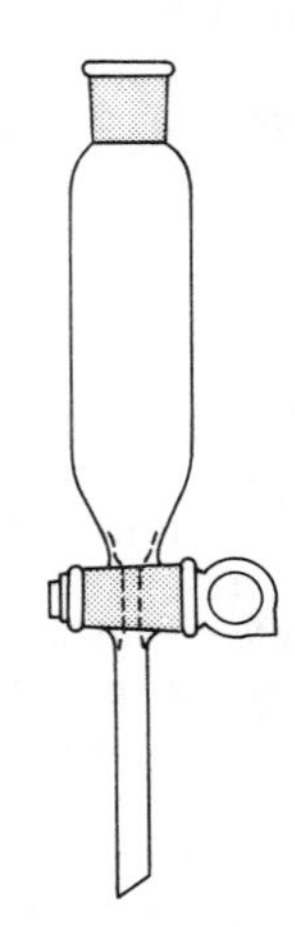

Fig. 56 Dropping Funnel

by drop when needed, (b) separating two *immiscible* liquids. Also known as a tap funnel or a separating funnel.

drug A substance which acts on the living body to relieve pain or produce some other effect. In some cases a person may be unable to give up taking the drug.

dry cell A form of *Leclanché cell* in which the ammonium chloride is present as a paste and not a solution. It also has the zinc *cathode* as its outer case. It is very widely used in torches, radios and many other instruments.

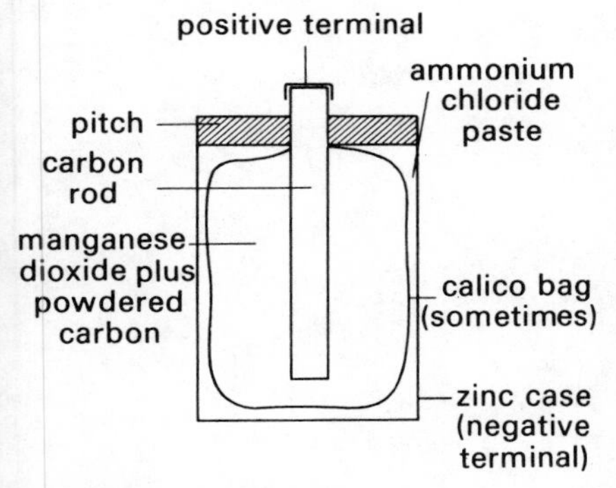

Fig. 57 The Dry Cell

dry ice Solid carbon dioxide, often used in the laboratory to produce low temperatures.

dry test A *chemical reaction* carried out on a *solid* as opposed to a *wet test* on a *solution*, e.g. heating the solid alone or with concentrated sulphuric acid.

drying agent A *deliquescent* or *hygroscopic* substance such as *anhydrous calcium chloride, silica gel* or *concentrated sulphuric acid* used for removing *moisture* from another substance or for keeping the other substance out of contact with damp air. See also *desiccant, desiccator.*

drying tower A glass vessel filled with a *drying agent* such as calcium oxide or silica gel through which a gas can be passed in order to dry it.

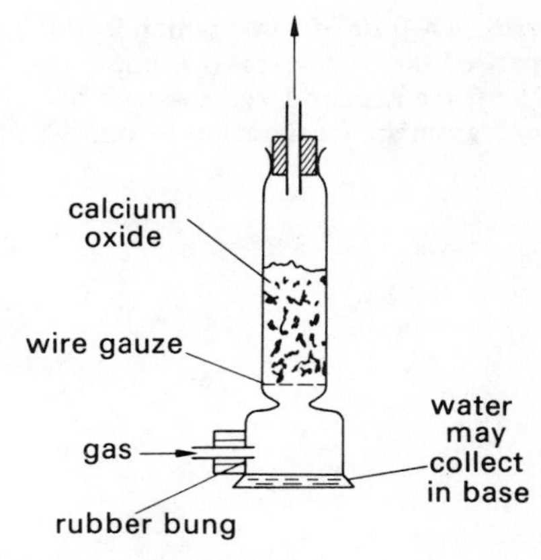

Fig. 58 Drying Tower

ductless gland Part of the body which makes an important chemical called a *hormone*, which is passed into the *blood* and travels round the body, e.g. the *thyroid gland.* The effect of the hormone may be shown in parts of the body distant from the gland.

duodenum Part of the *alimentary canal* (Fig. 5). It follows on from the *stomach.* It plays an important part in *digestion.* It *secretes* a liquid called the *succus entericus* which contains *enzymes.*

duralumin A strong yet light-weight *alloy* composed of *aluminium* with about 4% of *copper* and smaller amounts of magnesium, manganese and silicon. It is used in the construction of aeroplane parts.

Dutch metal Very thin sheets of a *copper-zinc alloy,* a variety of *brass.* Because of its large surface area compared with its mass, a sheet takes fire if dropped into a jar of chlorine. Also called *Dutch-gold.*

dye A coloured substance which, in *solution*, can be fixed firmly to the *fibres* of a material to be dyed, so that it is not affected by water, air, light or detergents.

dynamo An instrument for changing mechanical energy into electrical energy. See also *generator.*

dyne A *unit* of *force* which is not part of the *SI* system. It is equal to the force needed to give a *mass* of one gram an acceleration of one centimetre per second squared, and is therefore one hundred thousandth of a *newton* (or 1 newton = 10^5 dynes).

ear The *organ* of hearing and balance.

ear drum The tightly stretched piece of skin in the *ear* which shakes when *sound waves* hit against it. It is also called the tympanum.

ear ossicles The very small bones in the *middle ear.* They move as a result of sound waves shaking the *ear drum.*

earpiece An instrument which changes electric currents into sound *waves.* The current flows in a coil which is wound on a weak magnet and the changes in the current change the magnetic field. This makes an iron sheet (diaphragm) move in and out, so making the sound waves. See also *telephone.*

earth The metal cases of many electrical instruments are joined by the third of the connecting wires to the earth. If a fault happens so that the case touches the wire which leads in the electricity (the 'live' wire), then electricity flows safely through the earth wire and not through anyone who might touch the outside of the instrument. Such a flow of electricity should also melt the *fuse* in the circuit and cut off the current. Also called electrical earth.

earthenware Coarse baked clay: articles made from this.

earth's magnetism A free compass needle sets north-south because it lies along the *lines of force* of the earth's *magnetic field.* The origin of the earth's magnetic field is not yet completely explained. See also *declination, dip, meridian.*

ecdysis The breaking off of the outer skin of a young insect. Before a new skin forms the young insect can grow. When the skin hardens, the insect stops growing until the next ecdysis.

echo An echo is heard when a sound *wave* is reflected (*reflection*) from a hard surface like that of a building or a cliff.

eclipse The sun is eclipsed when the moon comes between it and the earth and so stops the light from the sun reaching the earth. A total eclipse happens when all the light is stopped and a partial eclipse when only part of the light is cut off. An eclipse of the moon happens when the earth comes between the sun and the moon. (Fig. 60.)

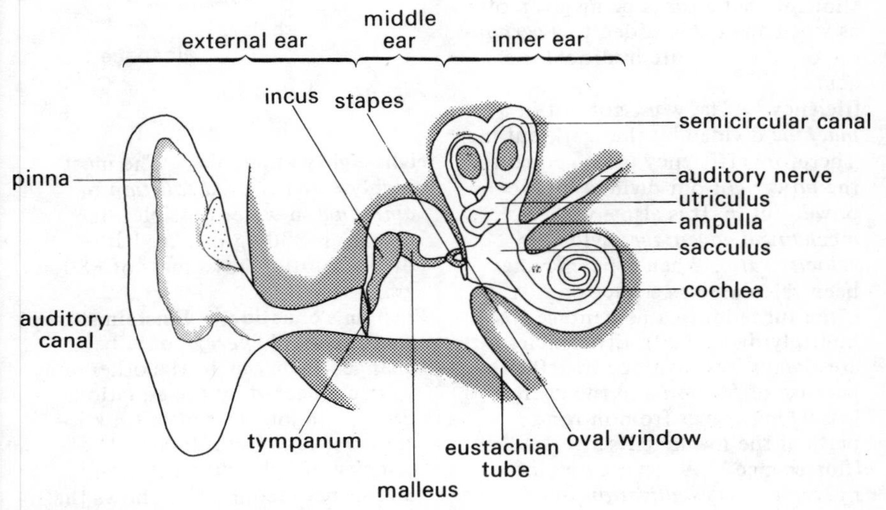

Fig. 59 Diagram of Human Ear in section

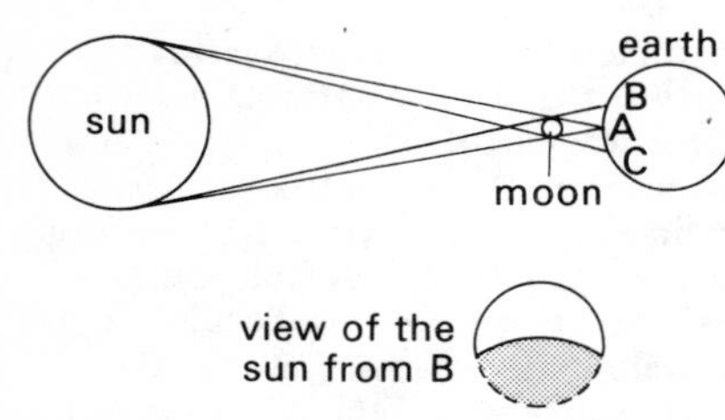

Fig. 60 The Eclipse of the Sun

ecology The science which studies plants and animals where they live and grow in nature.
ectoderm The outer skin.
ectoplasm The outer *cytoplasm* of a one-*celled* animal.
eddy currents Currents which are caused by *electromagnetic induction* and which flow in a block of metal rather than in a coil of wire. For example, eddy currents are set up in the metal round which the coils of a *transformer* are wound. They can be a cause of energy loss.
effector A *muscle* or *gland* which is brought into action after receiving a message from the *nervous system.*
efferent A *nerve* which carries messages from the *central nervous system* to a *muscle* or *gland.*
effervescence A bubbling or fizzing showing that a *gas* is being given off as when an *acid* is added to a *carbonate* or zinc to dilute hydrochloric acid.
efficiency The *work* got out of a *machine* divided by the work put in. Therefore efficiency is also equal to the *power* got out divided by the power put in. It is also equal to *mechanical advantage* divided by *velocity ratio.* When efficiency has been calculated as a fraction it is often turned into a percentage (by multiplying by 100). Efficiencies are always less than one (or 100%) because of *friction* and the energy loss which comes from moving parts of the machine itself.
efflorescence A 'drying-up'; loss of *water of crystallization* by *crystals* such as those of sodium carbonate on exposure to air. Such substances usually change to a white powder (the *anhydrous* solid) and are said to effloresce or to be efflorescent.
effort The *force* which is applied to a *machine* to make it do *work.*
effusion The escape of *gas* under pressure through an extremely fine hole.
egestion Removal of unused food.
egg The *female cell* from which the new young animal will grow. It must be *fertilized* by the male cell or *sperm.* Also used to describe the structure formed after *fertilization* in animals, where the young *embryo* grows inside a shell, as in a bird. See also *ovum.*

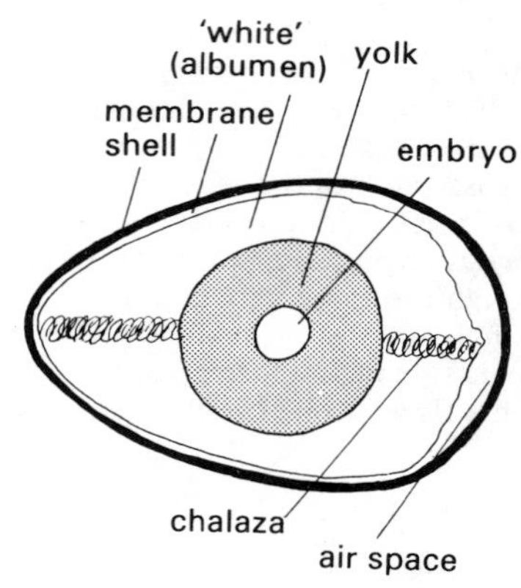

Fig. 61 A Bird's Egg

eight-eighty ammonia The most highly *concentrated solution* of *ammonia* in water possible: its density is 880 kg/m^3, i.e. 1 litre of the solution has a mass of 880 grams.
Einstein's equation Einstein showed that *mass* and *energy* could be changed from one to the other and were connected by the equation, Energy (in joules) = mass (in kilograms) x velocity of light^2. The velocity of light must be given in metres per second. This shows that a large amount of energy is released

when a very small amount of matter is used up.

elastic limit When a wire or spring is being stretched it will return to its first length when the stretching force is taken away unless it has been taken beyond the elastic limit. If it has been taken beyond this limit it will have a permanent increase in length (a permanent set) when the stretching force is taken away.

elasticity The property of an object which makes it return to its original size and shape after it has been stretched or twisted etc.

electric bell In this instrument an electric current flowing through a coil sets up a magnetic field which pulls a piece of iron towards it. This iron carries a hammer which hits the gong (the sounding part). The same movement also separates two connections and so switches off the current. The iron is fixed to a spring so that it moves back to make connection again, the current then flows once more and the whole pattern repeats itself.

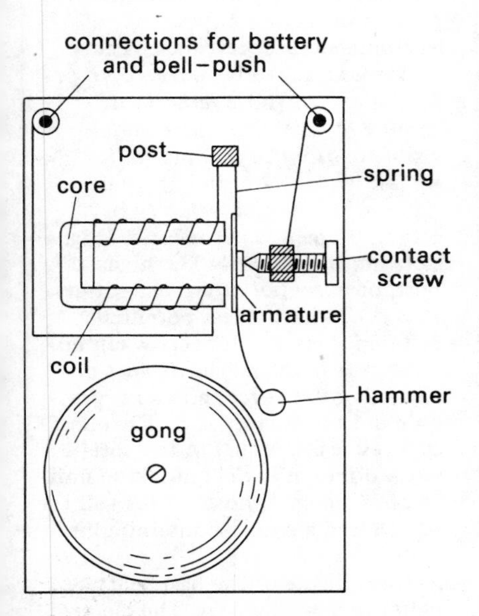

Fig. 62 The Electric Bell

electric charge Two of the basic particles of which matter is made carry what is called an electric charge. The *electron* has a negative charge while the *proton* has a positive charge. In ordinary matter the two are balanced and no effects are seen. However, if an excess of either particle is built up, for example by rubbing an *insulator* with a woollen cloth, then forces of attraction (between opposite charges) and of repulsion (between like charges) occur. If charged particles are made to move then an *electric current* is set up. The *SI unit* is the *coulomb* and the symbol is Q.

electric current A flow of electrically charged (*electric charge*) particles, e.g. *electrons* along a metal wire. An electric current can only be noticed by its effects. It sets up a *magnetic field,* causes heating of the *conductor* in which it flows, and can cause chemical action. See *electrolysis.* The *SI unit* is the *ampere* and the symbol is I.

electric field The space round an electrically charged (*electric charge*) object where the effects of that charge can be noticed.

electric motor An instrument for changing electrical energy to mechanical energy. Many electric motors contain a coil (*armature*) which turns in a *magnetic field* (see *Fleming's rule*). The direction of the current in the coil is changed every half turn by a *commutator* so that the coil continues to turn.

electric power The energy made available by an electric current in one second. It is equal to the *current* multiplied by the *potential difference* and therefore to current^2 multiplied by resistance or to (potential difference)^2 divided by resistance. As the symbol for power is P, this can be written entirely in symbols as $P = IV = I^2R = V^2/R$. The *SI unit* is the *watt.*

electrochemical equivalent The *mass* of a substance released at an *electrode* in a *voltameter,* by one *coulomb* of electricity. The *SI unit* is kilogram per coulomb.

electrochemical series A list of the *metals* arranged in order according to the ease with which they lose *electrons* and so form *ions.* When reversed, the metals are in the order in which they are deposited or liberated under normal conditions during *electrolysis.* The series very closely resembles the *activity series* and *hydrogen* can be included.

electrode The metal (or *carbon*) wire, rod etc., which leads an electric current into or out of a *cell, voltameter* or *valve* etc. See also *anode, cathode.*

electrolysis The *decomposition* of an *ionic compound,* either *molten* or in *solution* by the passage of electricity (*direct current*). The *ions* already present move towards the *electrode* which carries an opposite *electric charge* and are deposited or liberated there after giving up their electric charges.

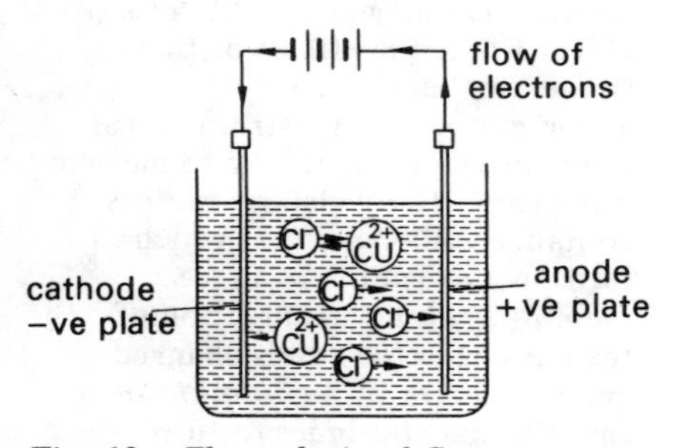

Fig. 63 Electrolysis of Copper Chloride Solution (showing movement of ions, Cu^{2+} and Cl

electrolysis, laws of See *Faraday's Laws of Electrolysis.*

electrolyte The liquid in a *voltameter* which is gradually broken up by the flow of the electric current. It may be a solution of a *salt* or a melted salt. The word was first used to describe the *compound* rather than its *solution.* Also the liquid in a *cell* which acts with the *electrodes* to set up the *electromotive force.*

electrolytic corrosion *Corrosion* occurring when two different *metals* are in contact in an *electrolyte,* e.g. iron and tin in rain water containing carbon dioxide as seen when opened food cans are left exposed to the atmosphere.

electrolyze To cause the *decomposition* of an *electrolyte* by the passage of an *electric current.*

electromagnet A coil of wire wound on to an iron frame. While an *electric current* flows through the coil, the iron becomes strongly magnetised and can be used for lifting other iron objects etc.

electromagnetic induction If a magnet is moved near a coil of wire (causing a changing *magnetic field*) a *current* is set up in the wire. This is called an *induced current* and the whole process is called electromagnetic induction. The changing magnetic field can also come from a changing current in a second coil near to the first one. See also *Faraday's Law of Electromagnetic Induction, Lenz's Law.*

electromagnetic spectrum The pattern, arranged by *wavelength* or *frequency*, of those *radiations*, *gamma rays*, *X-rays*, *ultra violet*, visible *light*, *infra red* and *radio waves*, which have similar properties, e.g. a common *velocity* of 3.0×10^8 metres per second. (Fig. 64.)

electromotive force The highest push or *potential difference* set up by a *cell.* This highest potential difference always appears when no current is flowing. The *SI unit* of electromotive force (e.m.f.) is the *volt* and the symbol is *E.* The e.m.f. of a cell is also equal to the total *work* done (in *joules*) in taking unit *electric charge* (one *coulomb*) all the way round a *circuit* containing the cell.

electron One of the basic particles contained in all *atoms.* The electron carries a negative *electric charge* and

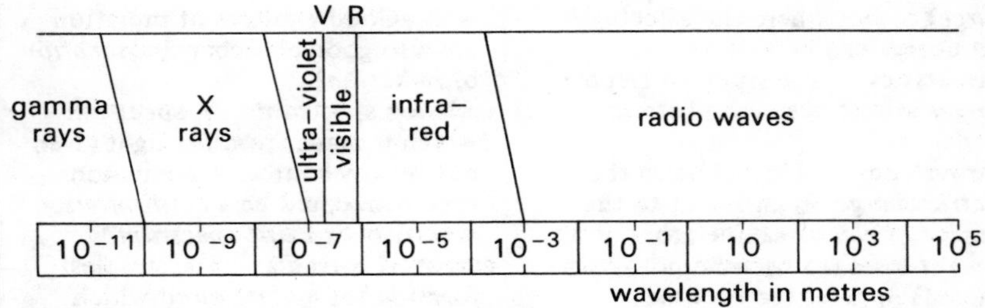

Fig. 64 Electromagnetic Spectrum Wavelengths

has a *mass* of about 1/1840 of that of a *proton*. Electrons move in orbits around the atomic *nucleus*, and in a neutral atom there are as many electrons as protons, equal to the *atomic number*. In a metal wire the flow of electrons forms an *electric current*. Electrons are shared by or transferred between atoms when *compounds* are formed from their elements. See also *beta particles*, *cathode rays*.

electron gun The first part of a *cathode ray tube* which produces a beam of *electrons*. It contains a heated *cathode* which gives out the electrons, *anodes* to accelerate (*acceleration*) them, a *grid* to control their number and a focusing cylinder to give a narrow beam.

electronic configuration The number of *electrons* in the different *shells* of an *atom* starting from the one nearest to the *nucleus*, e.g. *carbon* is 2.4; *sodium* is 2.8.1; *bromine* is 2.8.18.7.

electronic charge The *electric charge* on an *electron*; it is negative and equal to 1.6×10^{-19} *coulomb*. It is important because it is the basic unit of electric charge, no one has measured a charge smaller than this, nor one that is not equal to a whole number times this basic unit. See also *Millikan's experiment*.

electrophorus An *electrostatic* instrument for producing a series of equal *electric charges*. It consists of an ebonite (hard rubber) disc which is charged by rubbing with fur. A metal plate on an insulating (*insulator*) handle is placed on the ebonite, touched for a moment, and then taken away. The charge on the metal plate is then given to another *conductor* and the process repeated. See also *charging by induction*.

electroplating The deposition of a layer of a *metal* (e.g. copper) on an object forming the *cathode* of a *cell* during *electrolysis*. The metal to be deposited is made the *anode* of the cell in a *solution* of one of its *salts*. Electroplating may be used for improving the appearance of an object or for preventing *corrosion* (*rusting*), e.g. *chromium plating*, *tin plate*.

electroscope An instrument, of which there are several different types, which uses the movement of an electrically charged metal leaf, etc., to show the *potential difference* between two fixed, metal *conductors*. This potential difference is often caused by an *electric charge* being near to the electroscope and so the electroscope can be used to test for an electric charge. See also *gold leaf electroscope*, *Wulf electroscope*.

electrostatic attraction and repulsion Two positive *electric charges* or two negative charges will try to move away from (repel) each other but a negative and a positive charge will move towards (attract) each other. Thus 'like charges repel, unlike charges attract'.

electrostatic field The space round an electrically charged (*electric* *charge*) object where the effects of that charge can be noticed.

electrostatics The study of *electric charges* at rest, also called *static electricity*.

electrovalency This refers to the *electric charge* on an *ion* or to the number of *electrons* one atom of an *element* loses (to become positively-charged) or gains (to become negatively-charged) on forming an ion, e.g. Na^+, electrovalency = + 1; Cl^-, electrovalency = − 1; $SO_4{}^{2-}$, electrovalency = − 2.

electrovalent *Ionic*. Term used for a *compound* existing in the form of *ions* and not *molecules*, e.g. sodium chloride, Na^+Cl^-. The formulation of such a compound from its *elements* involves the transference of *electrons*

electrovalent bond An *ionic bond* in which *electrons* are transferred from one *atom* to another, e.g. in the formation of sodium chloride the sodium atom loses an electron to become a sodium *ion*, Na^+, and this electron is transferred to the chlorine atom $Cl + e^- \rightarrow Cl^-$; sodium chloride, Na^+Cl^- results.

element A substance which cannot be split into anything simpler in a chemical process. About 100 such substances are known. All the *atoms* of one element have the same atomic number. See also the table at the end of the book and *isotope*.

elytron The hard wing case of a beetle.

embryo The young plant or animal. The *seed* or *egg* before it starts to live on its own.

embryo sac The food container for the *embryo*.

embryology The study of the growth of the *embryo*.

embryonic membranes The skins around an *embryo*, used for protection or food supply. The *amnion* and chorion.

emergent ray See *refraction*.

emission of radiation Dull black surfaces give out (emit) *radiation* well, polished surfaces do not emit well. Good emitters of radiation are also good absorbers (*absorption of radiation*).

emission spectrum A *spectrum* which is made from the light given out by any source. An emission spectrum could be a *continuous spectrum* or a *line spectrum*.

empirical formula The simplest *formula* for a *compound* which shows only the proportions in which the *atoms* are joined together and not, necessarily, the actual number of each in the *molecule*, e.g. CH for ethyne, C_2H_2, or for benzene, C_6H_6; CH_2 for ethene, C_2H_4, and all the other alkenes (olefins).

emulsion A mixture of two liquids which, although *immiscible*, will not separate out after shaking: a *colloidal suspension* of one liquid in another.

endocrine organs Another name for the *ductless glands*.

endolymph The liquid inside the hearing and balance parts of the *inner ear*. It moves as a result of the sound waves hitting against the *ear drum*.

endoparasite A plant or animal which lives inside another plant or animal, e.g. tapeworm.

endoplasm The *cytoplasm* which is on the inside of a one-celled animal.

endoskeleton The bones of a *vertebrate* which are inside the *skin* and *muscles*.

endosperm An extra food supply found in some seeds, where the *cotyledon* is small, e.g. maize. The food is usually *starch*. Flour comes from *endosperms* of wheat and corn.

endothermic A *chemical reaction* which absorbs (*absorption*) *heat* from its surroundings while it is taking place, e.g. dissolving ammonium nitrate in water or precipitating (*precipitation*) calcium carbonate from solutions containing the necessary ions. The *energy* term is positive. $Ca^{2+}(aq) + CO_3^{2-}(aq) \rightarrow CaCO_3(s)$. $\Delta H = +10.5$ kJ per mole.

energy Anything that is able to do some *work* possesses energy. Several types of energy can be noted, e.g. *potential*, *kinetic*, wave (which includes light and sound etc.) heat, electrical, magnetic and nuclear. These types can be changed into one another without loss, and *mass* may also be changed into energy or energy into mass. The *SI unit* is the *joule* and the symbol is *E*. See also *conservation of mass and energy, Einstein's equation.*

energy levels The *electrons* which orbit the *nucleus* of an *atom* do so at different distances from it. The number of such distances (or *shells*) occupied in any particular atom is fixed, depending only on the number of electrons and the conditions. When an electron moves from one shell to another because of a change in conditions *energy* is given out or taken in. The different shells are called energy levels.

energy value A measure of the *heat energy* available by burning completely unit mass (1 g or 1 kg) of a food. It is expressed either in kilojoules or in kilocalories, the former being preferred.

enterokinase An *enzyme* used in *protein digestion* in the small *intestine.*

enteron The space inside the *alimentary canal.* A food cavity.

enthalpy The total *heat energy* contained in a substance, caused by vibration, rotation or some movement of its *atoms*, *ions*, *electrons* or *molecules;* this is difficult to measure and in *exothermic* or *endothermic* experiments it is the change in enthalpy which is determined.

environment The surroundings of plants and animals, e.g. the environment of a cactus is a desert.

enzyme A chemical substance produced by living *cells* which is able to change one substance into another without being changed itself. Enzymes are important in *digestion*, they are *proteins* and each enzyme affects only one type of chemical reaction, e.g. *amylase* changes *starch* to *sugar* and, in *respiration, oxidase oxidises* sugar. Enzymes are destroyed by heat. An enzyme is a biological *catalyst*. See also *yeast*.

epidermis The outer layer of a *leaf* or *skin.*

epididymis The tube which carries the *sperms* from the *testis* to the *vas deferens*. See *reproductive organs* (Fig. 141ii).

epigeal A kind of *germination* of seeds where the *cotyledons* come up above the ground, e.g. melon.

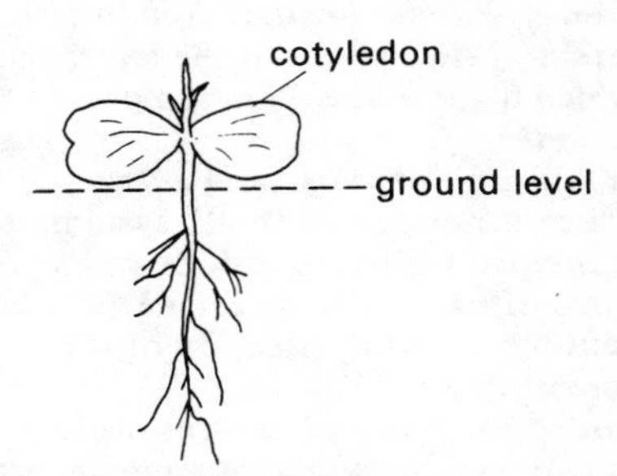

Fig. 65 *Epigeal germination*

epiglottis A *cartilage membrane* at the top of the *windpipe* which prevents food going down into the *lungs*.

epigynous Name given to a *flower* (Fig. 76ii) where the *ovary* is below the *petals*, e.g. orchid.

epiphyte A plant whose roots are above the soil, e.g. ferns growing on tree branches.

epithelium *Cells* on a surface, e.g. cells on the inside surface of the mouth.

Epsom salts *Hydrated* magnesium sulphate, $MgSO_4.7H_2O$. It is a colourless, crystalline solid, soluble in water and used as a purgative.

equation See *chemical equation.*

equations of motion The equations which describe the movement of a body. The simplest equation applies to an object moving with *constant speed*: distance travelled = speed × time. Another simple group of equations apply to a body moving with constant *acceleration* along a straight line. They connect first speed (u), last speed (v), acceleration (a), time (t) and distance travelled (s). They are: (1) $v = u + at$ (2) $s = \frac{1}{2}(u + v)t$ (3) $s = ut + \frac{1}{2}at^2$ (4) $v^2 = u^2 + 2as$ (5) $s = vt - \frac{1}{2}at^2$.

equilibriant A *force* which must be added to other forces to bring them into *equilibrium*.

equilibrium 1. A state of balance which happens although several *forces* are acting on a body. See also *stable*, *unstable* and *neutral equilibrium*. 2. See *chemical equilibrium*.

erepsin An enzyme in the *intestine* which helps in the *digestion* of *proteins*.

erg A *unit* of *work* (and *energy*) which is not part of the *SI* system. It is equal to the work done when a *force* of one *dyne* moves one centimetre in the direction of the force.

error of parallax The error made in reading the *volume* of a *liquid* in a *burette* or *graduated cylinder* by not having the eye in a direct line with the bottom of the *meniscus*. The diagram shows how different readings can be obtained as the eye is moved up or down.

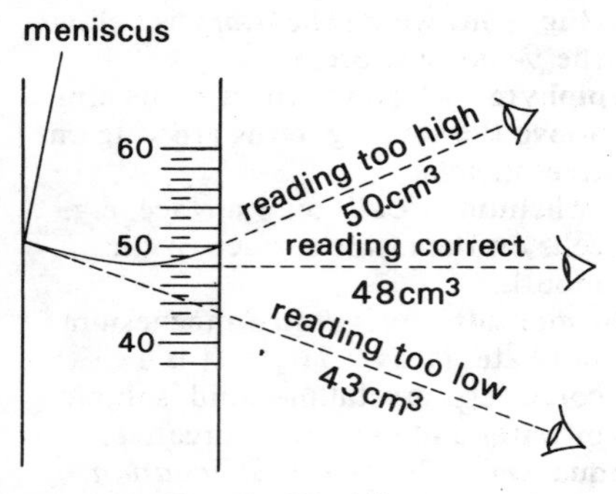

Fig. 66 *Error of Parallax*

erythrocyte A *red blood cell*.

ester A *compound* made by replacing the *hydrogen* of an *acid* with an *organic radical*; this is usually done by heating an organic acid with an *alcohol* in the presence of a little concentrated sulphuric acid, e.g. ethanol + ethanoic (acetic) acid→ ethyl ethanoate (acetate) + water. Ethyl ethanoate is the ester. Esters are usually sweet-smelling liquids present in perfumes, but may be solids such as those present in *vegetable* and *animal fats*.

ethane C_2H_6 The second member of the *alkane* (paraffin) series of *hydrocarbons*. It is a colourless, odourless gas which burns readily.

ethanoic acid CH_3COOH A *pungent*, colourless, *corrosive* liquid obtained in solution by *oxidation* of *ethanol* or from *ethyne* (acetylene) or ethanal (acetaldehyde). It solidifies at 16.7°C to '*glacial ethanoic acid*'. It is present in *vinegar* to the extent of 3–6% and is an important raw material in chemical industry. It used to be called acetic acid.

ethanol C_2H_5OH Also called 'spirits of wine', ethyl alcohol or just alcohol. The liquid obtained by the *fermentation* of *sugar* using *yeast*. In its purest form it is called absolute alcohol. With about 4% water present it is known as rectified spirit. Up to 10% of *methanol*, CH_3OH, may be added to produce industrial *methylated spirit* (I.M.S.) It is used in industry as a *solvent*. Often a *dye* is added to it before it is sold to the public as methylated spirit in order to make it unfit for drinking. It is a colourless, *flammable* liquid with a vinous smell and forms part of alcoholic drinks.

ethene C_2H_4 The first member of the *alkene* (olefin) series of *hydrocarbons*. It is a colourless sweet-smelling gas which burns with a yellow flame. It is used in the manufacture of *polythene*, *ethanol* and *glycol* (dihydroxyethane) and for controlling the growth of weeds. It used to be called ethylene.

ether $(C_2H_5)O$ A *volatile* organic liquid made by the *dehydration* of *ethanol* with concentrated sulphuric acid. It is a colourless, *flammable* liquid used as a *solvent* and as an *anaesthetic*. It is also called diethyl ether or ethoxyethane.

ethylene See *ethene*.

ethyne C_2H_2 A colourless, strong-smelling, poisonous gas formed by the action of water on *calcium carbide*, CaC_2. It is the first member of the *homologous series* of *hydrocarbons* called the alkynes. It is very *flammable* and is used for lighting, in welding and cutting metals (*oxy-acetylene torch*) and for the *synthesis* of *ethanoic acid* (acetic acid). It used to be called acetylene.

etiolation The state of a plant grown in the dark. It has long, thin stems and small yellow leaves.

eudiometer A long *graduated* glass tube used for (a) collecting a *gas* over water and measuring its *volume*, (b) measuring the changes occurring in the volumes of enclosed gases when they react together chemically.

eureka can A metal can with an overflow tube in the side. It is filled with a liquid until it just overflows. An object is then placed in the liquid and the *volume* of liquid which overflows is equal to the volume of the object. It is also called a displacement can.

eustachian tube The tube leading from the middle *ear* (Fig. 59) space to the throat. It relieves *pressure* inside the ear.

evaporate To change a *liquid* into a *vapour* by heating it or exposing it to the air.

evaporating basin A shallow dish in which solutions can be heated to remove some or all of the water. See *water-bath* (Fig. 171).

evaporation 1. A liquid will gradually turn to a *vapour* even at temperatures well below its *boiling point*. The liquid is said to evaporate and the process is called evaporation. Evaporation occurs because the faster moving *molecules* in the liquid have enough energy to break through the surface. If the space above the liquid is closed the vapour formed becomes *saturated* and although molecules still leave the liquid the same number return to it from the vapour to keep a balance. If there is an open space above the liquid *evaporation*. continues until all the liquid has gone. See also *cooling by evaporatiibn.* 2. Removal of some or all of a *solvent* from a *solution* by heating so as to *concentrate* the solution or to obtain a *residue* of the *solute*.

evergreen A plant or tree which does not lose its leaves in the cold season, e.g. rhododendron.

evolution 1. (Species) The idea that all living things developed from simpler *organisms* over a long period of time by changing gradually and becoming better suited to the surroundings. See *Darwin, natural selection*. 2. (Gaseous) Liberation of a *gas* by (a) *heating* a *compound* or (b) adding one compound to another, one usually being in solution, e.g. addition of an *acid* to a *carbonate* causes the evolution of carbon dioxide.

evolve A *gas* given off during a *chemical reaction* is said to be evolved.

excretion The removal of waste substances made in the body as a result of *metabolism*. (Fig. 67.)

exoskeleton The hard outer case around an animal, e.g. crab, turtle.

exothermic 1. Giving out *heat*. Applied to a *chemical reaction* in which *energy* in the form of heat is given out by the materials to the surroundings, e.g. burning coal or wood, adding concentrated sulphuric acid to water. 2. A compound formed by a reaction in which heat is given out.

expansion coefficient of gases See *Charles' Law*.

expansion of water Water shows two changes in *volume* as it is heated which do not happen in most liquids. (a) Ice, on melting, unlike

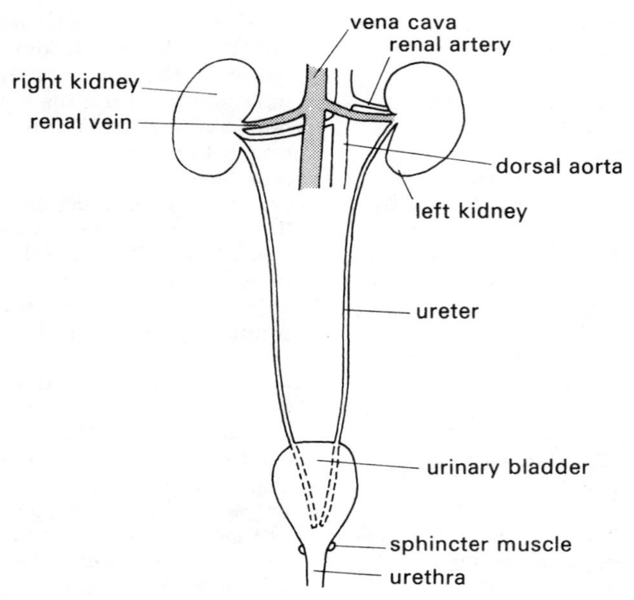

Fig. 67   Human Urinary System

most solids, makes a smaller volume of water than itself. This means that ice is less dense (*density*) than water (b) Water gets smaller in volume (contracts) as it is heated from 0°C to 4°C, this means that water has its highest density at 4°C.

expansivity   The increase in length (etc.) of unit length (etc.) of a material when it is heated through a change in temperature of one *kelvin*. In the simplest case of change in length (linear expansivity) change in length = length x linear expansivity x change in temperature. The *SI unit* is K^{-1} or 1/K and the symbol is α.

experiment   A practical exercise designed to illustrate some method or chemical reaction or scientific principle.

expiration   Breathing out.

explosion   A violent and rapid *chemical reaction* which is accompanied by sound. A large amount of *gas* at a high temperature is produced and this rapid increase in *pressure* in a restricted space may cause much damage.

explosive   1. A substance which, on heating or striking, produces a large quantity of gas in an extremely short time 2. Liable to explode, cause an explosion, i.e. to break up with much noise.

extensor muscle   A *muscle* which stretches a *joint*, e.g. the *triceps* which straightens the arm.

external gills   Breathing structures on the outside of the body as in young *tadpoles*.

extinguish   To put out a light or fire.

extracellular   On the outside of *cells*, usually liquid, e.g. the *plasma* of blood.

extract   To remove one substance from another usually by shaking the mixture with a *solvent* which *dissolves* only the required substance. *Filtration*, followed by *evaporation*, will leave the pure substance.

extraction   Removal of one substance from another by means of a suitable *solvent*, e.g. grease from fabrics with *petrol* or *kerosine*, or a *dye* dissolved in water may be removed by shaking with ether in which it is more soluble (*solution*). Many organic substances when prepared are impure and may be purified by shaking with ether.

eye The organ of sight.

eyepiece The *lens* (or group of lenses) in a *microscope* or *telescope*, etc., which is placed near to the eye.

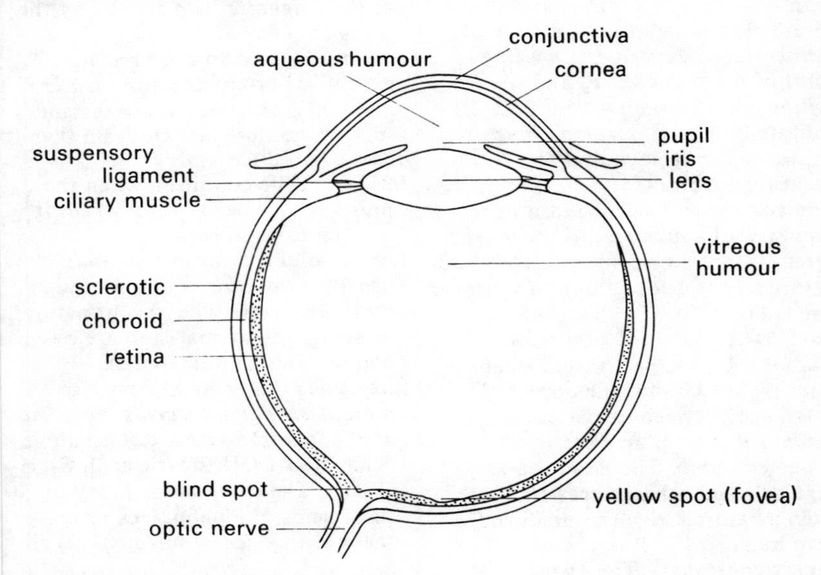

Fig. 68 The Human Eye in section

face Surface—name given to one of the flat bounding surfaces of a *crystal.*

facet A surface on a *bone* such as a *vertebra* (Fig. 166) where another bone *articulates.*

faeces The waste food material removed from the body through the *anus.*

Fahrenheit temperature A scale of *temperature* based on the *melting point* of pure ice (32 °F) and the *boiling point* of pure water (212 °F) both under one *atmosphere pressure.* This scale is not now used for scientific purposes.

fallopian tube The tube leading from the *ovary* of a *mammal* to the *uterus* (womb). In it an egg may be fertilized (*fertilization*). See *reproductive organs* (Fig. 141i).

farad The *SI unit* of *capacitance.* The symbol is F. A capacitor has a capacitance of one farad if the *potential difference* between its plates rises by one *volt* if it gains an *electric charge* of one *coulomb.* The farad is a very large unit and capacitances are more often measured in *micro*farads or *pico*farads.

Faraday constant The *quantity of electricity* which is needed to set free one mole (1 gram) of hydrogen during the *electrolysis* of a suitable *electrolyte.* One Faraday constant will also set free one mole of any *monovalent element* or *ion* under suitable conditions of electrolysis. The symbol is *F* and the *SI* value is 96 487 coulombs per mole.

Faraday's Laws of Electrolysis 1. The *mass* of substance set free at an *electrode* is *proportional* to the *current* flowing and to the time for which it flows. 2. The mass of an element set free by a given *quantity of electricity* (equals current multiplied by time) is proportional to the *relative atomic mass* of the element concerned divided by its *valency.* This second law may also be stated: The number of *Faraday constants* of electricity required to set free one atomic mass of an element during electrolysis is equal to the valency of the element.

Faraday's Law of Electromagnetic Induction The *electromotive force* or *current* produced by *electromagnetic induction* is *proportional* to the rate at which the *lines of force* of the *magnetic field* are cut by the turns of the coil.

fast breeder A *nuclear reactor* capable of producing more *fissile* material from non-fissile substances such as uranium and wolfram than it uses to produce *energy.*

fatigue The condition when the muscles have been overworked. It may give rise to *cramp.*

fats Solid *organic compounds,* usually a *mixture* of *esters* of high *molecular mass.* They occur naturally in plants and animals and are essential components of human diet.

fatty acids A general name for all *monobasic aliphatic acids,* both *saturated* and *unsaturated,* e.g. ethanoic acid, CH_3COOH, stearic acid, $C_{17}H_{35}COOH$. The *compounds* many of these acids of high molecular mass make with *glycerol* (an *alcohol*) are *fats.*

feather A structure which grows from the *skin* of a bird. Some feathers are used for flying, some for keeping the bird warm and some for special colouring to attract members of the opposite sex. (Figs. 14, 175.)

Fehling's solution A solution used for testing for *reductants.* It contains copper sulphate and *Rochelle salt* in *alkali* and forms red or yellow copper (I) oxide, Cu_2O, when warmed with simple sugars such as *glucose.*

femto- A prefix which means one thousand million millionth or 10^{-15}. It can be joined in front of any *SI unit,* e.g. 1 femtogram = 10^{-15} gram. The symbol is f.

femur Part of the *skeleton* (Fig. 148). The bone in the upper leg above the knee.

fenestra A thin place, as in *fenestra ovalis* in the *ear,* separating the middle ear from the inner ear.

fermentation A process in which *organic substances* are slowly broken down into simpler ones by means of *micro-organisms* or *enzymes.* The term usually refers to the action of *yeast* on *sugar* to produce *ethanol* and carbon dioxide, but milk turning sour is another example.

ferric Fe^{3+} The name formerly used for iron *compounds* containing the *trivalent* Fe^{3+} *ion,* e.g. $FeCl_3$. These are now called *iron(III)* compounds and are usually yellow or reddish-brown in colour, especially when in *solution.*

ferrous Fe^{2+} The name formerly used for iron *compounds* containing the *divalent* Fe^{2+} *ion,* e.g. $FeSO_4$. These are now called *iron(II)* compounds and are usually pale green in colour.

fertilization The joining together of a male cell (*sperm*) and a female cell (*ovum*) to make a *zygote,* which will grow into a new plant or animal.

fertilizer A chemical applied to the ground to replace substances removed from it by the growing of crops. Fertilizers supply elements essential to plant life such as *nitrogen* (in nitrates and ammonium salts), *phosphorus* (in phosphates) and *potassium* (in its salts).

fibre A thread or a number of threads spun together.

fibrin The solid part of *blood* which forms a cover over cuts. It is formed from *fibrinogen.*

fibrinogen The substance in the *blood* which makes it clot (to harden) when damage is done, e.g. a cut.

fibula Part of the *skeleton* (Fig. 148). The thin bone in the lower part of the leg.

field The space round an *electric charge,* a magnet or a *mass* where the effects can be noticed.

field lines Another name for the *lines of force* used to describe a *magnetic* or *electric field.*

field windings Fixed coils in an electric motor etc., which produce the *magnetic field* when an electric current flows through them.

filament 1. The fine wire (often made of tungsten) in an electric light, *valve, cathode ray tube,* etc., which becomes red-hot when an *electric current* is passed through it. 2. A very fine single fibre. 3. The stem of a *stamen.*

filter 1. Any material (cloth, sand, *porous* paper, gravel) capable of separating particles from a liquid in which they are suspended. See also *filter paper.* 2. To remove the *liquid* from a *suspension* by passing through a suitable material. To remove small solid particles from a gas.

filter flask A thick conical flask with a side-arm for attachment to a *filter pump.* A *funnel* with a *filter paper* is placed through a cork in the mouth of the flask and suction applied by running water through the pump. *Filtration* is speeded up. Usually a *Buchner funnel* is used as the paper is then sufficiently supported to withstand the strong suction.

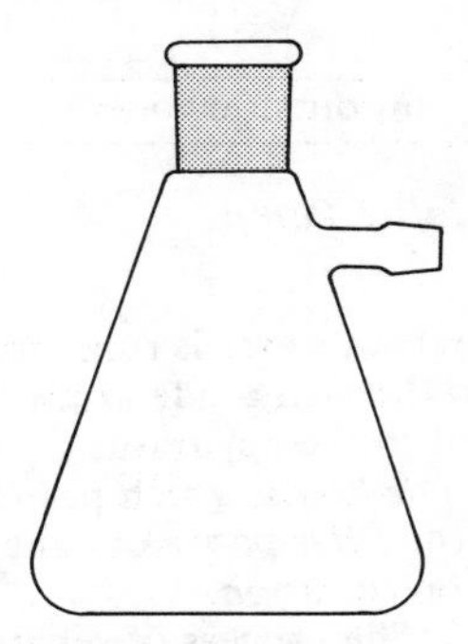

Fig. 69 Filter Flask

filter funnel A funnel used to support a *filter-paper* during *filtration.*

filter paper A *porous* paper made of *cellulose* (usually white, but sometimes black) which keeps back *solids* when a *suspension* is poured through it.

filter pump A device for drawing air through apparatus or for producing a *vacuum* in a closed flask or tube. It consists of a metal or glass pipe.

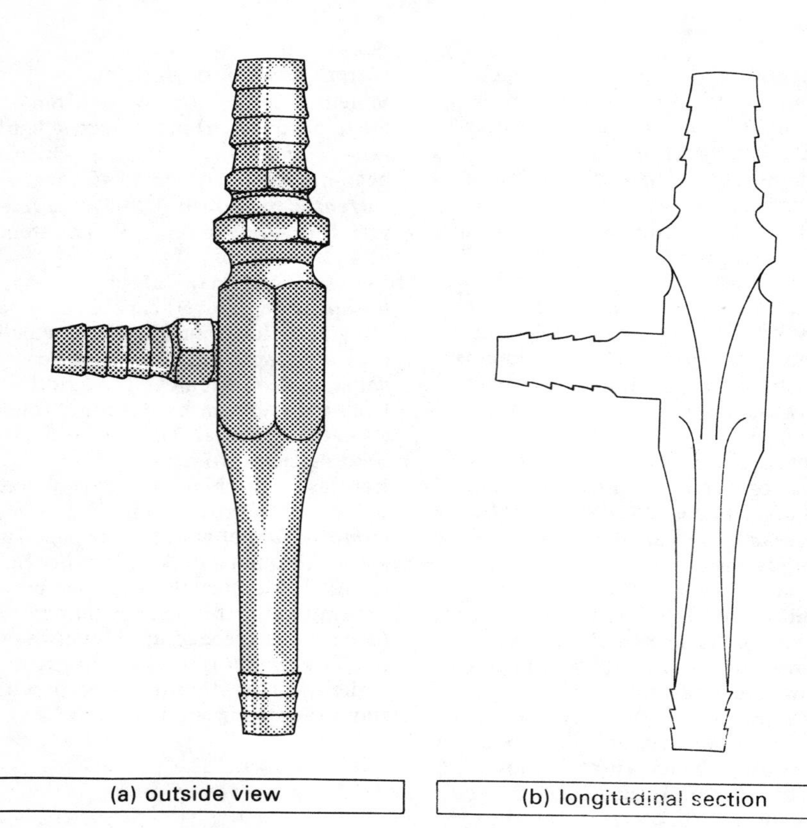

Fig. 70 Filter Pump

through which water is run, air being sucked in through a side-arm which is connected to the apparatus.

filtrate The *liquid* which passes through the *filter paper* or other material on *filtration.*

filtration The process of separating a *liquid* from an undissolved solid in *suspension* by pouring it through a *filter.* The term is also used if a gas containing small particles of solid is passed through a filter such as cotton wool.

fins The broad limbs of fish which help in movement and balance.

fibreclay A material obtained by heating *clay* (mainly *silica* and aluminium oxide) at a high temperature. It is used as a *furnace* lining and for making coarse *crucibles.*

fire extinguisher A steel container holding chemicals capable of putting out fires, either alone or on mixing together. There are several types: (*a*) *acid-soda:* in this *sulphuric acid* and *sodium hydrogencarbonate* react to produce *carbon* dioxide which forces out a jet of water; (*b*) foam: here the acid is replaced by a solution of aluminium sulphate which, on

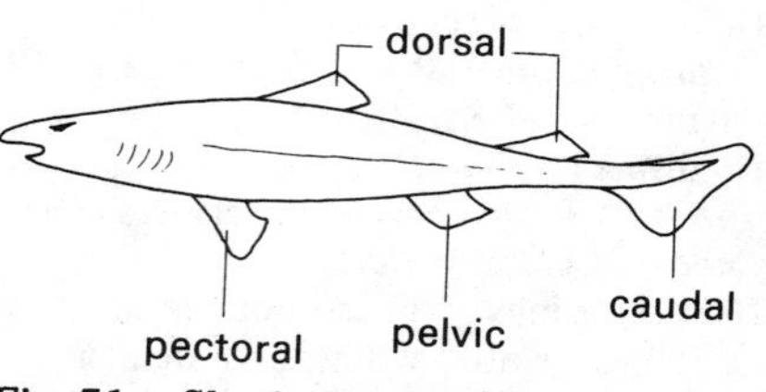

Fig. 71 Shark showing Fins

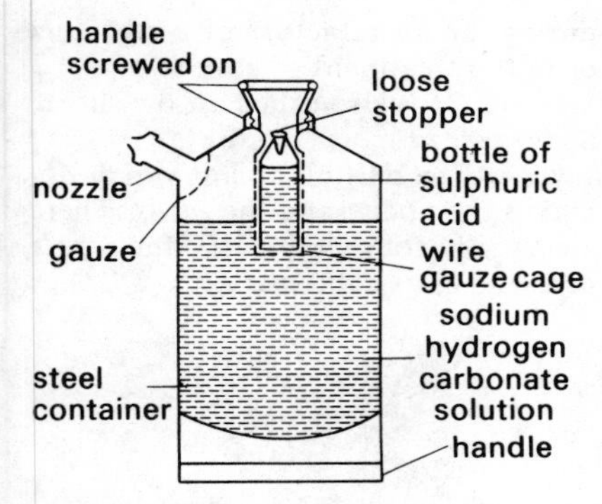

Fig. 72 *Acid-soda Type of Fire Extinguisher*

mixing with the sodium salt produces a foam of aluminium hydroxide, water and carbon dioxide and this cuts off air from the fire; (*c*) carbon dioxide: these contain this gas under pressure; (*d*) *tetrachloromethane,* CCl_4: this liquid is forced out when a knob is pressed; (*e*) dry powder: usually sodium hydrogencarbonate which *decomposes* in contact with the fire to produce carbon dioxide.

fission See *nuclear fission, binary fission.*

fissile Capable of splitting or being split, e.g. uranium.

fixed nitrogen *Nitrogen* available for use by plants in the form of nitrogen *compounds,* e.g. ammonium sulphate, sodium nitrate, calcium *cyanamide,* etc., as distinct from elemental or free nitrogen in the atmosphere.

fixed points *Temperatures* whose values are agreed and which can be used as a basis for a scale of temperature or for calibrating *thermometers.* For example, the melting point of ice and the boiling point of water under definite conditions.

flagellum A very thin living thread growing out of a living animal or cell, e.g. the tail of a sperm.

flame A region or zone between two *gases* or *vapours* where a *chemical reaction* is taking place with the production of heat and light.

flame test A method of identifying the *cation* in certain *salts* by the colour given to a *flame* when the substance is placed on a *platinum* wire, moistened with concentrated hydrochloric acid, and held in the flame of a Bunsen burner.

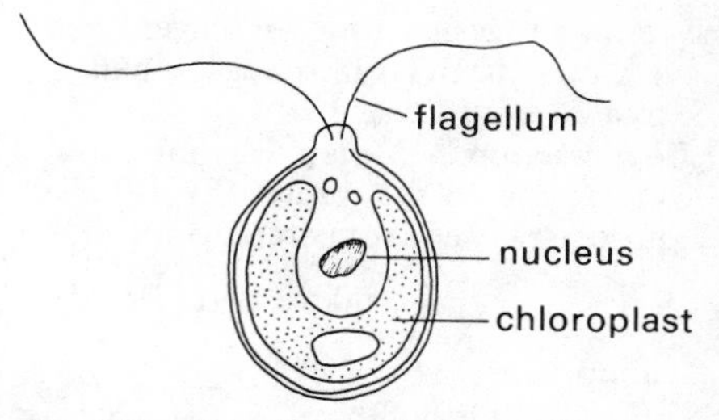

Fig. 73 *Flagellum of Chlamydomonas*

flammable (preferred to inflammable). Capable of burning readily.

Fleming's Rule This rule, also known as the left hand rule, connects the direction of *current, magnetic field* and the *force* which is set up by the field on the current, as in an electric motor. Stretch out the thumb, first and second fingers of the left hand so that they are at right angles to each other. Point the first finger in the direction of the magnetic field (north to south), the second finger in the direction of the current and then the thumb points in the direction of the force or movement.

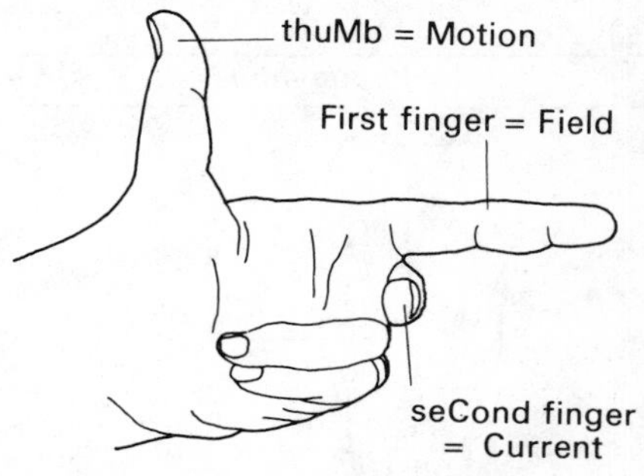

Fig. 74 *Fleming's Left Hand Rule*

flexor muscle A muscle which bends a *joint*, e.g. the *biceps* tightens to bend the elbow.

flocculation Very small particles

joining together to form bigger ones, e.g. *clay* particles in the *soil* when treated with *lime*.

flow diagram A plan drawing showing, in the correct order, the different processes used from start to finish during the manufacture of a substance or in the treatment of a piece of material. It is also called a flow sheet. See *Haber process* (Fig. 88).

dilute HNO_3 in
ammonia
nitrogen dioxide
excess NO_2 out
catalyst vessel
nitric oxide
conc. HNO_3 out
air or oxygen

Fig. 75 Flow Diagram for Manufacture of Nitric Acid

flower The part of a plant which makes the *pollen* and the *seeds*. There are usually four main parts, the *sepals, petals, stamens* and *pistil*.

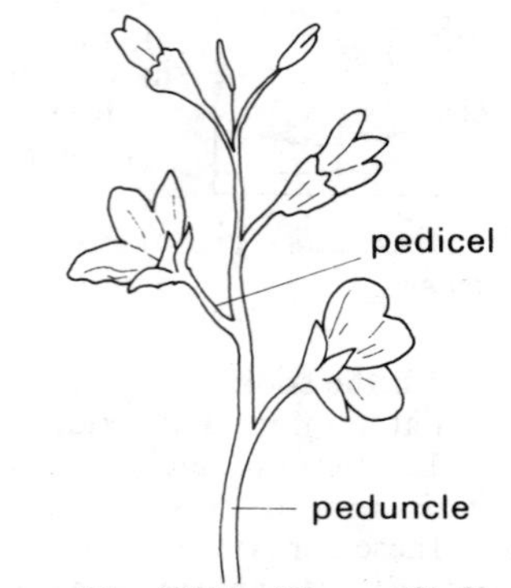

Fig. 76(i) Flower Stalks

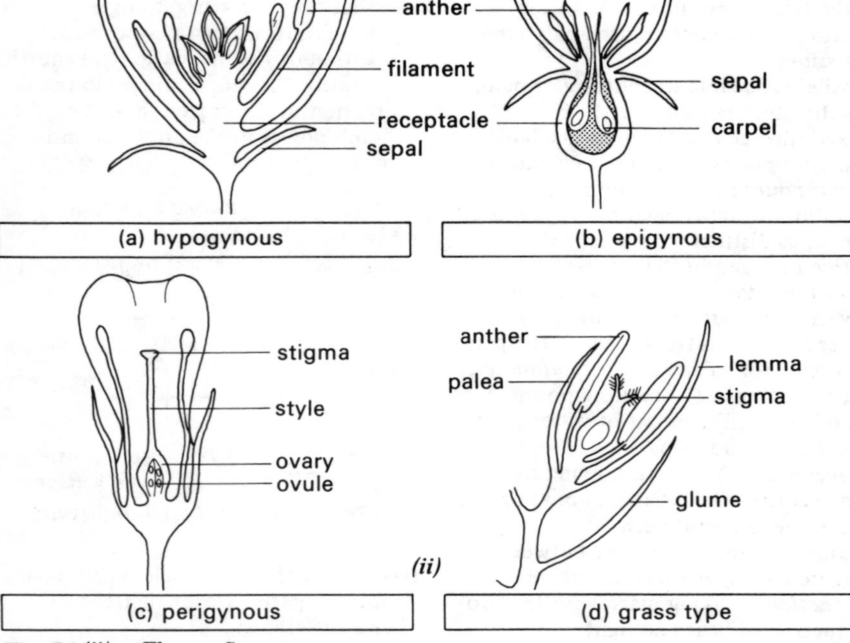

Fig. 76 (ii) Flower Structure

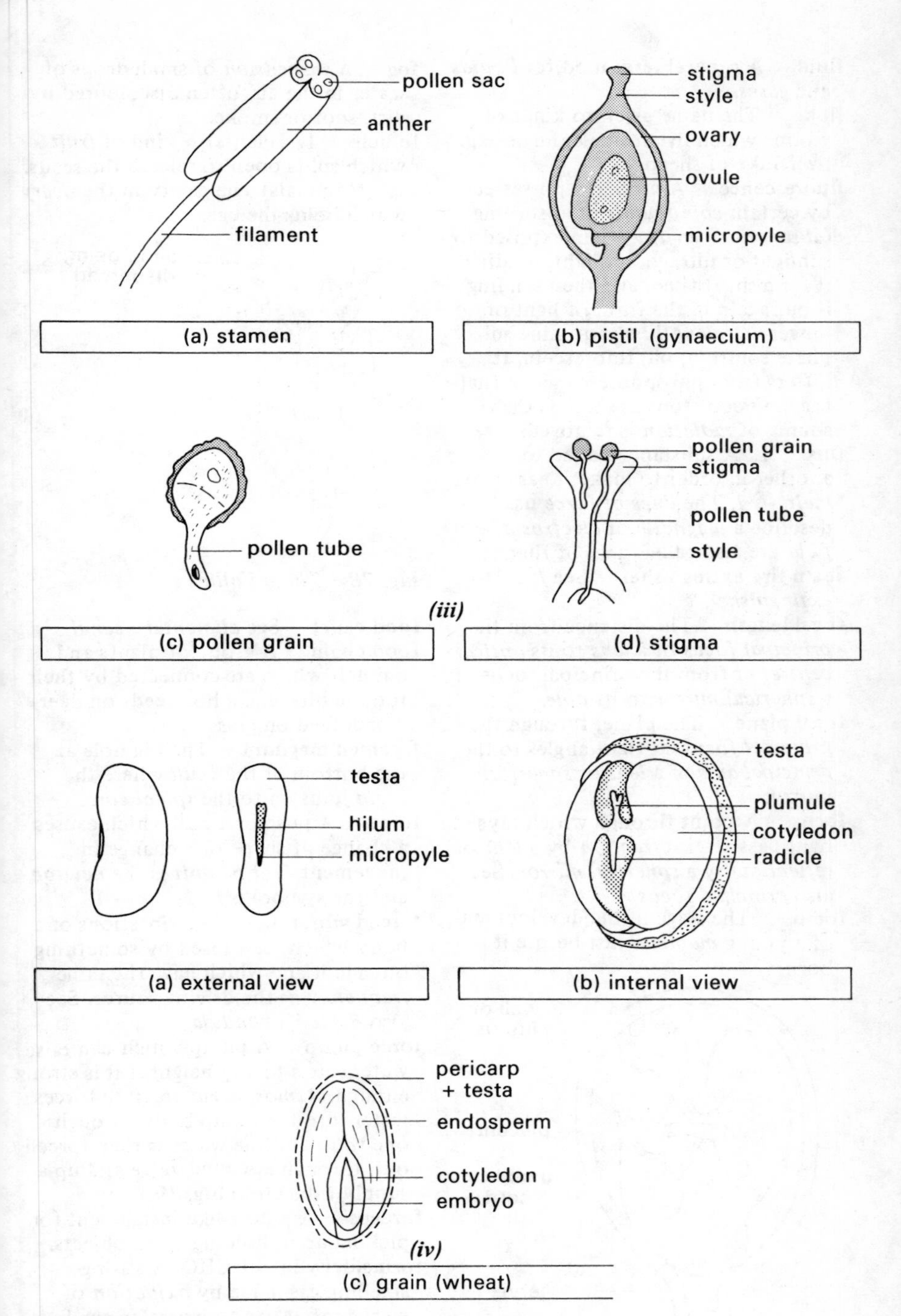

Fig. 76 (iii) Flower Structure (iv) Seed Structure

fluid A general term used for *liquids* and *gases.*
fluke The name given to kinds of worms which live inside animals, e.g. liverfluke of sheep.
fluorescence A *property* possessed by certain *compounds* of absorbing (*absorption*) *energy* when exposed to sunlight or ultra-violet light, holding it for a short time, and then sending it out again in the form of light of longer *wavelength;* e.g. quinine sulphate solution, oil, fluorescein. It differs from phosphorescence in that the emission stops as soon as the source of *radiation* is removed.
flux 1. A substance added to another in order to make it easier to *melt.* 2. The *lines of force* used to describe a *magnetic* or *electrostatic field* are also called lines of flux.
foam fire extinguisher See *fire extinguisher.*
focal length The distance from the *principal focus* of a *lens* to its *optical centre;* or from the principal focus of a *spherical mirror* to its *pole.*
focal plane The plane, through the *principal focus,* at right angles to the *principal axis* of a *lens* or *spherical mirror.*
focus A point through which rays of light pass after *refraction* by a *lens* or *reflection* by a *spherical mirror.* See also *principal focus.*
foetus The stage in the development of a young *mammal* just before it is born.

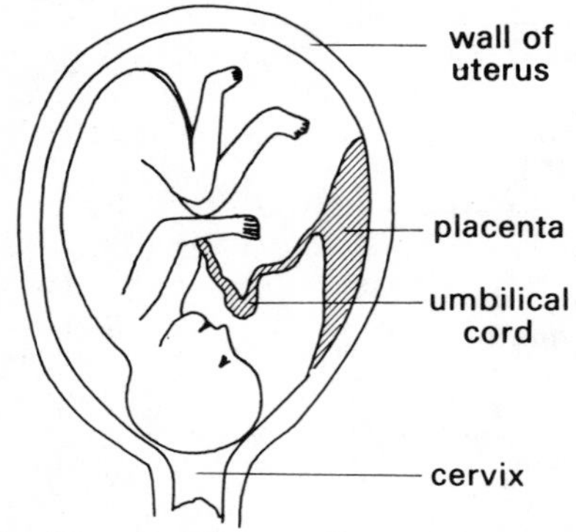

Fig. 77 The Human Foetus

fog A *suspension* of small drops of water in the air, often discoloured by dirt, soot or smoke.
follicle 1. (Plants) A kind of *fruit* which splits open to release the seeds. 2. (Mammals) The cavity in the *ovary* which holds the egg.

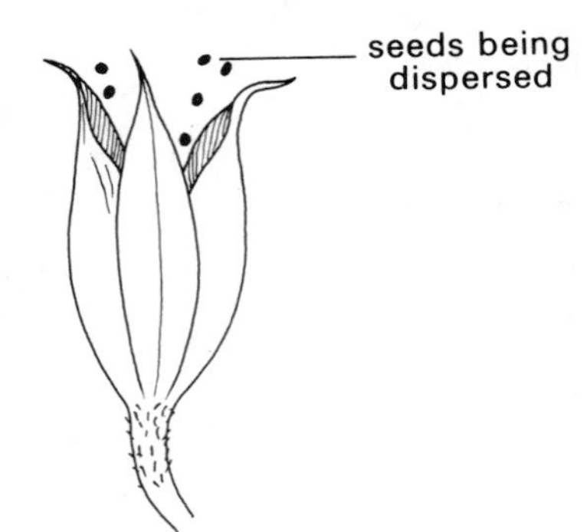

Fig. 78 Three Follicles

food canal See *alimentary canal.*
food chain A group of plants and animals which are connected by their food habits, e.g. a lion feeds on deer which feed on grass.
foramen magnum The big hole at the bottom of the *skull* where the *brain* joins on to the *spinal cord.*
force A push or a pull which causes a change of shape or a change in movement. The *SI unit* is the *newton* and the symbol is *F*.
forced vibration The vibrations of a body which are caused by something outside it and which have the same *frequency* as the driving source. See also *Barton's pendula.*
force pump A pump which can raise water (etc.) to any height if it is strong enough. *Atmospheric pressure* forces water into the pump body through one *valve* and the water is then forced out, through a second valve and up a supply tube etc. (Fig. 79.)
forceps A pincer-like instrument for picking up or holding small objects.
formaldehyde HCHO A strong-smelling gas made by *oxidation* of *methanol.* It has an irritating smell, is very soluble in water to give a solution

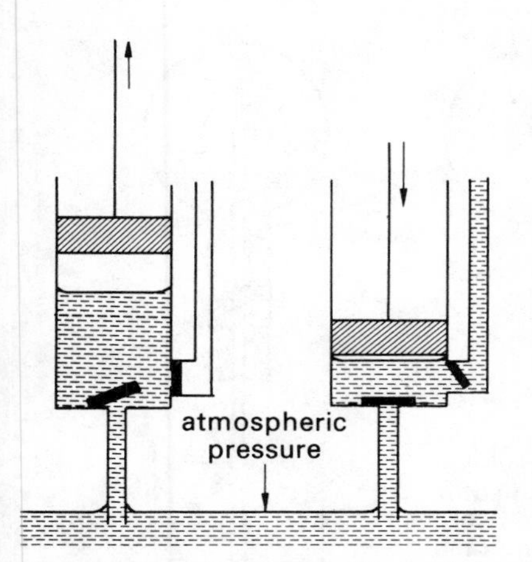

Fig. 79 The Action of a Force Pump

called formalin, and is used in the manufacture of some *plastics*. It is also called methanal.

formalin A 40% *solution* of *formaldehyde* in water, used as a *disinfectant.*

formic acid HCOOH The *acid* obtained by the complete *oxidation* of *methanol* or of *formaldehyde* (methanal). Industrially it is prepared by a catalytic action from hydrogen and carbon monoxide. It is colourless, corrosive and fuming and has a pungent smell. It is present in the sting of ants and is used in dyeing, tanning, electroplating and the manufacture of certain plastics. Its systematic name is methanoic acid.

formula An arrangement of *symbols* to represent the different *atoms* making up one *molecule* of a *compound* and showing the number of each present, e.g. H_2O–water; H_2SO_4 –sulphuric acid. The formula C_2H_4 shows that the smallest portion of the gas *ethene* which can exist as such contains two atoms of *carbon* and four atoms of *hydrogen.* A formula may also be written for an *element* if the molecule contains more than one atom, e.g. H_2–hydrogen; P_4–phosphorus: or for an *ion* such as NH_4^+, NO_3^-. A formula may be taken to represent either one molecule of the substance or one *mole,* e.g. H_2O in an equation can stand for 18 g of water. With *ionic* compounds such as *sodium chloride* there are no molecules and the formula NaCl indicates that there are equal numbers of sodium and chloride ions, Na^+ and Cl^-, in the compound.

formula mass (weight) This is the ratio of the total *mass* of all the *elements* making up one *molecule* of a *compound* to the mass of one *atom* of *hydrogen* on the scale where the *atomic mass* of *carbon* = 12. It is really a number, but is often expressed in *grams* (gram formula mass). It is easily worked out from a table of atomic masses, e.g. H_2O – F.M. = $(2 \times 1) + 16 = 18$; H_2SO_4 – F.M = $(2 \times 1) + 32 + (4 \times 16) = 98$; $Ca(OH)_2$ – F.M. = $40 + 2(1 + 16) = 74$.

Fortin barometer An accurate *barometer* which has the bottom mercury container made so that the mercury can be moved (by a screw) to touch a fixed pointer which is the zero of the scale. It may also have a *vernier* to help find the position of the upper mercury level more accurately.

fossil The preserved remains of plants or animals, usually in rock.

fountain experiment An experiment to show the great *solubility* of some gases. A flask is filled with a very soluble gas such as ammonia and fitted with a bung carrying a long glass tube as shown. The end of the tube dips into a beaker of water. As the gas *dissolves* the water is drawn up the tube and sprays out into the flask like a fountain. (Fig. 80.)

fovea Part of the *eye* (Fig. 68). The yellow spot at the back of the eye. It is very sensitive to colour.

fractional crystallization A method of separating a mixture of solids which have different *solubilities* in water. If two compounds such as potassium nitrate and sodium chloride are heated together in water to make a *saturated solution* of both, much more of the nitrate will *dissolve.* If this solution is allowed to cool to room temperature the nitrate will

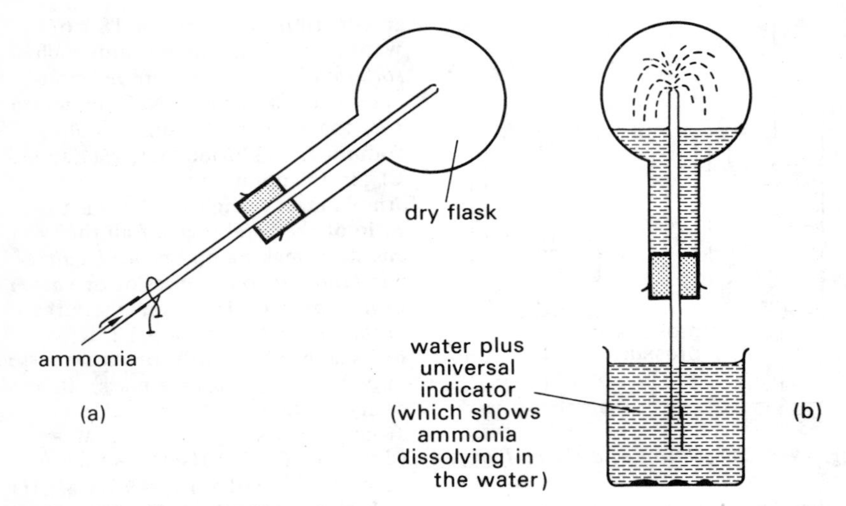

Fig. 80 The Fountain Experiment

crystallize out first as its solubility decreases very rapidly as the temperature falls.

fractional distillation A method of separating a mixture of liquids which have different *boiling points.* If a mixture of ethanol (b.p. 78.5 °C) and water (b.p. 100 °C) is heated in a flask fitted with a *fractionating column* and a *condenser,* most of the ethanol will distil over first at about 78 °C and when the temperature rises to 100 °C the water distils over. If several liquids are present each can be collected as a separate fraction at its own boiling point. See *petroleum, refine* (Fig. 135).

fractionating column A tall tube which can be filled with beads or rings or broken glass (see diagram). It is fitted into the neck of a flask in which *fractional distillation* is taking place. With a mixture of liquids in the flask, the one with the lowest *boiling point vaporizes* first and goes up the column to distil over; the other liquids with higher boiling points are hindered by the beads and *condense* back into the flask. As the temperature increases these liquids rise further up the tube and eventually each distils over as its own boiling point is reached.

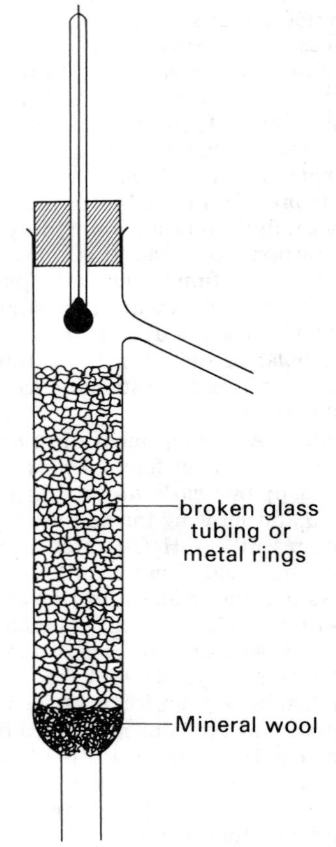

Fig. 81 Fractionating Column

Frasch process An arrangement of pipes around each other used in extracting sulphur from far below the ground. *Superheated* water under pressure is forced down the outside tube and melts the sulphur in the rocks at the bottom (see diagram). Compressed air down the central tube mixes with the sulphur/water *slurry* and forces it up the space between these two pipes and out at the top into large wooden vats where it solidifies. It is over 99% pure.

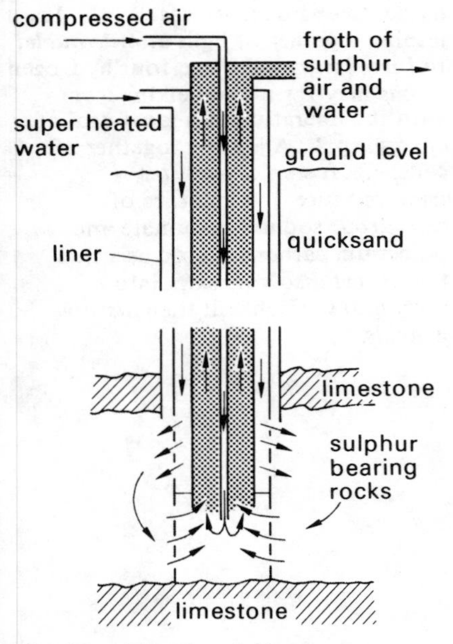

Fig. 82 The Frasch Process

free vibration The natural vibration of a body when it has been set into vibration and then left alone.

freezing mixture A mixture, usually of crushed ice and sodium chloride, which produces a temperature below 0 °C.

freezing point The *temperature* at which a liquid turns to its solid form, which is usually given for one *atmosphere pressure.*

freons Organic compounds containing carbon, fluorine and possibly chlorine; used in some *refrigerators* and freezing machines and in some fire extinguishers. They are not poisonous, do not take fire and have low boiling points. They are gases and are often used in *aerosols* for forcing out the liquid in the container.

frequency The number of *cycles* of any repeating process which occur in one second, e.g. the number of *waves* passing a point in one second. The *SI unit* is the *hertz* (Hz) and the symbol is f or ν.

friction The *force* opposing movement which happens when one surface rubs on another.

friction coefficient To obtain this, divide the greatest frictional force between two surfaces (the limiting friction, F_1) by the force pushing the surfaces together (this force is at right angles to the surfaces) F_2. I.e.

Friction Coefficient $= \frac{F_1}{F_2}$. It is just a number and has no *units* but depends on the nature of the surfaces.

fructose $C_6H_{12}O_6$ A simple *sugar* present in honey, in some ripe fruits and in the *nectar* of flowers. It is an *isomer* of *glucose.*

fruit The part of the plant which contains the *seed* or seeds. There are many kinds. Some are dry and split open, some are fleshy and have to be eaten by animals to set free the seeds. (Figs. 3, 78.)

fuel 1. A substance capable of burning at a controllable rate and of producing a considerable amount of *heat energy.* Fuels may be solid (coal, coke, wood), liquid (motor spirit, kerosine, oil) or gaseous (methane, coal gas, calor gas). 2. *Fissile* material used in a *nuclear reactor.*

fuel oil The *residue* left behind after the *fractional distillation* of crude oil or *petroleum.* After further treatment it can be used as *diesel* oil and also for domestic heating.

fulcrum A pivot or bearing about which a body can turn.

fumes Visible *vapours* consisting of

small airborne particles given off by a substance or appearing in a reaction, e.g. *ammonium chloride* produced in the reaction between gaseous ammonia and hydrogen chloride.

function Work done by a plant or animal, e.g. the function of the eye is to see.

fundamental note The basic *frequency* given out by a source of sound. See also *overtone, quality of sound.*

fungus Simple plants which are not green because they do not contain *chlorophyll.* They cannot make food for themselves and so are either *parasites* or *saprophytes.*

funnel See *filter funnel, thistle funnel, dropping funnel.*

fur The name given to the hard *deposit* left in kettles, boilers or pipes in which *hard water* has been boiled or through which it has passed. It is mainly calcium and magnesium carbonates. It is also called scale.

furnace An enclosed container, usually made of iron lined with firebrick, in which a very high temperature can be produced. Heating may be by gas or by electricity. See also *blast furnace, retort.*

fuse A thin piece of wire which protects an electric *circuit.* It is made so that it melts if too great a *current* flows.

fused 1. A substance in the molten state. 2. A substance which has been melted and allowed to solidify, e.g. fused (*anhydrous*) *calcium chloride.*

fusion 1. Changing a *solid* into a *liquid.* See also *latent heat.* 2. The joining together of light *atomic nuclei* to form larger ones, e.g. four hydrogen atoms may form one helium atom with the liberation of a great deal of *energy.* 3. A joining together of cells, e.g. fusion of *gametes.*

fusion mixture A mixture of *anhydrous* sodium carbonate and potassium carbonate used as a flux to convert *insoluble salts* into *carbonates* which will then *dissolve* in *acids.*

galena Lead sulphide, PbS, the commonest *ore* of lead.

Galilean telescope A *telescope* with a *converging lens* as an *objective* and a *diverging lens* as an *eyepiece*. It gives an upright *image*. Two such telescopes, fixed side by side, are called opera glasses.

gall bladder The place in the *liver* where the *bile* is stored. See *alimentary canal* (Fig. 5).

galvanize To coat with zinc. *Galvanized* iron is made by dipping sheets of iron into molten zinc in order to protect the iron from damp conditions likely to cause *rusting*.

galvanometer A sensitive instrument for showing the presence of, or measuring, a very small *electric current*.

gamete A *cell* (*sperm* or *ovum*) which has to join with the other kind of cell before it will develop into a new plant or animal.

gamma rays Certain *radioactive* materials give out gamma rays. These are of very short *wavelength* and are members of the electromagnetic spectrum. The *nucleus* which gives out the gamma rays is not changed except for a loss of *energy*.

ganglion A group of nerve *cells*. See *solar plexus, central nervous system*. group of nerve *cells*.

gas 1. An *element* or *compound* in such a physical *state* that its *molecules* can move about freely throughout the whole of the space occupied. It has no definite shape or *volume* as it fills any space into which it is put. A gas has a very low *density* and is easily compressed; gases that do not react together mix in all proportions. 2. An abbreviation (used especially in the U.S.A.) for gasoline, *petrol* or motor spirit.

gas carbon The *deposit* left on the walls of *retorts* in which *coal* has been heated in the absence of air. It is a good *conductor* of electricity and is used in making carbon *electrodes*.

gaseous Referring to the *gas* or vapour *state*. $H_2O(g)$ stands for steam.

gas laws The laws which describe the way in which gases behave. See also *Boyle's law, Charles' law, law of pressures, perfect gas*.

gas liquor When *coal* is heated in retorts the main products are coke coal gas and a watery liquid which contains ammonia and ammonium compounds.

gas oil One of the distillates (*distil*) obtained during the *fractional distillation* of *petroleum*. It is used in the preparation of *carburetted water gas*.

gasoline See *petrol*.

gas syringe A glass or plastic graduated cylinder in which a piston slides (see diagram); it is used for collecting a gas and measuring its *volume*.

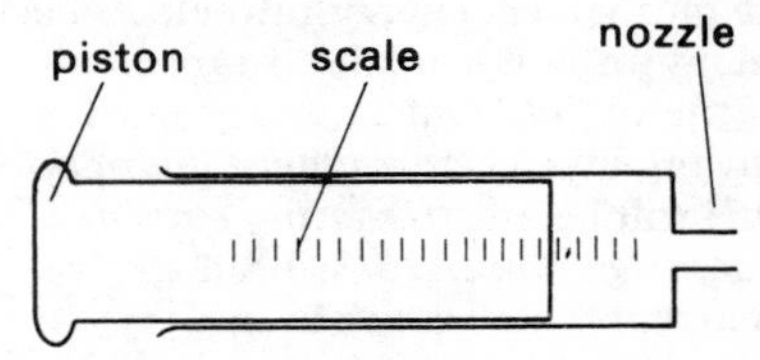

Fig. 83 *Gas Syringe*

gas tar See *coal tar*.

gas thermometer See *constant volume air thermometer*.

gastric Anything to do with the *stomach*, e.g. gastric juice.

gauze See *wire gauze*.

Gay-Lussac's law of combining volumes A law of chemistry concerned with the proportions in which *gases combine* by *volume*. It can be stated: When gases combine together they do so in proportions which bear a simple relationship to each other and to the product if that is a gas, all measurements being taken under the same conditions of *temperature* and *pressure*. Examples are 1 volume of hydrogen + 1 volume of chlorine form 2 volumes of hydrogen chloride; 3 volumes of hydrogen + 1 volume of nitrogen form 2 volumes of ammonia.

Geiger-Muller tube A metal or glass tube containing two *electrodes* and filled with a gas like argon, with a little bromine, at low *pressure*. When an *alpha* or a *beta particle* or *gamma radiation* enters the tube *ions* are formed. These move between the *anode* and the *cathode* so that there is a flow of electricity for a very short time (an electric pulse). These pulses are then counted by an electronic counter. These tubes are made in very many different forms. See also *ratemeter*, *scaler*.

gel A jelly-like substance where the particles stick together. Usually formed when a *sol* is heated.

gene Part of a *chromosome* which carries a particular *character*, e.g. gene for blue eyes. Genes are passed on from parents to children.

generator An instrument for changing mechanical energy into electrical energy. A coil is made to turn in a *magnetic field* and a current is set flowing by *electromagnetic induction*. This will be an *alternating current* unless a *commutator* is used to give a current which flows in one direction only. See also *dynamo*.

genetics The study of how *characters* are passed on from parents to the young.

genotype The set of *genes* present in a plant or animal. They are passed on in *heredity*.

genus One of the groups of plants or animals, e.g. the brown rat and the black rat belong to the same genus.

geotropism The effect that *gravity* has on the way stems and roots grow, e.g. the roots grow towards the centre of the earth.

germ The name given to any very small *organism* which causes disease, e.g. *bacteria*.

germination The growth of a *seed* into a young plant.

giga- A prefix which means one thousand million times or 10^9. It can be joined in front of any *SI units*, e.g. 1 gigahertz = 10^9 hertz. The symbol is G.

gill A thin piece of *skin* with *blood* in it, which is part of the breathing system of water living animals.

girdle 1. Part of the *skeleton*. A ring of bones, e.g. *pectoral* girdle at the top of the chest to which the arms are joined, and pelvic girdle at the bottom of the trunk to which the legs are joined. 2 Part of a *twig* (Fig. 162). Girdle *scars* show where *buds* have grown each year.

gizzard Part of the *alimentary canal* in earthworms and birds. Hard food is crushed in the gizzard.

glacial ethanoic acid Pure, concentrated *ethanoic acid*. As its freezing point, 16.6°C, is not much below normal room temperature it often forms ice-like *crystals* in cold weather. It used to be called glacial acetic acid.

gland A structure in the body which makes a chemical substance, e.g. *sweat gland* in skin, *thyroid gland* in neck.

glass A hard, *amorphous*, *brittle* material, usually *transparent* or *translucent*, made by *fusing* together an *acidic oxide* such as that of silicon (sand) or boron and one or more *basic oxides*, e.g. those of sodium, potassium or lead, and cooling the product quickly.

glasswool Material, something like cotton-wool (cotton), but consisting of very fine threads of glass. Used for *filtering*, for absorbing *corrosive* liquids and as an *insulator*.

Glauber's salt Crystalline (*crystal*) sodium sulphate, $Na_2SO_4.10H_2O$. Used as a purgative.

glomerulus A part of the *kidney* where the liquid which will become the *urine* is taken out of the blood. (Fig. 20.)

glottis The upper part of the *windpipe* where the *vocal cords* are found and sounds are made.

glucose $C_6H_{12}O_6$ A simple *sugar* (grape sugar) found in honey, in sweet fruits and in the blood. It is made in green leaves in the light from *carbon dioxide* and water. *Oxygen* is also made and passed out into the air. In this way *energy* from sunlight is stored in the plant and can be used for *growth* and *respiration*. Other *carbohydrates* taken into the body as food are converted into glucose before they can be used to provide energy. Glucose is used in brewing, in jam making and in confectionary.

glycerol $C_3H_8O_3$ Propane–1, 2, 3 triol, $CH_2(OH).CH(OH).CH_2OH$. An *alcohol* obtained as a *by-product* during the manufacture of *soap*. Fat + alkali→soap + glycerol. It can also be made from propane, C_3H_6. A thick, sweetish syrupy liquid, it is used in the manufacture of plastics and in some explosives such as nitroglycerin. It used to be called glycerin(e).

glycogen A kind of *starch*, found in many animals and a few plants.

glycol $CH_2.CH.CH_2OH$ An *alcohol* made from *ethene* and used as in *antifreeze* in motor car radiators because of its low *freezing point*. It is a colourless *viscous* liquid with a sweet taste. It is also called ethane 1, 2, diol.

goitre The swelling of the *thyroid gland* in the neck. Exophthalmic goitre is a disease caused by a large thyroid gland.

gold At. No. 79 R.A.M. 197 Rather soft yellow metal resistant to attack by air or water and by most acids.

gold leaf electroscope An *electroscope* in which a piece of gold leaf, which is supported by a strip of metal, moves in the *electric field* between the support and the case of the instrument.

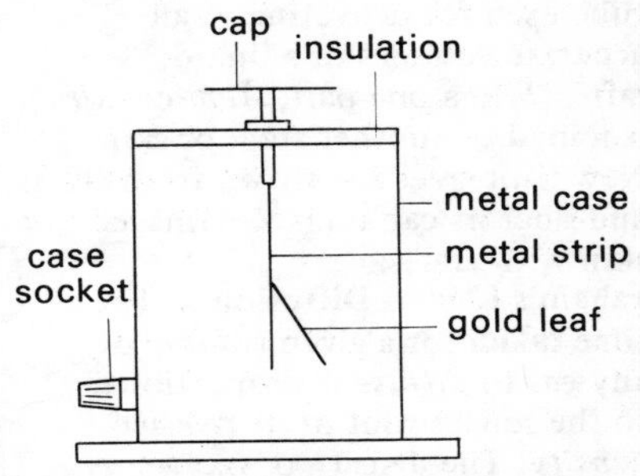

Fig. 84 The Gold Leaf Electroscope

gonad The general word used for the *reproductive organ*, male or female, in animals.

gradient The rate at which a slope, or a line on a *graph*, rises. It is usual in science to measure the gradient as vertical distance divided by horizontal distance, but gradients can be given as vertical distances divided by distance along the slope. If the graph is a curve, a straight line just touching the curve (a tangent) is first drawn at the point where the gradient is to be measured.

graduate To put a scale on an instrument.

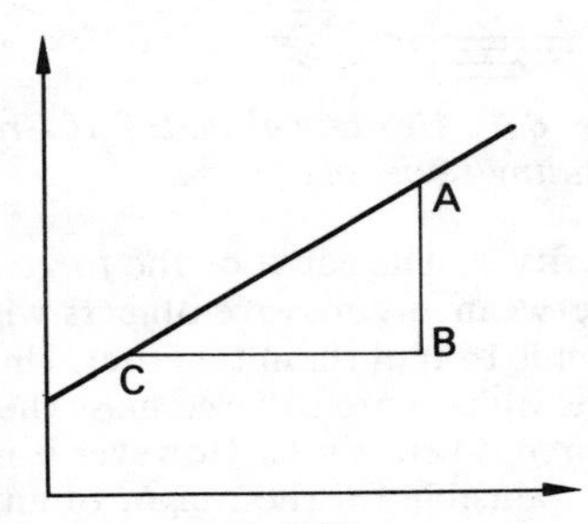

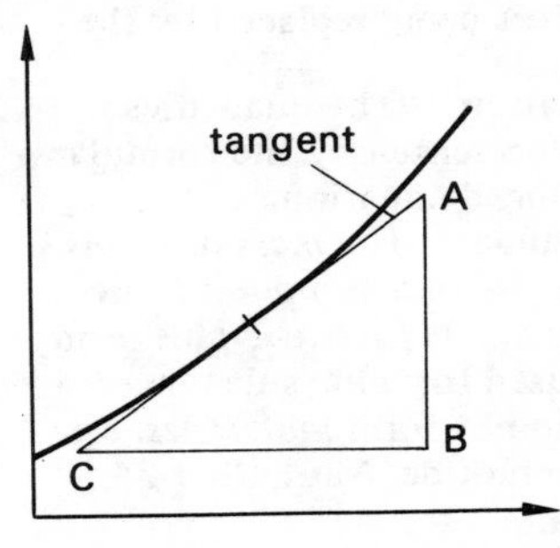

Fig. 85 Gradient $= \frac{AB}{BC}$

Fig. 86 Graduated Pipette

graduated cylinder A tall narrow glass or plastic vessel in which *volumes* of liquids can be measured. See also *measuring cylinder.*

graduated pipette A graduated tube used for delivering small accurate volumes of a liquid.

graft When one part, *stem* or *skin*, is joined to another stem or skin. New fruit trees are grown from grafts and doctors can replace damaged skin with grafts.

Graham's Law of Diffusion The time taken for a given *volume* of any *gas* to *diffuse* is proportional to the square-root of its *relative density*. The density of oxygen is 16 times that of hydrogen, so 100 cm^3 of oxygen will take 4 times as long to diffuse as 100 cm^3 of hydrogen.

gram One-thousandth part of the *SI unit* of *mass*, the *kilogram*. It is frequently used as the unit where small masses are concerned.

gram-atom 1. The *relative atomic mass* (atomic weight) of an *element* expressed in *grams*. 2. That *mass* of an element which contains *Avogadro's constant* of *atoms* (6.02×10^{23}) i.e. the mass of an element which contains one *mole* of atoms, e.g. 16 g of oxygen, 32 g of sulphur. As the term 'gram atom' is only one particular meaning of 'mole', it is being replaced by the word 'mole'.

gram-equation The quantities in grams represented by the formulae in a balanced equation.

gram-formula The *mass* in *grams* of one *mole* of a *compound* calculated from its *formula*. This term can be used for *ionic* substances which do not form *molecules*, e.g. sodium chloride, NaCl, 23 + 35.5 = 58.5 g.

gram molecular volume See *molar volume.*

gram-molecule 1. The *relative molecular mass* (molecular weight) of an *element* or *compound* expressed in *grams*. 2. That *mass* of an element or compound which contains *Avogadro's constant* of *molecules* (6.02×10^{23}), i.e. the mass of one *mole* of molecules of an element or compound, e.g. 32 g of oxygen, 17 g of ammonia, NH_3.

grape sugar *Glucose.*

graph A drawing, based on two *calibrated* reference lines (axes) at right angles to each other, which show how two quantities are connected. The matching values of the quantities are used to mark (plot) a series of points. These are joined, usually by a straight line or a smooth curve to complete the graph. See also *gradient.*

graphite A naturally-occurring allotrope (*polymorph*) of *carbon* in which the *atoms* are arranged in flat *planes* of six-atom rings (hexagons). As these planes can easily slide over each other, graphite is a good *lubricant*, either alone or mixed with oil or water. Mixed with clay it forms the 'lead' in pencils. It is a good conductor of electricity. It is also called black lead or plumbago.

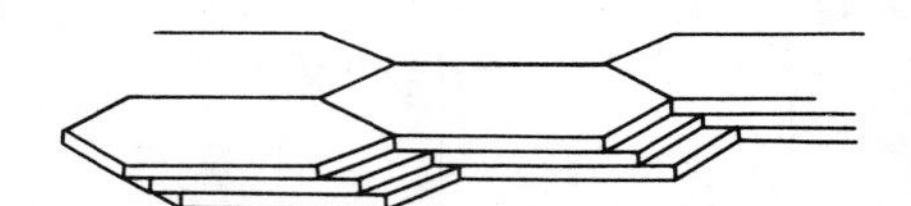

Fig. 87 Hexagonal Plates of Graphite "sliding" over each other

gravity The cause of the *force* between any massive objects which tends to pull them together. Unless the objects are of large mass the force is very small. However it is responsible for the *weight* of any object on earth, and for the paths of the planets in their movement round the sun, etc.

green vitriol *Crystalline iron(II) sulphate*, $FeSO_4.7H_2O$.

grey matter The outside part of the *brain* and the inside part of the *spinal cord*. It is where the *nerve cell bodies* are found. See *central nervous system* (Fig. 34ii).

grid 1. The open wire network which forms the third *electrode* in a *triode*, etc. 2. The national system which links *generators* and users of electricity together by overhead (or sometimes underground) *high-tension* wires.

gristle The soft but tough part of the *skeleton*. See *cartilage*.

group 1. A collection, usually of similar substances. Used particularly in referring to the different families of elements grouped vertically in the *periodic table*. 2. A group of *atoms* which stays as a unit in a series of similar compounds, e.g. $=CO_3^{2-}$, the carbonate group or radical; the —COOH group in organic acids.

growth A condition of healthy plants and animals. Growth is caused by an increase in *cells* which leads to an increase in size.

guard cell One of the two *cells* which surround the *stoma* in the skin of a leaf. The size of the opening is controlled by the guard cells. The guard cells contain *chlorophyll*.

gullet Another name for the *oesophagus*.

gums General name for *organic* substances obtained from trees such as accacias; they react with or dissolve in water and are used as adhesives for sticking things together.

gunpowder An *explosive* mixture of 75% potassium nitrate, 15% charcoal and 10% sulphur. When it is exploded the *volume* of *gas* produced is many times that of the original solid.

gut Another name for the *alimentary canal*.

gynaecium The female part of the flower which contains the *ovules*. It consists of *ovary*, *style* and *stigma*. The *pollen grains* land on the stigma leading to *fertilization* of the ovules. (Fig. 76.)

gypsum $CaSO_4.2H_2O$ A *crystalline* form of calcium sulphate. It occurs naturally and, although only slightly *soluble*, it causes *permanent hardness* in water which passes through it. When it is heated it loses three-quarters of its *water of crystallization* and forms Plaster of Paris.

Haber process The industrial method of making *ammonia*. *Nitrogen* and *hydrogen* in the proportion of 1 volume to 3 are passed under high *pressure* over a *catalyst* (usually containing iron) at a *temperature* of about 450 °C. $N_2 + 3H_2 \xrightleftharpoons{\Delta,\ iron} 2NH_3$ The ammonia is liquefied or dissolved in water. See also *flow diagram*.

habitat Where a plant or animal a actually lives, e.g. the habitat of the deer is grassland.

haematite Naturally-occurring red *iron(III) oxide*, Fe_2O_3, the main *ore* of iron.

haemoglobin The red colouring matter present in *blood*, consisting of haem and globin (a *protein*). It combines readily with oxygen and also with the poisonous carbon monoxide. It carries the oxygen round the body.

haemophilia A *disease* where *blood* keeps on flowing from a wound for a long time. It is usually found in men and boys, but can be handed on through the mother. It is an example of *sex linkage*.

hair Living growths from the *skin* (Fig. 149) of *mammals*. The live part stays below the skin. Hair may be of different kinds and colours in different mammals. See also *root hair*.

half-life The time it takes for one half of any quantity of a *radioactive nuclide* to *decay*. It is therefore also

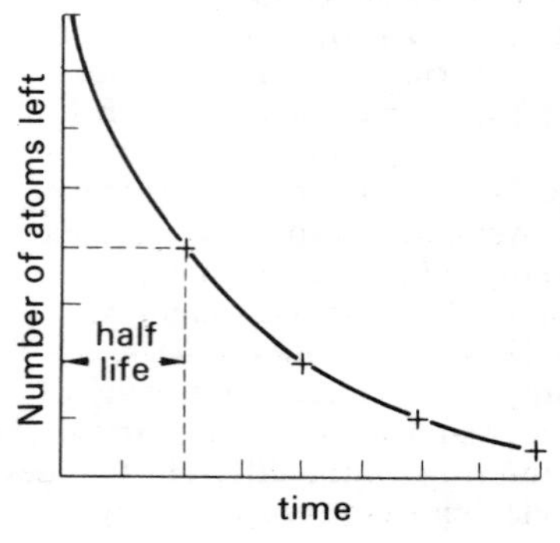

Fig. 89 *Half-life*

equal to the time for the amount of the radioactive nuclide present to fall from one half to one quarter, from one quarter to one eighth, etc., of the first amount. Half-lives can vary from

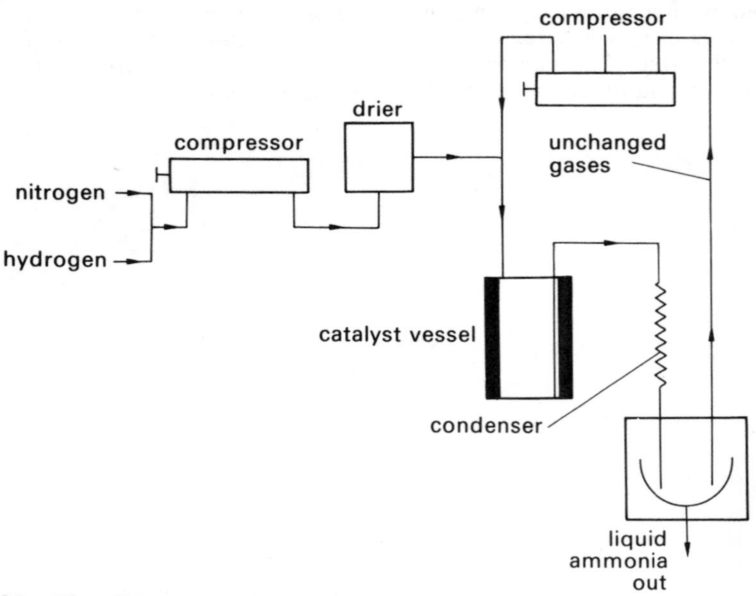

Fig. 88 *Flow Diagram for Haber Process*

millions of years to millionths of a second or less. The value suggests which nuclide is present.

halide A *binary compound* containing one of the *halogens* combined (*combination*) with one other element.

halogens The elements fluorine, *chlorine, bromine, iodine* and astatine which form Group VII of the *periodic table.* They have closely related *properties* and all have seven *electrons* in their outer *shell.* They are non-metals and their *reactivity* decreases in the order shown above.

hard water Water which will not readily form a *lather* with *soap.* This hardness is caused by the presence in *solution* of calcium or magnesium *ions.* The hardness can be removed by the addition of *zeolite* or of *sodium carbonate.* There are two kinds of hardness; *temporary hardness* and *permanent hardness.*

Hare's apparatus Two vertical glass tubes are joined at the top to a third tube. The lower end of each of the vertical tubes is placed in a container of liquid. When the *pressure* inside the tubes is reduced by sucking at the third tube, the liquids rise up the vertical tubes. The height risen is smaller the denser (*density*) the liquid and the ratio of the heights risen in the two tubes is equal to 1/ratio of the densities of the two liquids.

haltere The very small back wings of a fly which it uses for balance.

hammer One of the *ear ossicles,* the *malleus.*

haustoria Growths from a *fungus* which lives inside another plant. The fungus takes in its food through the haustoria.

health salts A mixture of *sodium hydrogencarbonate, citric acid* and sugar. When water is added the solid acid dissolves and reacts with the carbonate to liberate *carbon dioxide* and the mixture *effervesces.*

heart The *blood* pump of the body. It is different in different groups of animals. In a *mammal* it is a double pump, sending the blood first to the lungs and then around the body. (Fig. 91.)

heart wood The hard wood in the middle of a tree trunk.

heat 1. (*vb.*) To raise the *temperature* of a body by supplying *energy* by means of electricity, by *burning* a *fuel* or by the transference of some other kind of energy. 2. (*n.*) See *heat energy.*

heat energy The internal *energy* possessed by a substance because of the motion of its *atoms, molecules* or *ions.* It is usually measured in *joules* (formerly calories) and is transmitted to other bodies by *conduction, convection* or *radiation.* As the *temperature* is raised the motion gets faster and stronger and the heat energy of the substance increases. If a solid the substance usually melts; if a liquid it will vaporize. The term 'heat' is often wrongly used instead of temperature.

heat of combustion The amount of *heat energy,* measured in *joules,* which is given out when a definite *mass* of a substance is burnt completely in oxygen. If the amount burnt is one *mole* the term 'molar heat of combustion' is used.

heat of displacement The *heat energy* given out or taken in for every *mole* of a particular reactant (*react*) completely used up when a *displacement reaction* occurs, e.g. $Zn(s) + CuSO_4(aq) \rightarrow ZnSO_4(aq) + Cu(s)$; H = −218 kJ, i.e. 218 kJ of heat energy are given out for every mole of zinc or copper sulphate involved.

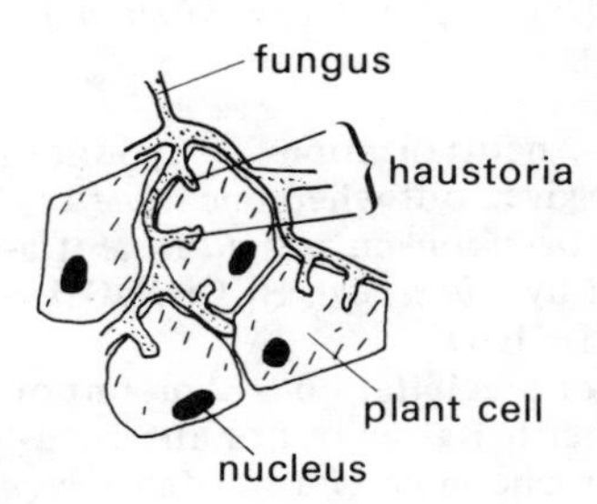

Fig. 90 Haustoria

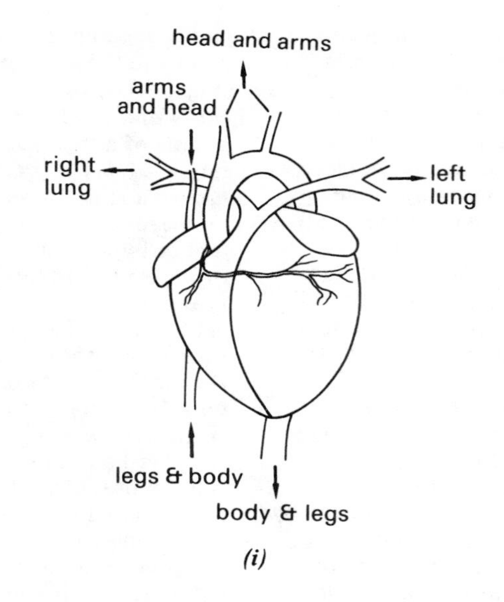

(i)

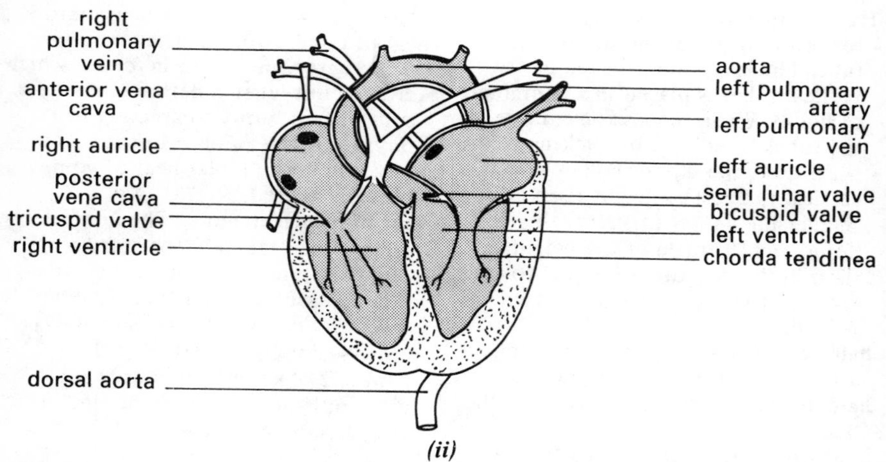

(ii)

***Fig. 91** (i) Front view of Human Heart (ii) Internal view of Heart and Main Blood Vessels*

heat of neutralization Amount of *heat* given out when one *mole* of *hydrogen ions* in an *acid* is neutralized by a *base*, e.g. $HCl + NaOH \rightarrow NaCl + H_2O$.

Heat of precipitation Amount of heat either given out or absorbed when one mole of a substance is precipitated (*precipitation*) during a *chemical reaction.*

heat of reaction The heat change occurring during a *chemical reaction*, which may be *exothermic* or *endothermic.*

heat of solution The amount of *heat* given out or taken in when one *mole* of a substance is *dissolved* in a large *volume* of water.

heavy water A form of water in which the two hydrogen *atoms* are replaced by *deuterium* atoms. It is found (about 1 part in 5000) in natural water and the name is also used for mixtures of ordinary water and deuterium oxide if there is much more of the heavy form than usual. (*Boiling point* 101.42 °C, *freezing point* 3.82 °C, *density* 1 105 kg/m^3).

helium He At. No. 2, R.A.M. = 4 The second lightest of the *elements,* and the first member of Group O of the *periodic table.* It is a *monatomic gas* and occurs in the mineral, cleveite.

hepatic Anything to do with the *liver,* e.g. the hepatic *artery* takes blood to the liver.

herbaceous Not woody, e.g. grass.

herbivorous An animal which feeds only on plants, e.g. deer.

heredity (*adj.* Hereditary) The passing on of *characters* from parent to the young in plants and animals.

hermaphrodite An animal which is both *male* and *female,* e.g. an earthworm.

hertz The *SI unit* of *frequency.* The symbol is Hz. A frequency of one hertz means that a complete *cycle* of events repeats itself once every second

heterogeneous Not of uniform composition throughout, e.g. a mixture of sand and sugar or benzene and water.

heterozygous A word used in *heredity.* One *gene* inherited from the father and the other from the mother are slightly different, e.g. a child may receive the gene which causes brown eyes from one parent and the gene which causes blue eyes from the other parent. One of the pair is usually *dominant* and this is the one which shows. The child would have brown eyes, the blue being *recessive.*

hexagonal Six-sided; used in referring to the flat 6-atom rings of *graphite.*

hexamethylene diamine $H_2N(CH_2)_6NH_2$ A *reagent* used in the preparation of *nylon,* also called 1.6. diaminohexane.

high tension A high voltage or *potential difference.* This is a common, rather than a scientific, use of words.

hilum The mark on a *seed* where it was joined to the *fruit* wall.

histamine The substance which causes the red mark and swelling when the *skin* is damaged by, for example, an insect bite.

Hofmann voltameter A three-limbed glass or plastic vessel (see diagram) used in *electrolysis.* The volume of any *gases* liberated can be measured.

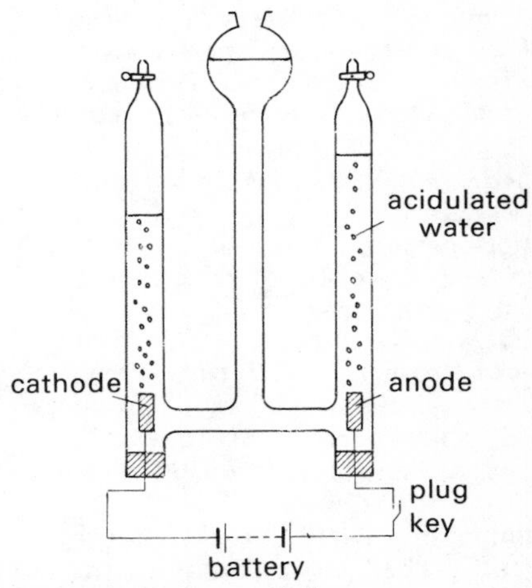

Fig. 92 Hofmann Voltameter

homeostasis A state of steadiness in the body, e.g. *blood* temperature in a human remains steady at about 37 °C in a healthy person.

homo The scientific name for man—Homo sapiens.

homogeneous Of one kind only—of uniform *composition* throughout; e.g. a mixture of ethanol and water.

homoiothermic Warm blooded, as in birds and *mammals.*

homologous series Series of *organic compounds* of similar chemical *properties* and physical properties which gradually change with increase in *mass.* All compounds in any one series have the same general *formula,* each member differing from the one before by a constant group, CH_2—, of *atomic mass* 14. *Alkanes,* C_nH_{2n+2} *alcohols,* $C_nH_{2n+1}OH$; *alkenes,* C_nH_{2n}; *alkynes,* C_nH_{2n-2}; *arenes,* C_nH_{2n-6}; etc.

homozygous A word used in *heredity.* The pair of *genes* for a *character* inherited from the two parents are exactly the same.

honey The sugary liquid made by a bee from the *nectar* of flowers.

Hooke's Law The increase in length of a spring or wire under *tension* is *proportional* to the stretching *force.* See also *limit of proportionality.*

hormone A substance produced by a *ductless gland* and passed around the body in the *blood.* Different hormones have different uses in the body, e.g. *thyroxin* influences growth; *adrenalin* influences the *nervous system.*

horsepower A *unit* of *power* which is not part of the *SI* system. It is equal to 745.7 *watts.*

hot wire instrument An instrument for measuring an *electric current* through the heat energy it produces. This heat energy causes a wire to expand and this movement is made to turn a pointer.

humerus Part of the *skeleton* (Fig. 148). The *bone* in the top part of the arm.

humid Damp and warm.

humidity The amount of water *vapour* in the air.

humus That part of the *soil* which comes from dead plants or animals or the excreta (*excretion*) of animals. It gives food in the soil and holds water.

hybrid An animal or plant which has come from parents of different sorts, e.g. the mule is a hybrid between a horse and a donkey.

hydrate A *salt* having a fixed percentage of *water of crystallization,* e.g. copper sulphate pentahydrate, $CuSO_4.5H_2O$.

hydrated 1. The opposite of *anhydrous;* possessing *water of crystallization* as hydrated sodium carbonate, $Na_2CO_3.10H_2O$. 2. Used of an *ion* joined to one or more *molecules* of water, e.g. the hydrated *hydrogen* (*hydronium*) ion, H_3O^+, or the hydrated copper(I) ion, $Cu(H_2O)_4^{2+}$.

hydraulic press Two cylinders of different diameters are connected together. Both are fitted with pistons and the space below the pistons is full of oil or water. A small *force* applied to the piston in the smaller cylinder needs a large force applied to the piston in the larger cylinder for balance as the *pressure* (force/area) must be the same on both sides. This large force can be used in many forms of press, etc.

hydride A *binary compound* of *hydrogen.* The hydrides of *non-metals* (e.g. ammonia, NH_3) are *covalent* compounds, but those of *metals* (e.g. sodium hydride, NaH) are salt-like and *ionic* and on *electrolysis* of the *fused* salt the hydrogen comes off at the *anode.*

hydrocarbons A general name for *compounds* containing *carbon* and *hydrogen* only. They may have open *chains* (*straight* or *branched*) as butane, C_4H_{10}, or closed chains (rings) as benzene, C_6H_6. See also *alkanes, alkenes, alkynes, arenes.*

hydrochloric acid HCl An *aqueous solution* of *hydrogen chloride.* The *concentrated* acid is colourless, *fuming* and *corrosive.* The acid forms *salts* called *chlorides* by reactions with metals, metal oxides and metal carbonates. It used to be called spirit of salt.

hydrogen H At. No. 1 R.A.M. 1 The lightest *element* there is. A colourless, *flammable* gas which combines with all *non-metals* other than the *argonons* to form *covalent compounds* and also with most active *metals* (Groups I and II of the *periodic table*) to form salt-like *ionic* hydrides. It is a very good *reductant* and is liberated at the *cathode* during the *electrolysis* of many solutions. Manufactured by the *Bosch process,* from *water gas* or from *methane,* CH_4, it burns readily to form water and is used in the manufacture of *ammonia,* and for converting oils to fats (hydrogenation). It occurs naturally combined in water and oil, in most organic compounds and in all living things. The hydrogen *atom* contains one *proton* and one *electron;* an *isotope,* deuterium, has in addition one neutron; a *radioactive* isotope, tritium, has one proton, two neutrons and one electron. Originally, hydrogen was the standard on which all relative *atomic masses* were based.

hydrogenation Combination with *hydrogen* to increase the amount of hydrogen already present, e.g. conversion of *alkenes* such as *ethane,* C_2H_4, to an *alkane,* ethane, C_2H_6. Industrially this is done in the preparation of margarine and involves the converting of an *unsaturated oil* to a *saturated fat* by *vaporizing* it in the presence of a heated *nickel catalyst.*

hydrogencarbonate ion HCO_3^- The *acid salts* of *carbonic acid,* H_2CO_3, contain the HCO_3^- ion (formerly called the bicarbonate ion). Only a few of these salts are known, some are only known in *solution.* Sodium hydrogencarbonate is used in *baking-powder* and calcium hydrogencarbonate in solution causes water to be *hard.* All hydrogencarbonates give *alkaline* solutions and when treated with a dilute *acid evolve carbon dioxide.*

hydrogen chloride HCl A *gas* which is colourless when dry, but *fumes* strongly in moist air. It is made industrially by the controlled burning of hydrogen in chlorine or as a by-product during certain organic reactions. It is a *covalent compound, soluble* in toluene without any change in this *property,* but when it dissolves in water it forms an *ionic solution* called *hydrochloric acid.*

hydrogen ion H^+ A *positively-charged* hydrogen *atom;* a *proton,* which is present in aqueous *solutions,* from many of which it is liberated during *electrolysis* as hydrogen gas. It is usually *hydrated* in the solution the *hydronium* ion, H_3O^+, in which it is joined to one *molecule* of water. The general *properties* of *acids* in solution are due to the presence of hydrogen ions. In the metallic *hydrides,* the hydrogen carries a *negative charge* and is liberated at the *anode* if the *fused salt* is *electrolyzed.*

hydrogen ion concentration A measure of the *acidity* of a solution; it is expressed in *grams* or *moles* per litre of solution. It is usually represented by the term pH which is a scale running from 1 to 14, i.e. from very *acidic* to very *alkaline.* Excess of H^+ *ions* renders a solution *acidic* ($pH < 7$), but excess of OH^- ions make it *alkaline* ($pH > 7$). Water contains equal numbers of both ions and so is *neutral* and has a pH of 7.

hydrogen peroxide H_2O_2 A *compound* of hydrogen and oxygen which is *unstable* and therefore a good *oxidant* as it liberates oxygen. It is a syrupy liquid which in *concentrated* form is used for supplying oxygen to some rockets, torpedoes and submarines. When dilute it is a good *bleach* for silk and hair.

hydrogensulphate ion HSO_4^- The *acid salts* of *sulphuric acid,* H_2SO_4, contain the HSO_4^- *ion* (formerly called the bisulphate ion).

hydrogen sulphide H_2S A colourless, poisonous gas, smelling like bad eggs and produced by the action of dilute *acids* on certain metal *sulphides,* usually iron(II) sulphide, FeS. It is a strong *reductant* and is much used in *qualitative analysis.*

hydrogensulphite ion HSO_3^- The *acid salts* of *sulphurous acid,* H_2SO_3, contain the HSO_3^- *ion* (formerly called the bisulphite ion). When these *salts* are treated with a dilute *acid* they *evolve sulphur dioxide.*

hydrolysis 1. The chemical *decomposition* of a compound by water, often with the formation of an *acid* and a *base,* but not necessarily so. *Sodium carbonate* in water has a strongly *alkaline* reaction due to the presence of *hydroxyl ions* OH^-. Phosphorus pentachloride with water forms a mixture of hydrochloric and phosphoric acids. $PCl_5 + 4H_2O \rightarrow$

H_3PO_4 + 5HCl. 2. The conversion of *esters* into *alcohols* and acids by the action of water. A little concentrated sulphuric acid is also added. Ethyl ethanoate (acetate) + water→ *ethanol* + ethanoic (acetic) acid.

hydrometer An instrument which is floated in a liquid to measure its *density*. The hydrometer has a weighted bulb at the bottom and a long stem containing a scale. The denser the liquid in which it floats the less of the stem there is below the surface (*principle of flotation*), and so the scale can be *calibrated* to give densities directly.

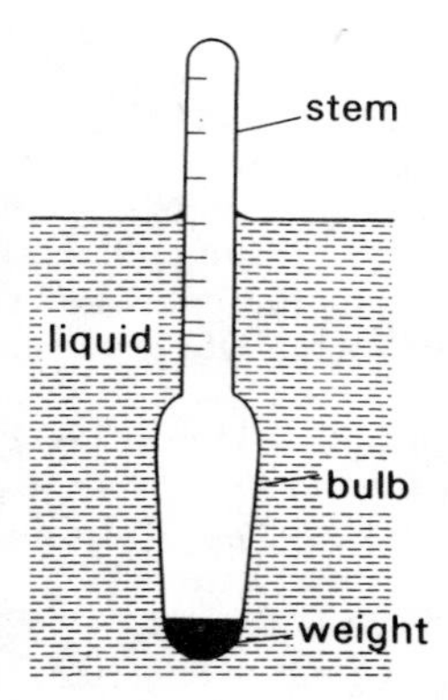

Fig. 93 Commercial Hydrometer

hydronium ion H_3O^+ The simple *hydrogen ion*, H^+, is thought not to exist as such in *solution*, but to be joined to a *molecule* of water to form the *hydrated* hydrogen ion, H_3O^+. It is also called the hydroxonium ion.

hydrophyte A water living plant, e.g. water lily.

hydroponics Growing plants in water without any soil.

hydrostatics The study of *fluids* at rest.

hydrotropism Growth of a *root* towards water.

hydroxides Same as hydrated oxides. *Basic compounds* containing a *cation* (usually a metal ion) joined to one or more hydroxyl (OH) groups, e.g. NaOH, sodium hydroxide; $Al(OH)_3$, aluminium hydroxide; NH_4OH, ammonium hydroxide. They may be regarded as obtained from water by replacement of one hydrogen ion by a metal. If *soluble* in water, hydroxides form *alkaline* solutions containing the positive metal ion and the negative hydroxyl ion. However, most hydroxides are *insoluble* in water and change to the *oxide* on strong heating. They neutralize *acids* to form *salts* and water only.

hydroxyl ion OH^- The negatively-charged OH group present in *hydroxides* and present in excess in *alkaline solutions;* these have a *pH* greater than 7. Hydroxyl *ions* are present in water to the same extent as *hydrogen ions* and are frequently set free at the *anode* during *electrolysis.* Immediate *decomposition* occurs to form water and oxygen: $4OH^- \rightarrow 2H_2O + O_2 + 4e^-$.

hygrometer An instrument, of which there are several types, for measuring *humidity*.

hygroscopic Having a tendency to *absorb* moisture, but not in sufficient amount to form a *solution;* e.g. calcium oxide, *silica gel.* Term also used for liquids such as concentrated sulphuric acid which have this property.

hypermetropia Long sight. A fault in the eye which causes a person to be unable to see near objects clearly. It is corrected by a *converging lens* used in spectacles.

hypha The very thin thread of a *fungus.*

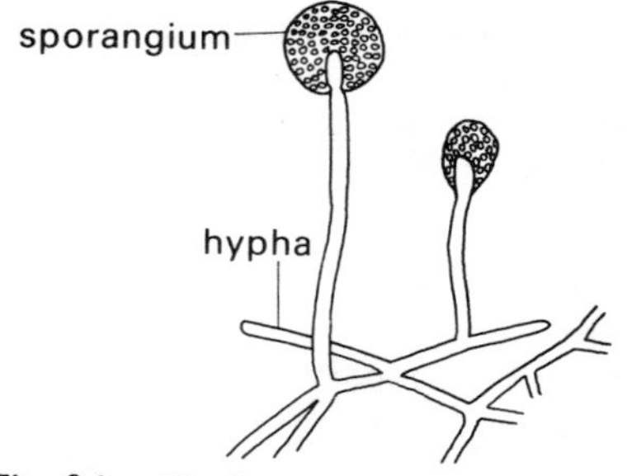

Fig. 94 Hyphae of mould

hypo Photographers' name for *sodium thiosulphate crystals,* $Na_2S_2O_3.5H_2O$, used by them as a 'fixing' agent for a developed film by dissolving away the silver bromide which has not been changed by the light.

hypochlorous acid HClO or HOCl A very *unstable acid* which is present in *chlorine solution,* in household *bleaches* and in some *disinfectants.* It readily breaks down into *hydrochloric acid* and *oxygen.* It forms salts called hypochlorates.

hypogeal When the *cotyledons* of a *seed* remain below the ground, e.g. peas.

hypogynous When the *ovary* of a *flower* (Fig. 76ii) is above the *petals.*

hypothalamus Part of the *brain.* It controls simple behaviour. It is joined to the *pituitary gland.*

hypothesis An idea put forward to explain the way in which certain substances act under certain conditions, but which needs to be tested by experiment to show that it is true.

hypsometer A special container used for calibrating (*calibrate*) *thermometers* at the *boiling point* of water.

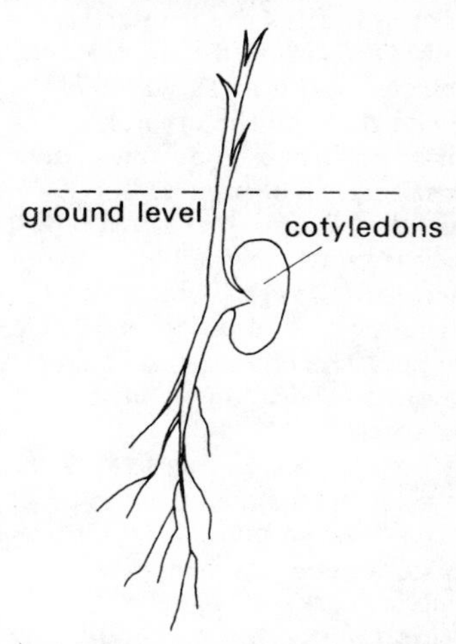

Fig. 95 Hypogeal germination

ice H_2O Water in the solid *state*. As it is less dense (*density*) than water it floats on the surface.

Iceland spar $CaCO_3$ A *crystalline* form of *calcite* (*calcium carbonate*) which possesses the property of double *refraction*, i.e. if a *crystal* is placed over a mark which is then viewed through the crystal, the mark appears to be double. Used in making Nicol prisms.

ideal gas See *perfect gas*.

identical twins When a fertilized (*fertilization*) *egg* divides into two so that two children are made from one egg. The children have exactly the same inherited (*heredity*) characters.

ignition 1. Setting on fire. 2. Heating very strongly. 3. Refers especially to the *burning* of a *petrol*-air mixture in a car engine by means of an electric spark.

ignition temperature The *temperature* below which a substance will not take fire. It is relatively low for most liquid *fuels*, but high for most inorganic solids and liquids.

ileum Part of the *alimentary canal* (Fig. 5). The small *intestine*.

illumination The amount of light which falls per second on unit *area* of a surface. The *SI unit* is the *lux* (*lumen* per square metre) and the symbol is E.

image The picture of an object produced by a *lens* or *spherical mirror*. See also *real image*, *virtual image*.

imago The fully formed stage of an *insect*.

immiscible A term applied to *liquids* which do not mix together, e.g. oil and water.

immunity The ability of a plant or animal to resist attacks of *disease*, e.g. some men are immune to *malaria*.

impermeable membrane A skin which will not let any liquids pass through it, e.g. *cork* in plants, or the *cuticle* of an insect.

impervious A term used for a material or rock through which water or other *liquids* cannot flow.

impulse 1. When a *force* acts for a short time, the impulse is equal to the force acting multiplied by the time for which it acts. 2. A message carried by the *nerves*.

impure spectrum A *spectrum* in which the colours partly mix again after having been separated by the prism etc. See also *pure spectrum*.

incidence An incident ray is one which strikes a lens, mirror or block of glass, etc. The angle of incidence is the angle between this ray and the *normal* at the point where it strikes. See *critical ray* (Fig. 47).

incisor One of the sharp front teeth. Used for cutting food. See *tooth* (Fig. 159i).

inclined plane A simple *machine* made from a straight board of wood, etc., raised at one end and placed so that objects may be slid or rolled up it rather than lifted. The *velocity ratio* is equal to the vertical height risen divided by the length of the board.

incubation 1. The development of an *egg*. 2. Incubation period – the time taken for a *disease* to show itself after it has been caught.

incus One of the *ear* (Fig. 59) *ossicles*. Also called the *anvil*.

indehiscent A kind of *fruit* which does not split open, e.g. coconut.

indian ink A dense, black permanent ink which is a *suspension* of *lamp-black* ground up with water containing a little *gum*.

indicator 1. A substance which is one colour in *acid solution* and another in *alkaline*, e.g. *litmus* (red→blue); *phenolphthalein* (colourless→red). Some indicators such as *universal indicator* show a whole range of colours and give a close approximation of the *pH* of the solution. The indicator may be used in solution or absorbed by pieces of filter paper which are then dried. 2. other kinds of indicators are used for *reactions* not involving acids or alkalis, the end of the reaction being indicated by a particular change of colour.

induced current An *electric current* produced by *electromagnetic induction*.

induction coil This instrument is partly like an *electric bell* for an electric current flowing through a coil on an iron frame sets up a *magnetic field* which pulls another piece of iron towards it to separate two connections and so switch off the current. The moving iron is fixed to a spring so that it moves back to make the connection again, the current then flows once more and the whole pattern repeats itself. This also means that the magnetic field is building up and dying away many times a second. So a second coil, with many turns of thin wire, is wound over the first. A very high *potential difference* is set up across this coil by *electromagnetic induction* and if the wires from the ends of the second coil are brought near to each other, sparks will jump across the gap.

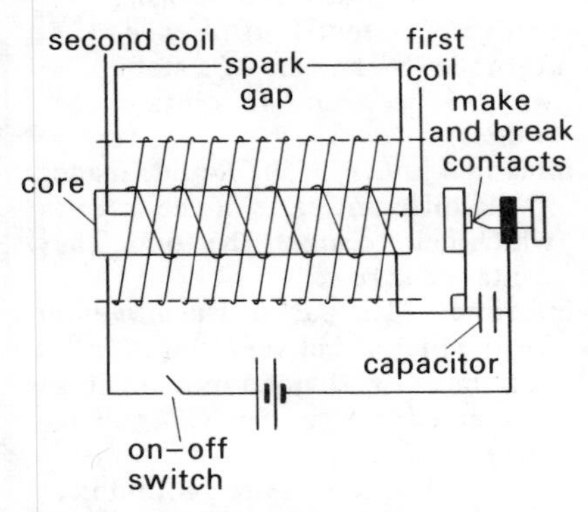

Fig. 96 The Induction Coil

indusium The cover over the sporangia (*sporangium*) on a fern leaf.

industrial methylated spirit (*abbr.* I.M.S.) A variety of *methylated spirit* free from the dye, consisting of 95% *ethanol* and 5% *methanol*.

inert Either not entering into any *chemical reaction* at all or entering into very few chemical reactions and then only with difficulty, i.e. not readily changed by chemical means.

inert gases See *argonons*.

inertia An object remains at rest or moves at *constant speed* in a straight line unless a force acts. This is said to be due to the inertia of the object.

inferior ovary See *epigynous flower*.

inflammable See *flammable*.

inflorescence A group of *flowers* growing on the same plant.

influenza A *disease* caused by a *virus*.

infra red radiation *Radiation* which is part of the *electromagnetic spectrum* but has a longer *wavelength* than visible light. It cannot be seen by the eye but is found by the heating it can produce. It is used for heating in industry and medicine and also for photography using special films.

ingestion The taking in of food through the mouth.

ingot A mass of *metal* poured into a mould or suitable shape ready for further treatment.

inks Deeply-coloured *liquids* which often contain organic compounds of *iron* (black or blue) or suitable *dyes*.

inner ear The part of the *ear* structure which contains the actual hearing and balance parts.

inoculation Putting into the body a weak form of a *disease*, e.g. cow pox, which will cause the blood to make *antibodies* and these will protect the body against the more serious form of the disease.

inorganic This term refers to the compounds of metals and non-metals, but excludes those of carbon other than the *oxides* and the metal *carbonates*. The elements themselves come into this classification.

Insecta A group of *Arthropoda*.

All insects have six legs, either one or two pairs of wings (except some *parasitic* kinds) and the body divided into head, *thorax* and *abdomen*. Many insects have *metamorphosis*. There are many kinds of insects which have good or harmful connections with man.

insecticide A chemical, often in powder form, capable of destroying insect pests on food crops. In solution it may be used as a spray.

insemination Placing *sperms* into a female, either in *mating* or artificially by man, as used in producing many calves from one bull by putting stored sperms into different cows. Artificial insemination is abbreviated to AI.

insertion The end of a *muscle* which is joined to a fixed bone, e.g. muscles fixed to the *skull* at one end and to *vertebrae* at the other, which cause the neck *joint* to bend.

insoluble Incapable of being *dissolved* by a particular liquid. When an insoluble substance is mixed with a liquid there are still two distinctly separate substances, e.g. chalk and water, benzene and water. This is a relative term as most substances do dissolve in water even though it is to a very tiny extent.

inspiration Breathing in.

instinct Something which is in an animal at birth which makes it behave in a special way without having to learn the method, e.g. a spider has an instinct for making its web.

insulator A substance which does not allow heat or electricity to flow easily through it. Non-metals and especially plastics, are good insulators (or very bad *conductors*).

insulin A *hormone* made in the *pancreas* which controls the amount of *sugar* in the *blood*. If there is not enough insulin made, *diabetes* is caused.

intelligence The ability to reason and to understand.

intercellular Between *cells*, e.g. the air spaces in a *leaf*.

intercostal muscles The *muscles* which cause the *ribs* to move in breathing.

interference When two *waves* of the same type cross, it is possible for the top (or crest) of one wave to be in the same place as the bottom (or trough) of the second wave. If this happens the effect of the two waves is zero at that point. However, at other points crests or troughs will come together and the effect of the wave is increased. This pattern of zero effect at some points and increased effects at others is called an interference pattern and the two waves are said to interfere with each other.

internal combustion engine An engine, like that of a car, in which a fuel is burnt inside a cylinder closed by a piston, so changing the energy of the fuel into mechanical energy.

internal resistance The electrical *resistance* inside a cell or other source of *potential difference*.

internode The part of a stem which is between two joints (*nodes*).

intestinal juices The liquids made in the different parts of the *intestine* which help to *digest* the food. They contain *enzymes*.

intestine The part of the *alimentary canal* which is between the *stomach* and the *anus*. Divided into small and large intestine. See *villus* (Fig. 167).

inulin A kind of *starch* found in some plants.

invar An *alloy* of iron and nickel (36%) which hardly changes in length on heating. It is used in watches and other accurate instruments which must not be troubles by changes in temperature.

inversely proportional Changing in the opposite way, but at the same rate, e.g. when the *pressure* on a certain *volume* of *gas* is increased the *molecules* are pressed closer together so the volume must decrease. The product of the two quantities concerned, here Pressure x Volume (PV), remains constant.

invertase An *enzyme* in the *intestine* which changes *sucrose* to *glucose*.

invertebrate An animal which has no backbone (*vertebral column*), e.g. worm, snail.

investigate To discover, or find out what happens when certain things are done, e.g. when a substance is heated or solutions are mixed together.

involuntary action Some action performed which is not under the control of the will power, e.g. the *pupil* of the eye gets smaller in bright light.

iodides *Binary compounds* of *iodine* which contain the iodide *ion*, I^-, e.g. *potassium iodide*, KI.

iodine I At. No. 53 R.A.M. 127 A *solid element* of the *halogen* family in Group VII of the periodic table. It *crystallizes* in small silvery plates and on heating *sublimes* to a violet *vapour*. *Dissolved* in *alcohol* (it is only slightly soluble in water) it forms a brown solution, 'tincture' of iodine, used on small cuts and wounds. A certain amount of iodine is essential for the correct working of the *thyroid gland* in humans or they may suffer from *goitre*. It is used as a test for starch, a blue colour being positive.

ion 1. An *atom* which has become electrically charged by gaining or losing *electrons*. *Metal* atoms form positive ions by loss of electrons, Na^+, Cu^{2+}, while *non-metal* atoms gain electrons to form negative ions, Cl^-, O^{2-}. 2. *Acid radicals* and some other groups of atoms also exist as ions, e.g. sulphate ion, SO_4^{2-}, nitrate ion, NO_3^-, phosphate ion, PO_4^{2-}, ammonium ion, NH_4^+. Ions carry electricity through a solution during *electrolysis*. See also *anion, cation*.

ion-exchange A reaction in which *metal ions* in a *compound* are replaced by others. A chemical *water softener* contains a natural *mineral*, *zeolite*, or a similar *synthetic* one. Zeolite is a complex sodium compound which removes the calcium *ions* present in *hard water* and replaces them by sodium ions, forming calcium zeolite at the same time: thus the water is softened. See also *soft water*.

ionic See *electrovalent*.

ionic bond Term used to describe the *electrostatic* attraction existing between oppositely-charged *ions* in a *compound*, e.g. between the *positive* sodium ions and the *negative* chloride ions in solid sodium chloride. In order to 'break' the bond and allow the ions to move about one of two things must be done: (a) *heat energy* must be supplied in order to *melt* the substance or (b) the compound must be placed in water which has the power of weakening the bond. It is also called an *electrovalent bond*.

ionic dissociation See *ionization*.

ionic equation A *chemical equation* written in such a way as to show that the *reaction* occurring involves *ions*, e.g. any barium compound in solution will react with any soluble sulphate in solution to form *insoluble* barium sulphate as a precipitate. This may be written $Ba^{2+}(aq) + SO_4^{2-}(aq) \rightarrow BaSO_4(s)$. Also, any ammonium salt heated with any alkali will liberate ammonia; $NH_4^+ + OH^- \xrightarrow{\Delta} NH_3(g) + H_2O$

ionization 1. The production of *ions*, as when *covalent hydrogen chloride* gas *dissolves* in water to produce *hydrochloric acid* which contains hydrogen and chloride ions. It does not correctly describe the change produced when a salt such as sodium chloride is placed in water; here the ions are already present in the solid, but separate when water is added.

Ionization in many solutions increases as they are made more *dilute*. 2. The production of *ions*, for example, by the effect of sparks, particles from *radioactive* materials, *X-rays* or *gamma rays* on gases.

ionization chamber A container in which the amount of *ionization* caused by a *radioactive* material, etc., may be studied. A *potential difference* is set up between two *electrodes* and the *ions* move, making a very small *electric current* which is measured by sensitive instruments.

ionize To form ions.

iris The coloured part of the *eye* (Fig. 68) as seen from the outside.

iron Fe At. No. 26 R.A.M. 56 A greyish, magnetic *metal* which is fairly *reactive* towards acids and steam. It burns in oxygen to form *iron(II) iron(III) oxide*, Fe_3O_4 and tends to *rust* (oxidize) when in contact with both air and water at the same time. It occurs naturally as haematite, Fe_2O_3, and magnetite or lodestone, Fe_3O_4, from which it is obtained by heating the *ore* with coke and limestone in a *blast furnace* (Fig. 18). It is used in the manufacture of *steel* for industrial purposes. In the body it is used in the making of *haemoglobin*, the red colour of *blood*. In plants, it helps to make *chlorophyll*, the green colour of leaves.

iron(II) iron(III) oxide Fe_3O_4 A 'mixed' oxide of iron containing the metal in *valency* states 2 and 3. It behaves in *chemical reactions* as though it is a mixture of iron (II) oxide and iron(III) oxide (FeO and Fe_2O_3). With *acids* it forms a mixture of *salts* in which the metal exerts its different valencies. It is also known as tri-iron tetroxide, black iron oxide and magnetic iron oxide. It occurs naturally as magnetite and lodestone.

iron(III) oxide Fe_2O_3 A reddish-brown solid occurring naturally as *haematite*; it can be made in the laboratory by heating *iron (II) sulphate* strongly. It is used as a polishing powder (rouge), for colouring glass and as a paint for ironwork. It used to be called ferric oxide.

iron (II) sulphate $FeSO_4$ A light green powder which forms crystals of green vitriol, $FeSO_4.7H_2O$. It is used as an *insecticide* and in the making of inks and *pigments*. When strongly heated it gives off sulphur dioxide and sulphur trioxide and leaves iron (III) oxide. It used to be called ferrous sulphate.

irritability Another name for *sensitivity*.

ischium Part of the *skeleton*. One of the bones of the *pelvis*.

islets of Langerhans *Cells* in the *pancreas* which make *insulin*.

isobar A line on a map joining points having equal *atmospheric pressure*.

isoclinic line A line on a map joining points having equal magnetic *dip*.

isogametes Two *cells* which are alike and which join together to form a new plant or animal. See *gamete*.

isogonal line A line on a map joining points having equal magnetic *declination*.

isomers *Compounds* in which the *molecules* have the same number of the same kinds of *atoms*, but the atoms are arranged differently, e.g. C_2H_6O may be CH_3OCH_3, methyl methanoate (dimethyl ether) or C_2H_5OH, ethanol. The arrangement of the atoms is

```
    H       H
    |       |
H - C - O - C - H          or
    |       |
    H       H

    H   H
    |   |
H - C - C - O - H
    |   |
    H   H
```

Butane, C_4H_{10}, exists with four different structural arrangements of the atoms and so shows isomerism. Because of the different arrangements of the atoms, isomers have different properties.

isomorphous Term used for minerals which have a similar composition and crystallize in a similar form.

isotopes All the *atoms* in any one *element* have the same number of *protons* (this determines the *atomic number*), but the number of *neutrons* (and therefore the *relative atomic mass* and the *mass number*) may vary. These different kinds of atoms are called isotopes and many naturally-occurring elements are a mixture of isotopes. Most hydrogen atoms have no neutrons, but a few have one giving an isotope, deuterium, of mass number 2 instead of 1. The average atomic mass of hydrogen as usually prepared is 1.008, taking into account the number of each kind of atom present. Chlorine has two isotopes with 18 and 20 neutrons (17 protons) respectively, giving mass numbers of 35 and 37. Since there are 3 atoms of the lighter isotope to each one of the other the average atomic mass is 35.5. Isotopes have identical chemical properties (except for hydrogen and deuterium), but differ in those physical properties which depend on the mass of the atom, e.g. density and melting or boiling point.

jaw The bones surrounding the mouth. Teeth (*tooth*) are fixed into the jaws of *vertebrates*. Some insects have jaws not made of bone. See *mandible*.

Jenner, Edward In 1796 he did the first *inoculation* to prevent *smallpox*.

joint Where two *bones* come together so that one can move and cause bending or stretching. Also found in the boneless legs of *invertebrates*, e.g. insects.

joule The *SI unit* of *work* and *energy*, the symbol is J. It is equal to the work done when a *force* of one *newton* moves one metre in the direction of the force.

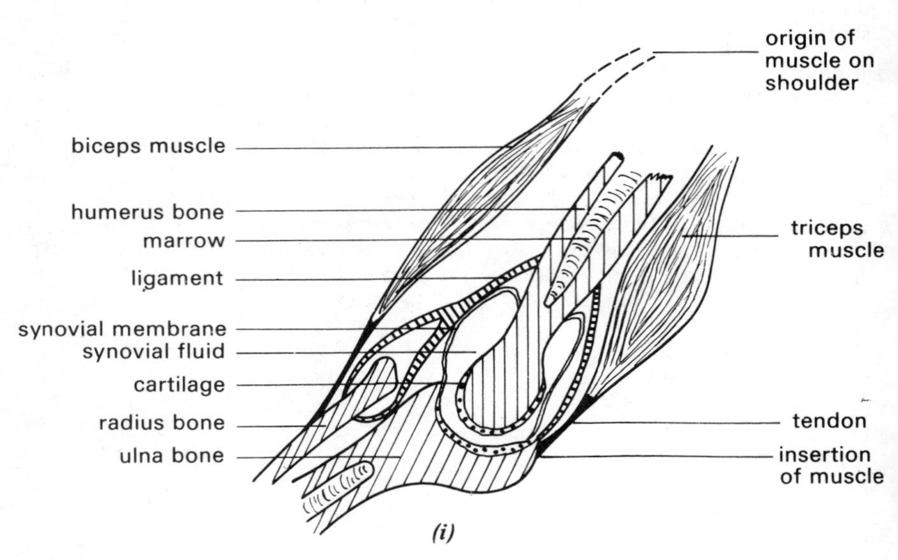

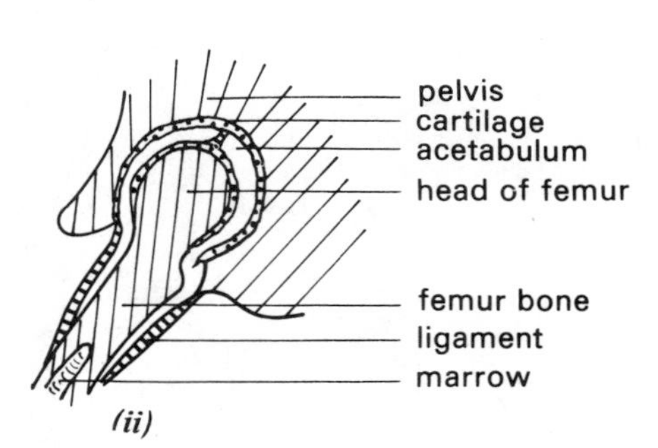

Fig. 97(i) Hinge Joint of the Elbow
(ii) Ball and Socket Joint of Hip

kaolin A hydrated aluminium silicate used in making pottery and *porcelain.* Also called china-clay.

keel 1. The lower *petal* of a bean *flower.* 2. The narrow, sharp edge of the *breast bone* of a bird.

keeper A piece of soft iron placed across the ends of a horse-shoe magnet or between the ends of two bar magnets put side by side with opposite *magnetic poles* at each end. The keeper reduces the loss of magnetism from the magnet.

kelvin The *SI unit* of *temperature*. It is a basic SI unit and the symbol is K. The *absolute temperature* when ice, water and water vapour are in equilibrium (balance) with each other is stated to be 273.16 K. The kelvin is the same size as the *Celsius* degree.

keratin The *protein* substance of *hair* and nails.

kerosine One of the lighter fractions (boiling point range 150–300 °C) obtained by the *fractional distillation* of crude *petroleum.* Also called paraffin oil. It is a mixture of *hydrocarbons* used in domestic heaters and in oil-burning engines.

kidney An *organ* in the body which has the work of taking out water and useless substances from the blood. It is made of different kinds of *cells* and is joined to the *urinary bladder* by a tube, the *ureter.*

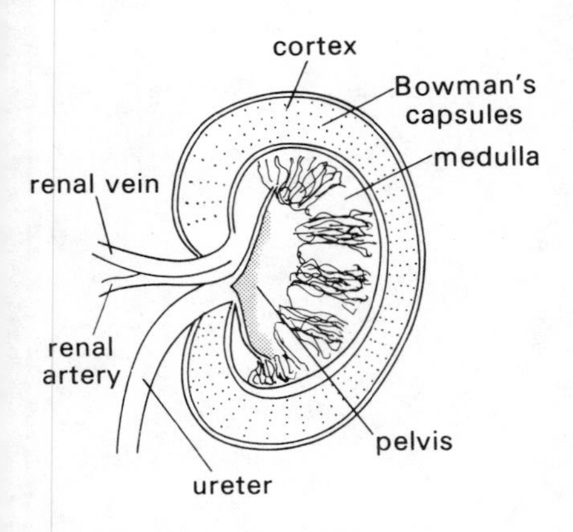

Fig. 98 *Section of Kidney*

killed spirit *Hydrochloric acid* which has been treated with *zinc* until no more *hydrogen* is liberated. The resulting zinc chloride solution is used as a *flux* in *soldering.*

kilo- A prefix which means one thousand times or 10^3. It can be joined in front of any *SI unit,* e.g. 1 kilometre = 10^3 metres. The symbol is k.

kilogram The *SI unit* of *mass.* It is a basic SI unit and the symbol is kg. The kilogram is the mass of a special piece of platinum-iridium *alloy* kept by the Bureau International des Poids et Mesures near Paris.

kilogram-force A *unit* of *force* which is not part of the *SI* system. It is equal to the *weight* of a *mass* of one *kilogram,* and is about 9.81 *newtons.*

kilogram-weight An older name for *kilogram-force.*

kilowatt-hour A *unit* of electrical *energy* which is not part of the *SI* system. It is mainly used in finding the cost of electrical energy. It is equal to a *power* of 1 kilowatt (1000 *watts*) used for one hour.

kinetic energy The *energy* which a body has because it is moving. It is equal to one half the *mass* of the body multiplied by its *velocity* squared. The *SI unit* is the *joule* and the symbol is E_k, T or K.

kinetic theory The theory which explains various ways in which matter behaves, e.g. *evaporation, diffusion, latent heat, Brownian movement* and the *gas laws.* It states that matter is made up of *atoms* or *molecules* which are moving. The amount of movement depends on whether the substance is a solid, liquid or gas. In a solid the particles can only vibrate, in a liquid they are free to move except that they cannot pass easily through the surface but in a gas they can move with complete freedom and at high speed.

Kipp's generator A semi-automatic *gas* generator used in cases where a gas can be produced in the cold from a *solid* and a *liquid*, e.g. hydrogen

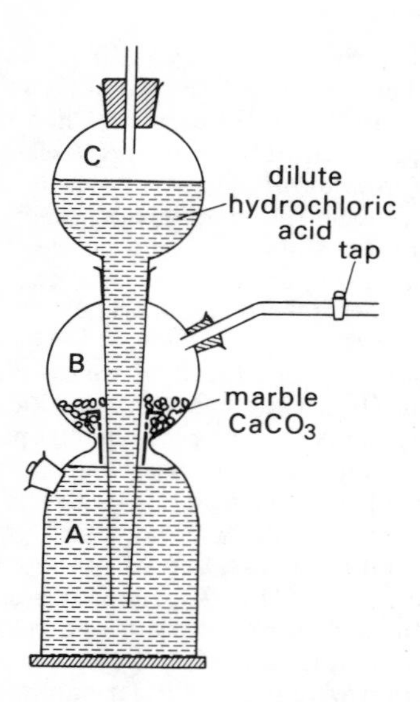

Fig. 99 Kipp's Apparatus for Generation of Carbon Dioxide

sulphide, carbon dioxide, hydrogen. The solid, possibly marble, $CaCO_3$, is placed in the middle compartment B and dilute hydrochloric acid poured into C. When the tap is opened the acid rises into B where it reacts with the marble. When the tap is closed the gas continues to come off for a time, but its increasing pressure soon forces the liquid back into A and C out of contact with the solid. The action then stops.

knee jerk reflex One of the *involuntary actions* of the body. When a tap is given below the knee, the foot gives a quick movement which cannot be stopped.

knocking A tapping noise occurring in *petrol* engines when the spark *explodes* the mixture of *fuel* and air just before the correct moment, often because of too high a *pressure* at that moment. It can be caused by using a *motor spirit* of too low an *octane number*. Also called pinking.

krypton Kr At. No. 36 R.A.M. 84 One of the *argonons* present in the *atmosphere* (1 part in 1 000 000). Like the other gases in this *group* it is *monatomic*. It is used in filling some electric lamps.

labellum The flat end of the mouth tube of a housefly. Liquids are sucked up and pass to the *stomach*.

labium Part of the mouth of *insects*—the lower lip.

labrum The upper lip of an *insect's* mouth.

lachrymal gland The structure at the back of the *eye* which makes the tears.

lacteal A very small tube found in the *villus* (Fig. 167). Fatty foods after *digestion* pass into the lacteals and then are carried away by the *lymph*.

lactic acid $C_3H_6O_3$ (a) The acid produced in milk by the action of certain *bacteria* which cause it to turn sour. Chemically it is 2-hydroxypropanoic acid, $CH_3CH(OH)COOH$ and is a colourless crystalline solid. (b) A substance made in *muscles* during *energy* release. It is formed in the absence of *oxygen*. It is important in the working of muscles Too much lactic acid causes the muscles to tighten up.

Lamarck, J. B. A Frenchman (1728–1777) who had the idea that if an animal tried hard to become strong, its young would start life a little bit stronger. The idea is not now thought correct.

lamina 1. A thin sheet of material. 2. The flat green part of a *leaf* (Fig. 101 i).

lamp black A form of impure *graphite* or *soot* made by burning substances rich in *carbon* (e.g. oil) in a limited supply of air. It is used in the production of rubber for motor tyres, of *Indian ink* and black paint and in the manufacture of gramophone records. It is also called carbon black.

lard An *animal fat* containing a mixture of *esters*.

larva The young stage of an *insect*, e.g. the *caterpillar* of a *butterfly*.

larynx The part inside the throat where the voice sounds are made. See *lung* (Fig. 109 i).

latent heat The heat *energy* needed to change a solid to a liquid (latent heat of fusion) or a liquid to a *vapour* (latent heat of vaporisation). Latent heat is supplied at *constant temperature* and is given up when the vapour turns back to a liquid or the liquid to a solid. See also *specific latent heat*.

lateral displacement The sideways movement of a ray of light when it passes through a rectangular block of glass, etc.

lateral inversion The turning of an *image* seen in a plane mirror so that the right hand side of the object becomes the left hand side of the image.

lateral lines On the sides of a fish. Shallow tubes which tell the fish about movements in the water.

lather The froth (bubbles) produced when *soap* is stirred with *soft water*.

lattice See *crystal lattice*.

laughing gas Former name for *dinitrogen oxide* (nitrous oxide), N_2O.

Law of Constant Composition See *Constant Composition, Law of*.

Law of Constant Proportions See *Constant Composition, Law of*.

Law of Definite Composition See *Constant Composition, Law of*.

Law of Definite Proportions See *Constant Composition, Law of*.

Law of Multiple Proportions See *Multiple Proportions, Law of*.

Laws of Electrolysis See *Faraday's Laws of Electrolysis*.

Law of Pressures The *pressure* of a fixed *mass* of gas at *constant* volume is directly *proportional* to the *absolute temperature*. The pressure therefore increases by by $\frac{1}{273}$ of the pressure at 273 K (0°C) for every kelvin rise in temperature. In equation form

$$\frac{\text{pressure}_1}{\text{absolute temperature}_1}$$

$$\frac{\text{pressure}_2}{\text{absolute temperature}_2}$$

laxative A substance which causes a loosening of the bowels.

L-D process A *steel*-making process in which *oxygen* is directed on to the molten *iron* in a *Bessemer* converter through a water-cooled tube. Also called the Ling-Donawitz process.

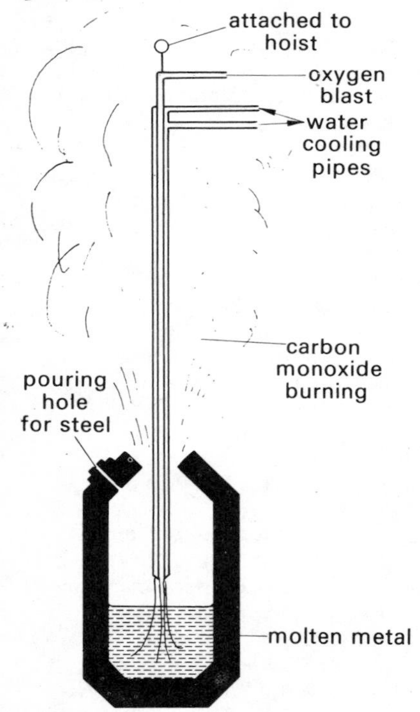

Fig. 100 Bessemer Converter: L–D process

lead Pb .At.No. 87 R.A.M. 207 A bluish-grey heavy *metal*, low in the *electrochemical series*. It resists *corrosion* and is used in plumbing, in accumulators and as a shield in X-ray and nuclear work.

lead accumulator An *accumulator* which has two sets of lead plates in sulphuric acid. Usually six lead accumulators are connected together to make a car battery.

lead acetate paper Strips of filter paper soaked in a *solution* of lead acetate (ethanoate), $(CH_3COO)_2Pb$, and allowed to dry. When exposed to *hydrogen sulphide* gas a brownish-black stain of lead sulphide is produced.

lead (II) oxide PbO Powder obtained by burning *lead* or heating lead carbonate or nitrate. It exists in a yellow form (massicot) and a red-orange variety (litharge). It is used in the manufacture of some paints and glass.

lead tetraethyl $Pb(C_2H_5)_4$ A colourless liquid added to *petrol* in internal combustion engines to prevent the petrol-air mixture igniting (*ignition*) too soon. Commonly called 'anti-knock'.

leaf The flat green structure which grows out from the stem. It is made of different parts. There are small holes (*stomata*) in the skin of the leaf, through which gases pass in and out. The leaf contains *chlorophyll* and this makes it able to make food (see *photosynthesis*). Water passes out of the leaf as a gas. Some leaves are changed so that they can store food or water.

leaf fall The loss of leaves from a tree in the cold season. See also *deciduous*.

leaf scar The mark on a *twig* left when a leaf falls from a stem.

leaflet The little leaves formed when a big leaf is divided into parts, e.g. pea.

Leclanché cell A *cell* in which the negative *electrode*, a zinc rod, is placed in a *solution* of ammonium chloride kept in a glass jar. The positive electrode, a carbon rod, is surrounded by manganese dioxide and carbon powder and is fixed in the middle of a porous pot which is also put in the ammonium chloride solution. See also *dry cell*. (Fig. 102.)

legume The *fruit* of the pea/bean group of plants. It splits open to scatter the *seeds*. (Fig. 103.)

lens A piece of glass (etc.) with one or usually two curved surfaces. It is used to make rays of light come

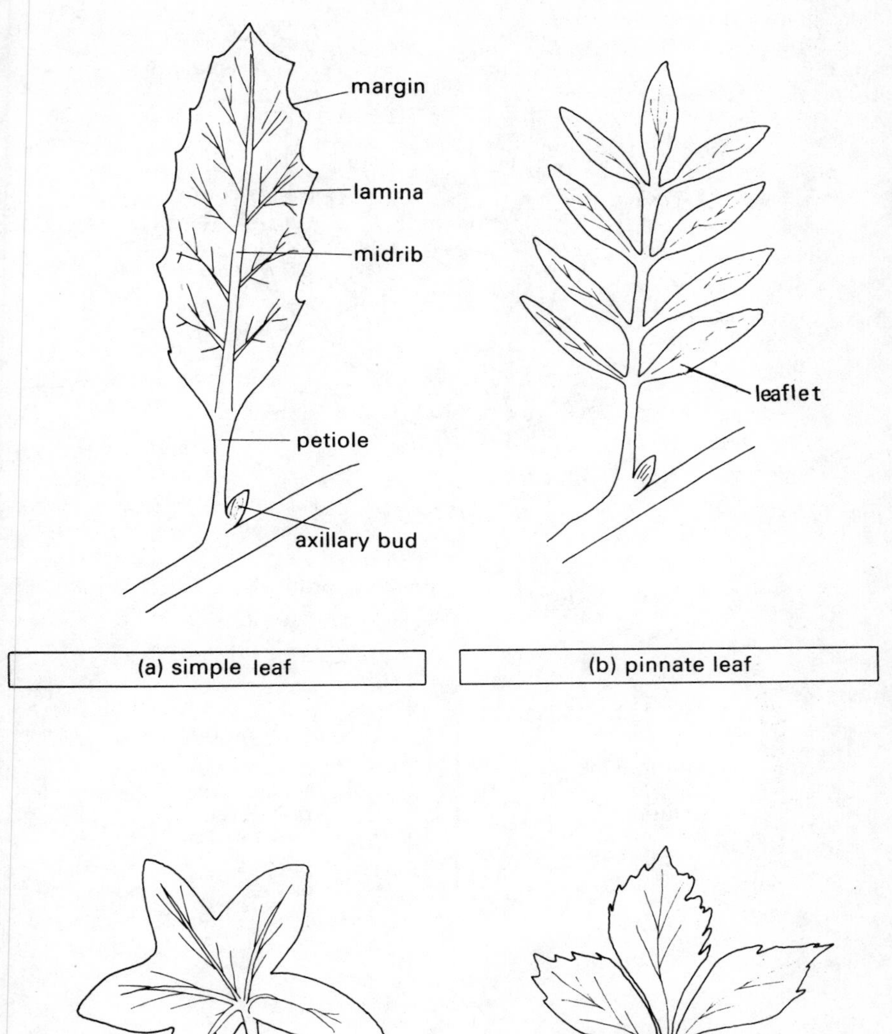

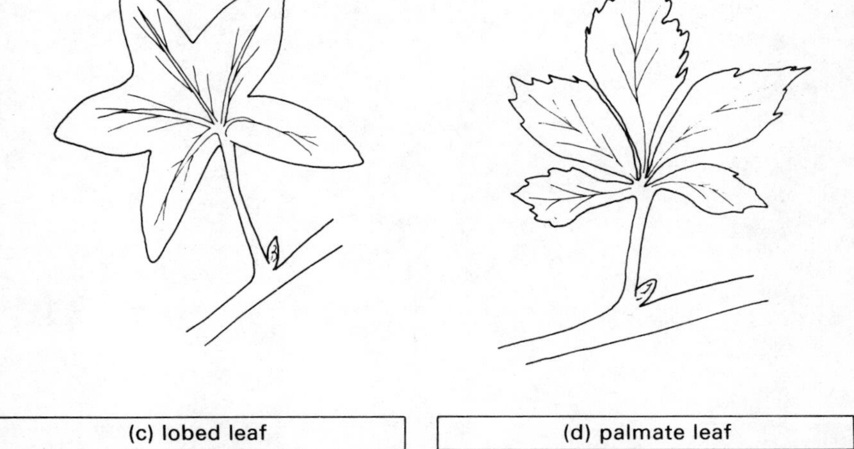

Fig. 101*(i)* *The Different Types of Leaves*

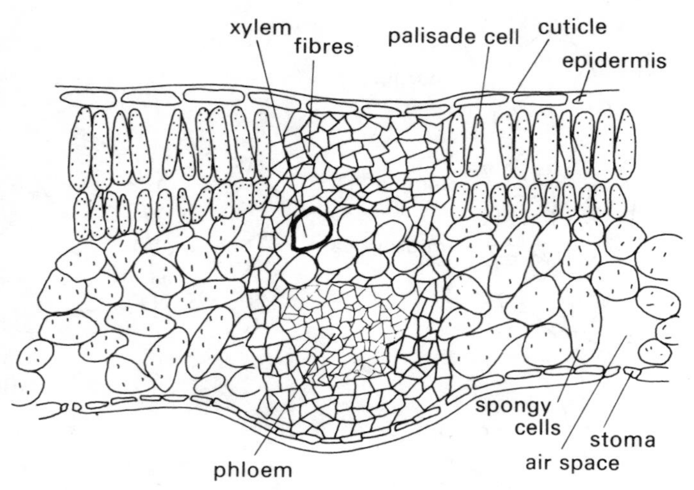

Fig. 101 (ii) *Magnified cross-section of Leaf*

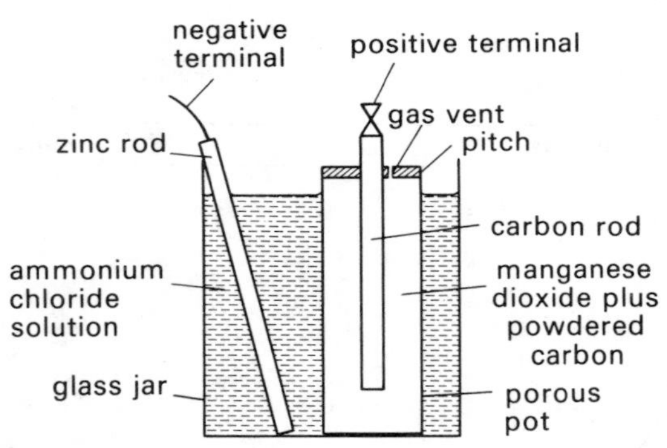

Fig. 102 *The Leclanché Cell*

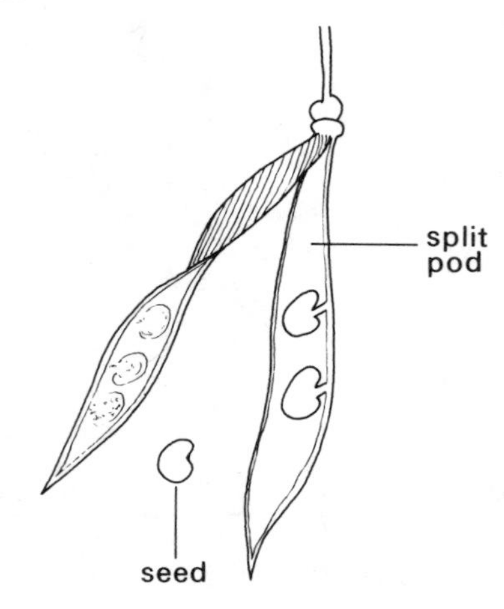

Fig. 103 *Legume*

together or spread out. Lenses are found in *cameras,* projectors, spectacles, *telescopes,* etc. See also *converging lens, diverging lens.*

lens formula An equation which connects the distance (u) of an object from the *optical centre* of a *lens*, the distance (v) of the *image* from the optical centre and the *focal length* (f) of the lens:

$$\frac{1}{u} + \frac{1}{v} = \frac{1}{f}$$

lenticel The breathing openings on a *twig* (Fig. 162). They are filled with loose *corky cells.*

Lenz's law The current produced by *electromagnetic induction* always flows in the direction which opposes the change that is happening. E.g. if a magnet is being pushed into a coil, the *induced current* sets up a *magnetic field* which tries to push the magnet away.

lepidoptera The scientific name for the group of butterflies and moths.

Leslie's cube A metal can with square sides, each side having a different surface finish. One may be polished, another painted so that it has a rough black surface, etc. It is used to study the *emission of radiation.*

leucocyte A *white blood cell.* There are different kinds of leucocytes, having different *functions.* See *cell* (Fig. 33 ii).

leucoplast A colourless small structure in a *cell.* It usually turns green if light is present, e.g. in the skin of a potato.

lever A bar which can turn about a pivot or *fulcrum.* Usually a force applied to one point on the lever causes another force to be set up at another point. See also *principle of moments.*

Leyden jar An old form of *capacitor.* A glass jar has the lower part covered, inside and outside, with metal foil (very thin sheet). An insulating (*insulator*) lid carries a metal rod, ending in a chain knob on the end of this rod is one connection, the outside foil is the other.

lichen A simple plant, half *fungus* and half *alga.* Lichens are able to live in very difficult places, on rocks and where it is very cold and dry.

Liebig condenser A water-cooled tube used for condensing (*condensation*) vapours produced during distillation.

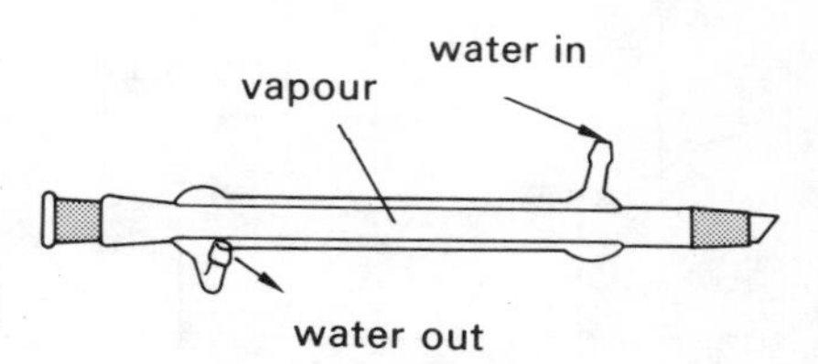

Fig. 104 Liebig Condenser

life history The different stages in the life of a plant or animal, e.g. an *insect.*

lift pump A pump which can only raise water (etc.) through a limited height. *Atmospheric pressure* forces water into the pump body through one *valve* and the water then passes through a second valve and overflows as the piston is raised. (Fig. 105.)

ligament A strong-string-like structure used to join *bone* to bone. See *joint* (Fig. 97).

light 1. (Physics) That part of the *electromagnetic spectrum* which can be seen by the eye. 2. (Biology) The *energy* from the sun which is used to make *sugar* in a *leaf.* This will happen only if *chlorophyll* is present.

lighting-back In a burner where air is drawn into the *gas* stream near the base of the barrel, the *pressure* of the gas keeps the *flame* at the top of the barrel. If the gas supply is slowly reduced, the flame tends to travel back down the tube and eventually it will do so and burn at the jet at the base. This tendency is greater with *coal gas* than with *natural gas.*

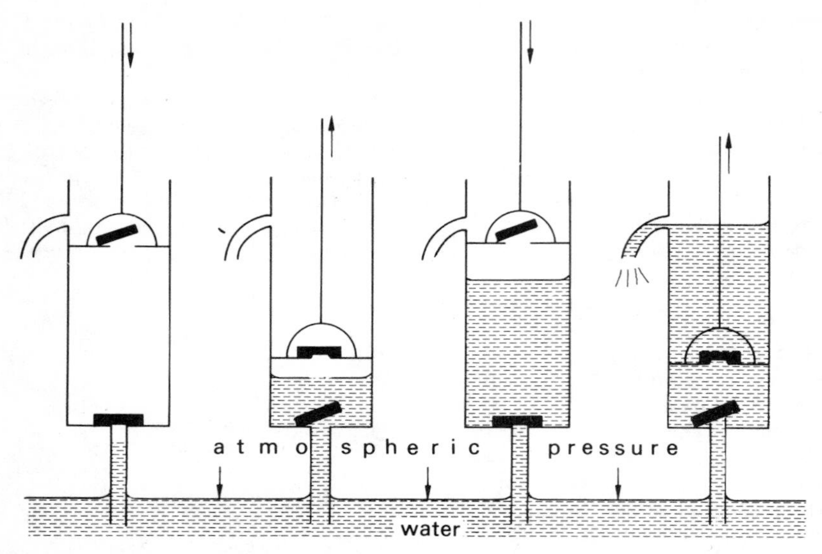

Fig. 105 The Action of a Lift Pump

lightning A large spark which passes between two *electrically charged* clouds or from one such cloud to the earth.

lignin One of the *constituents* of wood where it is often combined with *cellulose*. It is removed by treatment with a solution of calcium hydrogensulphite, $Ca(HSO_3)_2$, in the preparation of pulp for the paper and *rayon* industries. Lignifying means changing to wood.

lignite A fossil *fuel*—brown coal—of more recent origin than *coal* itself.

lime *Calcium* containing substance found in animal *bones* and shells, and used by plants for good growth. It can be added to *soil* to improve it when it is very heavy or full of *disease* causing *organisms*. See also *calcium oxide*.

lime-kiln A tall *furnace* in which *limestone* is heated above 825°C to convert it into the *oxide*. Heating is usually by means of producer gas.
$CaCO_3 \xrightarrow{\Delta} CaO + CO_2$

limelight Intense white light obtained by *heating lime* in an *oxy-hydrogen* flame.

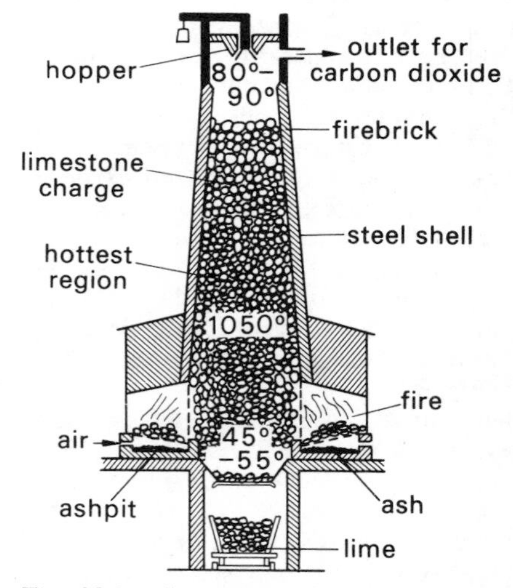

Fig. 106 Continuous Vertical Lime-kiln

limestone A form of *calcium carbonate* which occurs in many districts. It is brown or grey in colour and was originally laid down under the sea. It is fairly hard and is used as a building material and for the production of *quicklime*.

lime water See *calcium hydroxide.*

limit of proportionality When a spring or wire is being stretched the increase in length is *proportional* to the force applied up to a certain limit, called the limit of proportionality. See *Hooke's law.*

limiting friction The highest value of the *frictional* force between two surfaces.

line of force A line used to show the direction and hence the pattern of an *electrostatic* or *magnetic field.* Lines of force run from positive *charge* to negative charge or from north *magnetic pole* to south, and are the paths of a free positive charge or north pole in the field.

line spectrum A *spectrum* formed by *light* from a discharge lamp or a flame containing a trace of a chemical compound. It has separate lines of one or more colours and can be used to see what substance was put in the lamp or flame.

linkage A term used in *genetics* when two characters are passed on together, e.g. *colour blindness* is usually linked with the male sex.

Linnaeus Carl A Swede, the first man to work out a system of classifying (*classification*) plants and animals. He gave the scientific names, e.g. man is *homo sapiens.*

lipase An *enzyme* which works on *fats* in *digestion*, changing them to *fatty acids* and *glycerine* which can be taken into the *lacteals.*

liquefy To turn to a liquid.

liquid A *state* of matter between a solid and a gas. A certain *mass* at a constant *temperature* has a fixed *volume*, but takes up the shape of the vessel in which it is contained, e.g. water in a flask, $H_2O(l)$. Individual particles are free to move about within the confines of the liquid.

litharge See *lead (II) oxide.*

lithium Li At. No. 3 R.A.M. 7 The lightest *metal element.* Silvery-white and one of the *alkali* metals: like the other elements in Group I of the periodic table it reacts with water to liberate hydrogen.

litmus A purple *dye* which acts as an *indicator*, becoming red in *acids* and blue in *alkalis.* It is obtained from a *lichen.*

litre A *unit* of *volume* which is not part of the *SI* system and should not be used in accurate work. It is equal to one cubic decimetre (10^{-3} m^3) and the symbol is l.

liver A large internal *organ* in the *abdomen*. It is joined to the *duodenum* by the *bile duct*. The liver has many different kinds of work to do. It makes *bile*, stores *carbohydrates*, breaks down *amino acids* and deals with worn out *blood cells.* See *alimentary canal* (Fig. 5).

liver fluke A kind of flat worm which is a *parasite* in the *liver* of sheep.

liverwort A kind of green plant which lives in damp places. It is usually flat without stem or leaves— a *thallus.*

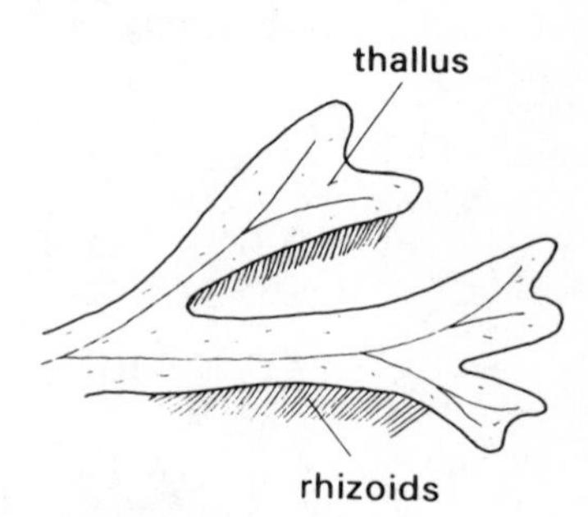

Fig. 107 Liverwort

load The *force* which comes from a *machine* to raise an object etc.

loam A kind of *soil* which contains *sand* and *clay*. The best sort of soil for growing plants.

local action The gradual dissolving of a metal *electrode* of a *cell*, even when no current is flowing. It is caused by other metals (impurities) in the electrodes.

locomotion The movement of the

whole animal from one place to another.

lodestone A magnetic form of natural iron oxide. Pieces may be used as simple *magnetic compasses.*

long sight See *hypermetropia.*

longitudinal wave A *wave* in which the to and fro movement of the particles is in the same direction as that in which the wave is travelling.

loudness The human way of describing the *amplitude* of a sound wave.

loudspeaker An instrument which changes *electric currents* into sound *waves.* A coil carrying the electric current moves in the *field* of a permanent magnet. A paper cone, fixed to the coil, moves with it and sets the air moving.

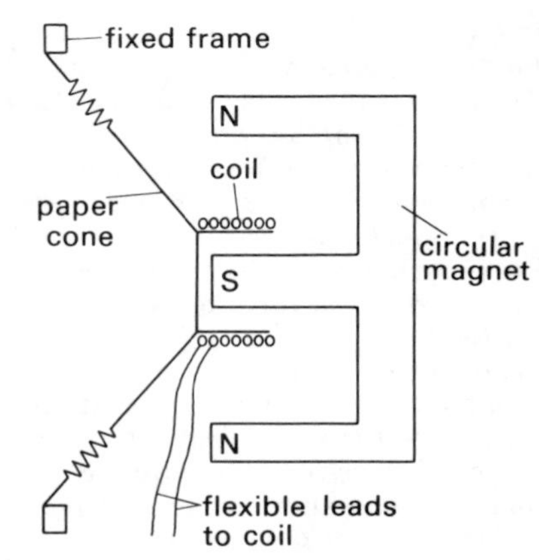

Fig. 108 The Moving Coil Loudspeaker

lubricant A substance such as oil or graphite which is applied to the moving parts of machinery in order to reduce *friction* between them.

lumbar To do with the back of the body, e.g. lumbar *vertebrae.*

lumen The *SI unit* of *luminous flux*. The symbol is lm. It is equal to the amount of light falling on to one square metre of the surface of a sphere (ball) of radius one metre when a source of one *candela* is placed at the centre of the sphere. Thus a source of one candela gives out 4π lumens.

luminous flame The yellow flame produced when *hydrocarbons* are burnt with little or no air mixed with the gas (e.g. a Bunsen burner

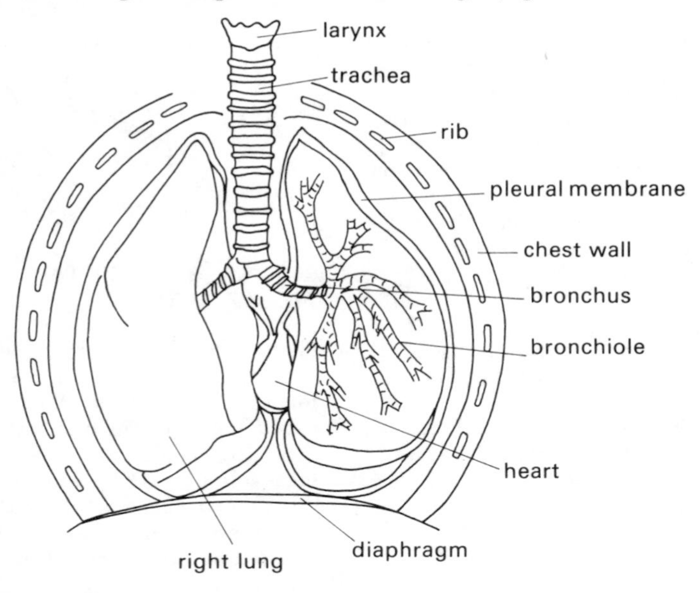

Fig 109*(i)* *The Respiratory System (left lung opened up)*

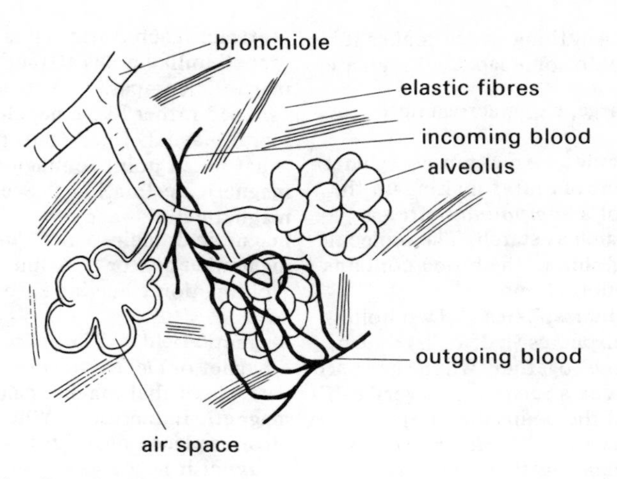

Fig. 109 (ii) *Detail of Lung*

with the air-hole closed). The yellow colour is caused by particles of hot *carbon*, *combustion* being incomplete, and a cold object placed in the flame will have *soot* deposited on it.

luminous flux The total amount of light passing through an area in one second. The *SI unit* is the *lumen* and the symbol is Φ.

luminous intensity The amount of light given out by a point source of light in one second and passing through one square metre of the surface of a sphere (ball) of radius one metre placed so that the source is at the centre of the sphere. The *SI unit* is the *candela* and the symbol is *I*.

lung The breathing organ of the body. It is made from elastic *tissue*, with many air tubes and air spaces.

lux The *SI unit* of *illumination*. The symbol is lx. It is the illumination of a surface receiving one *lumen* on each square metre and therefore of a surface one metre away from a source of one *candela*.

lymph A watery liquid in the body which comes from the *blood*. It carries food to the *cells* and later goes back into the blood.

lymph glands Produce *antibodies* which go into the *lymph* and *blood* and which are important in fighting *disease*. They are found in different parts of the body.

lymphocyte One kind of *white blood cell*.

machine Anything which makes it possible to do some *work* with greater ease.

macro Large, e.g. macrogamete—the big *egg*.

macromolecule An extensive (giant) arrangement of *atoms* making up the *molecule* of a *compound*—often a *polymer*, such as starch. The molecule of haemoglobin in the blood contains about 10 000 atoms.

Magdeburg hemispheres Two hollow metal hemispheres (half balls) which fit very close together. When the space inside is made a *vacuum* it is very difficult to pull the hemispheres apart because the *atmospheric pressure* is pushing them together.

maggot The young stage (*larva*) of kinds of flies.

magnalium A *metal alloy*, consisting mainly of *aluminium*, but containing some magnesium and copper. It is both light and strong and is used in the manufacture of aeroplane parts.

magnesia See *magnesium oxide.*

magnesium Mg At. No. 12 R.A.M. 24 A light grey *metal* which readily burns with a brilliant white flame. Used in lightweight *alloys*. One of the alkaline-earth metals of Group II of the Periodic Table. It is used to make *chlorophyll* in a plant.

magnesium hydroxide $Mg(OH)_2$ A white powder, slightly *soluble* in water to give an *alkaline solution*. A *suspension* in water is called 'milk of magnesia' and is used to relieve excess *acidity* in the stomach.

magnesium oxide MgO A white powder formed by burning magnesium or heating magnesium carbonate. It is very slightly *soluble* in water to give an *alkaline solution.*

magnesium sulphate $MgSO_4$ The *crystalline* form, $MgSO_4.7H_2O$, is known as Epsom salt and is used as a purgative.

magnetic attraction and repulsion Two north *magnetic poles* or two south poles will try to move away from (repel) each other, but a north and a south pole will move towards (attract) each other. Thus 'like poles repel, unlike poles attract'.

magnetic compass A magnet (usually shaped rather like a needle), or a card with magnets fixed to it, pivoted so that it can point to magnetic north.

magnetic declination See *declination.*

magnetic dip See *dip.*

magnetic domain See *domain theory.*

magnetic equator A line on a map joining points having zero magnetic *dip.*

magnetic field The space round a magnet or *electromagnet,* where the effects of that magnet can be noticed.

magnetic induction When a piece of iron is placed near a *permanent magnet* it becomes a magnet itself. This magnetism is said to be induced in the iron and only remains while the permanent magnet is near.

magnetic line of force A line used to show the direction and hence the pattern of a *magnetic field.* Magnetic lines of force run from north *pole* to south pole and are the paths of an imaginary free north pole in the field.

magnetic materials A magnetic field has some effect on all materials, but on iron, cobalt and nickel (ferromagnetic substances) and many *alloys* the effect is very great. Some materials, like soft iron, are easily magnetized but also lose their magnetism easily. Others, like certain steels, are hard to magnetize but keep their magnetism well.

magnetic pole The point in a magnet where the effects of the magnetism are strongest. If a magnet is freely pivoted so that it points north and south, the pole at the end pointing north is called a north pole, and the one at the other end is called a south pole. Magnetic poles always appear in twos, one being a north and the other a south pole.

magnetic screening If a magnet is surrounded by a sheet of iron no *lines of force* pass through the iron and the magnet has no effect outside the iron sheet. In the same way if the magnet is outside the iron sheet the space inside the sheet is free from the effects of the magnet.

magnetic variation Another name for magnetic *declination.*

magnetization The process of turning a piece of *magnetic material* into a magnet.

magnetite Fe_3O_4 Magnetic iron oxide, tri-iron tetroxide, a naturally-occurring *ore* of *iron*. It is attracted by a magnet, but only in the form of lodestone can it attract other objects to itself.

magnification The number of times the *image* produced by a *lens* or *spherical mirror* is bigger (in length) than the object. The magnification is also equal to the distance of the image from the lens or mirror divided by the distance of the object from the same lens or mirror.

magnifying glass A *converging lens* held so that the object being studied is nearer to the lens than its *principal focus.* An upright, *virtual image*, larger than the object, can be seen.

malachite A bright-green *mineral* consisting mainly of *basic* copper(II) carbonate, $Cu(OH)_2.CuCO_3$. It decomposes readily on heating to give black copper(II) oxide.

malaria A *disease* caused by a very small animal, plasmodium, which gets into the *blood* as a result of a bite from a *mosquito.* It multiplies very quickly and causes a fever, which may end in death.

malleable Said of a *metal* capable of being beaten or rolled into thin sheets.

malleus The small hammer bone in the *ear* (Fig. 59).

malpighian body One of a very large number of small tube structures in the *kidney.* They are connected with the removal of substances from the *blood* which form the *urine*. See also *Bowman's capsule.*

malpighian layer The part of the *skin* (Fig. 149) which makes new *cells* as the old ones are worn away.

maltase An *enzyme* which changes a double *sugar* (*maltose*) to *glucose.*

maltese cross tube A special and simple type of *cathode-ray tube.* It contains a single *anode* in the form of a fancy (maltese) cross and no X or Y plates. A sharp shadow of the cross is formed on the fluorescent screen showing that the *cathode rays* (*electrons*) travel in straight lines.

maltose The *sugar* found in many *germinating seeds.* It is also formed during *starch digestion* in animals.

mammal The highest group of animals. Mammals have their young born, they make *milk* to feed the young and the bodies grow *hair,* e.g. goat, man.

mammary gland The parts of the body of a *mammal* which make *milk.* The number of gland outlet is different in different mammals, e.g. two in a human, four in a cow.

mandible The *bone* of the lower jaw or the jaw of an insect.

manganese(IV) oxide MnO_2 A black powder used in several laboratory reactions, (*a*) it *oxidizes* concentrated hydrochloric acid to chlorine; (*b*) it catalyzes (*catalyst*) the decomposition of hydrogen peroxide into water and oxygen; (*c*) it helps the liberation of oxygen from potassium chlorate on heating. It is used in the making of *glass* and in dry *cells.* It used to be called manganese dioxide.

manganin The trade name of an *alloy* containing about 84% copper, 4% nickel and 12% manganese. It is often made into a wire and used for making accurate *resistors* as its *resistance* does not change very much as its temperature changes.

manometer A glass (etc.) tube bent into the shape of a letter U and about half filled with a liquid (mercury, water, oil etc.) When it is joined to a container full of gas under a pressure different from *atmospheric pressure* the liquid comes to rest at a different height in each side of the U-tube. The difference in pressure may be calculated from the difference in height (= difference in height x *density* of liquid x *acceleration due to gravity*).

marble A *crystalline* form of *calcium carbonate* capable of taking a high polish.

marrow The fatty substance inside a hollow *bone. Red blood cells* are made in the marrow.

marsh gas See *methane.*

marsupial One of a group of *mammals,* e.g. kangaroo. The young live in a space (the pouch) formed by a loose piece of *skin* on the *abdomen.*

mass A measure of an object's resistance to a change in motion (i.e. to an *acceleration* or *deceleration*). This resistance depends on the amount of matter in the object. The *SI unit* is the *kilogram* and they symbol is *m.*

mass number The total number of *neutrons* and *protons* in the *nucleus* of an *atom.* If it is necessary to show a *nuclide* with its mass number, it is written at the top, e.g. ^{1}H, ^{12}C, or ^{16}O. See also *atomic number.*

mass spectrometer A complicated instrument which will separate *electrically charged* particles according to their *masses,* by making them pass through *magnetic* and *electrostatic fields.* The resulting pattern shows the composition of the material being tested.

massicot See *lead(II) oxide.*

mastication Mixing up the food in the *mouth,* using the *teeth* and the *tongue.* The food is mixed with *saliva* and made easier to swallow.

mate When a male and a female animal join together in *sexual reproduction.*

matrix The solid parts between *cells,* as in *bone.*

maxilla The upper jaw.

maximum thermometer A *thermometer* which will show the highest *temperature* it reaches. Often a small metal rod (index) is pushed in front of the mercury as the temperature rises and left behind as it falls.

mean free path The average distance travelled by a molecule (etc.) between collisions.

measuring cylinder A tall, cylindrical glass (or plastic) vessel, with marks down one side, used to measure *volumes* of *liquids* with moderate accuracy. See also *graduated cylinder.*

meatus The *ear* hole which leads to the *eardrum* (*tympanum*).

mechanical advantage In a *machine,* the *load* divided by the *effort.*

mechanical energy *Energy* due to position or movement. See also *kinetic energy, potential energy.*

mechanical equivalent of heat When heat and *work* were measured in different *units,* the mechanical equivalent of heat was equal to the number of work units which, when completely changed into heat, would give one heat unit. In *SI* units, where heat and work are both measured in *joules,* the mechanical equivalent of heat is not needed.

mechanism The route by which a *chemical reaction* takes place and the changes involved while it is going on.

median The middle line of the body, e.g. median *fins* of a fish.

medulla The inner part, e.g. the *pith* in the middle of a stem, the inner part of the *brain* or *kidney. Medulla oblongata* is the part of the brain which joins the *spinal cord.* (Fig. 137.)

medullary ray Lines of living *cells* in a woody stem. They come out from the middle towards the outside. They pass and store food.

mega- A prefix which means one million times or 10^{6}. It can be joined in front of any *SI unit,* e.g. 1 megahertz = 10^{6} hertz. The symbol is M.

meiosis The type of division of the *nucleus* which occurs when *gametes* are formed. The double set of *chromosomes* is halved, so that each gamete receives one set. When *fertilization* occurs, the double set is made up again.

Meker burner A Bunsen burner with the top of the barrel widened to fit in a small grid which breaks up the *flame* into a large number of smaller flames and produces a higher temperature as *combustion* is more nearly complete.

melt 1. To heat a *solid* until it becomes a *liquid.* 2. The liquid obtained by melting a solid. See *fused.*

melting point The *temperature* at which a *solid* becomes a *liquid* on heating at a given pressure (usually 760 mm of mercury). For pure crystalline substances this is quite definite, but many amorphous substances, such as wax, soften gradually over a range of temperature.

membrane A thin skin, e.g. the *cell* membrane.

Mendel, Gregor An Austrian monk (1822–1884) who did a great deal of the early work on the laws of *heredity.* He showed that inherited characters remain unchanged when *sexual reproduction* takes place. The character may not show up in the young, but may re-appear in future generations. See *allele, dominant, recessive.*

meniscus The curved surface of a liquid, most noticeable in a narrow tube. If a liquid wets the tube the meniscus is as shown (water); if it does not, then the surface is *convex* (mercury).

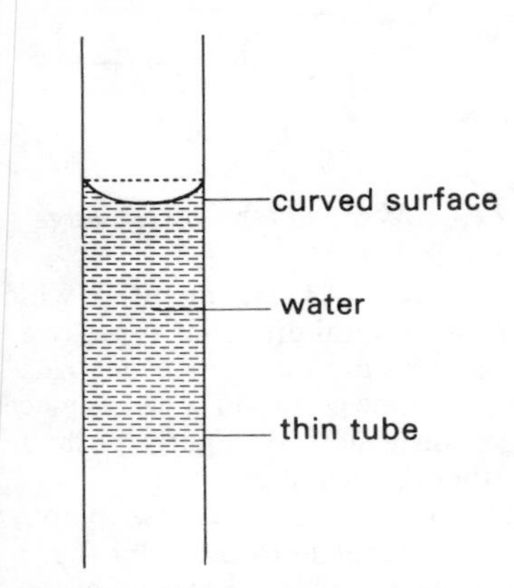

Fig. 110 Meniscus

meninges *Membranes* around the brain. When these are diseased, *meningitis* may occur.

menstruation The monthly flow of *blood* in a woman which keeps the *uterus* fresh and clean. When a baby is growing, the flow of blood stops until the baby is born.

mercury Hg At. No. 80 R.A.M. 200 A silvery-liquid metal used in *thermometers,* barometers and other scientific instruments. Its alloys with other metals are called *amalgams.*

meridian An imaginary line running north-south at any point. Either magnetic or true north can be used to fix either the magnetic or the geographic meridian.

meristem The part of a stem or *root* where active *growth* takes place, e.g. right at the tip of the stem and in buds. In woody stems, there is a meristem just under the *bark* and it is this which makes the trunk get thicker.

mesentery The thin *skin* which holds the different parts of the *alimentary canal* together.

mesophyll The *cells* (*palisade* and *spongy*) inside a *leaf.*

metabolism The name given to all the processes which happen in the life and growth of a living thing which concern chemical substances and *energy.* Divided into *anabolism* and *catabolism. Metabolic rate*—the speed at which usable energy is produced in the body.

metacarpus Part of the *skeleton* (Fig. 148). The bones of the hand.

metal An element which readily forms *positive ions.* Most *metals* are solid at room temperature and are hard, dense, brittle, malleable, good conductors of heat and electricity and have considerable tensile strength. There are exceptions to all these. Chemically, metals *burn* to form *basic oxides.* Most metals react with *acids* to form *salts.*

metamorphosis The changes which occur when a young animal becomes older and different in shape, e.g. tadpole to frog, caterpillar to butterfly.

metaphosphoric acid HPO_3 The acid formed when phosphorus pentoxide (phosphorus smoke), P_2O_5, dissolves in cold water. It forms *salts* called metaphosphates and these contain the PO_3^- *ion.*

metatarsus Part of the *skeleton* (Fig. 148). The bones of the foot.

methane CH_4 An odourless, invisible

gas, the first member of the *alkane* series of *saturated hydrocarbons.* It occurs naturally as a result of the decay of damp vegetation (marsh gas) and in coal mines (fire damp). It is often associated with *oil* deposits and burns readily with a *non-luminous flame* to form carbon dioxide and water. Also known as marsh gas, *natural gas,* North Sea gas and fire damp.

methanol CH_3OH The first member of the *alcohol* series formed from the *alkanes.* It is a colourless, poisonous liquid and is used in industry as a *solvent.* It is added to *methylated spirit* to render the latter unfit to drink. It used to be called methyl alcohol or wood spirit.

method of mixtures A way of making heat measurements. A hot object is mixed with a cold object and the steady *temperature* of the mixture is noted. When the heat *energy* lost by the hot object is put equal to that gained by the cold one the quantity which is not known can be calculated.

methylated spirit A liquid fuel made up of 90% ethanol, the rest being mainly methanol and a dye.

metre The *SI unit* of length. It is a basic SI unit and the symbol is m. It is equal to 1 650 763.73 times the *wavelength* of a certain line in the *spectrum* of krypton.

metre bridge An instrument for measuring electrical *resistance.* A *metre* length of resistance wire is stretched across a board and the ends are joined by copper strips to two *resistors.* A *galvanometer* and a *cell* are also used in the *circuit.* See also *Wheatstone network.*

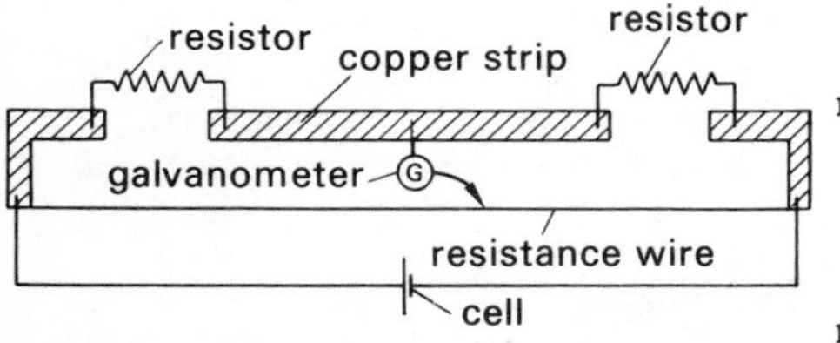

Fig. 111 The Metre Bridge

micro- 1. A prefix meaning 'small'. 2. A prefix which means one millionth or 10^{-6}. It can be joined in front of any *SI unit,* e.g. 1 microfarad $= 10^{-6}$ farad. The symbol is μ.

microbiology The study of the very small living things, e.g. *bacteria* and *algae.*

microcrystalline Consisting of such extremely small *crystals* that the substance may appear *amorphous,* e.g. some varieties of carbon and phosphorus.

micrometer An instrument used to measure small lengths accurately. The object whose length is to be measured is placed between the fixed and the moving part of the instrument, and the moving part is screwed up until it just touches the object. A scale, usually in two parts, on the screw shows the size of the object.

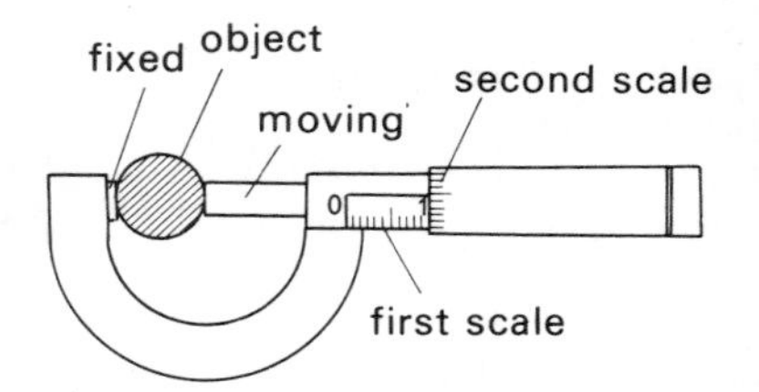

Fig. 112 Micrometer Screw Gauge

microphone An instrument, of which there are several different types, for changing sound *waves* into *electric currents.* One common type has pieces of carbon which are slightly pushed together or moved apart by and fro movement of a metal plate. The plate is made to move by the sound wave. This changes the *resistance* of the instrument and, if a *battery* is connected, the current flowing will also change.

micropyle The small hole in an *ovule* through which the *pollen tube* grows to take the male *cell* to the *egg.* The micropyle later becomes a hole in the *seed* through which water is taken in.

microscope An instrument which produces a highly magnified (*magni-*

fication) *image* of a very small object. It contains two *converging lenses* fixed so that the magnified image produced by the first lens (*objective*) is magnified even more by the second lens (*eyepiece*).

middle ear The space, filled with air, just inside the *eardrum*. The ear bones are found here. A tube joins the middle ear to the throat (*eustachian tube*).

midrib The main *vein* of a *leaf* (Fig. 101i).

mil See *millilitre*.

mild alkali The name given to *sodium* and potassium *carbonates* which have an *alkaline* reaction in *solution*.

mild steel An *alloy* of *iron* and *carbon* with less than 0.15% of carbon. Mild steel is used for structural engineering purposes.

milk The white watery liquid produced by the female *mammal*. It contains all the food needed by the young.

milk of lime A *suspension* of *calcium hydroxide* in water.

milk of magnesia See *magnesium hydroxide*.

milk teeth The first set of teeth of a young mammal. They drop out and the *permanent teeth* grow.

milli- A prefix which means one thousandth or 10^{-3}. It can be joined in front of any *SI unit*, e.g. 1 millimetre = 10^{-3} metre. The symbol is m.

Millikan's experiment A very important experiment which studied the movement of very small drops of oil in an *electric field*. It showed that *electric charge* was always found as a whole number times a basic amount, the *electronic charge*.

millilitre 1 thousandth of a *litre*. Usually shortened to 'mil' and this is abbreviated to ml. 1 ml = 1 cm^3.

Million's reagent The liquid used to test for *protein*. A white *precipitate* is made and this turns red when heated.

mineral A substance occurring naturally in the earth and not containing living matter, e.g. diamond, coal, clay, marble, oil.

mineral carbon *Diamond* and *graphite*.

mineral oil See *petroleum*.

mineral salts Substances in the soil water (plants) or in food (animals). They are needed for growth, e.g. *potassium nitrate, calcium phosphate*.

minimum deviation The smallest change in direction of a ray of light that is possible, as it passes through a *prism* (etc.). See also *deviation*.

minimum thermometer A *thermometer* which will show the lowest *temperature* it reaches. Often a small metal rod (index) is pulled back by the alcohol as the temperature falls and left behind as it rises.

mirage The appearance of something, in a particular place, that is not in that place due to the *refraction* of *light* by a cold or hot layer of air. For example, light from the sky, refracted by a layer of hot air just above the hot sand in a desert or the hot surface of a road, gives the appearance of a pool of water.

mirror A piece of glass (etc.) which supports a metal surface that reflects (*reflection*), in a regular way, most of the light which falls on to it.

miscibility The property possessed by some liquids of mixing in all proportions with others to form a *homogeneous* substance, e.g. ethanol and water.

mist A *suspension* in air of small drops of a *liquid*, usually water, but may be other substances such as *hydrochloric acid* (this is why the concentrated acid appears to *fume*) or sulphuric acid.

mitochondria Very small particles in a *cell* which control the chemistry of *respiration*.

mitosis The changes which take place when a *nucleus* divides in a growing part of a plant or animal. Each *chromosome* splits into two and the two halves separate, one half going into each new *cell*.

mitral valve The *bicuspid valve* found in the *heart*. It controls the flow of *blood* from the left *auricle* to the left *ventricle.*

mixture Two or more substances together forming a *heterogeneous* mass, e.g. iron and sulphur, or a *homogeneous* one such as hydrogen and oxygen. There is no new substance formed (*compound*), the *properties* are not fixed and separation by physical means is usually fairly simple.

mobile Thin and runny like water.

moderator A substance such as graphite used in *nuclear reactors* to slow down the *chain reaction* set up by the release of *neutrons* as a result of *fission.*

moisture Dampness; often caused by the *condensation* of *water vapour.*

molar Concerning the *mole;* some property or factor depending on the mole.

molar solution A *solution* containing one *mole* of a substance dissolved in one *litre* of solution, e.g. 98 g of sulphuric acid in one litre of *aqueous* solution.

molar volume The *volume* occupied by one *mole* of a substance under certain conditions, e.g. for gases at *standard temperature and pressure* this value is 22.4 litres or approximately 24 litres at *room temperature and pressure* (20 °C and 760 mm of mercury pressure).

molarity The number of *moles* of a substance contained in one *litre* of *aqueous solution.* One mole of sodium hydroxide has a *mass* of 40 g, so a solution containing 80 g per litre is two molar (2M) and a solution with 4 g per litre is one-tenth molar (0.1M).

molars The teeth at the back of the mouth. They grind up hard food. See *tooth* (Fig. 159i).

mole The *SI unit* of *amount of substance.* It is a basic SI unit and the symbol is mol. A mole of any substance contains as many particles (*atoms, molecules, ions, electrons* etc.) or groups of particles as there are atoms in 0.012 kilograms of the carbon *isotope* whose *mass number* is twelve, i.e. 6.02×10^{23} particles. It is always necessary to follow the word mole with the name of the particles, e.g. a mole of oxygen molecules. The mole replaces the older terms gram-atom, gram-molecule, etc. For any substance the mole corresponds to the *formula-mass* in grams.

molecular equation A *chemical equation* in which all the substances are put down as *atoms* or *molecules* and not as *ions,* e.g. $Zn + H_2SO_4 \rightarrow ZnSO_4 + H_2$.

molecular formula The *formula* of a *chemical compound* showing the number and kind of *atoms* present in the *molecule,* but not their actual arrangement, e.g. benzene, C_6H_6; ethanol, C_2H_5OH.

molecular mass See *relative molecular mass.*

molecular size The size of a *molecule* can be roughly found by placing a small drop of a substance like oleic acid on the surface of water. The *volume* of the drop and the *area* over which the oleic acid spreads are measured and then the thickness of the oleic acid film can be calculated. The film is thought to be one molecule thick and the experiment usually gives a result of about 10^{-9} m.

molecular weight See *relative molecular mass.*

molecule The smallest part of an *element* or *compound* which can exist on its own and still show all the properties of the element or compound. There may be one or more *atoms* in the molecule and they may be of the same kind as in an element (hydrogen, H_2) or of different kinds in a compound (water, H_2O).

Molisch test A test used in identifying a substance as a *carbohydrate.* The substance in solution is mixed with a little 1-naphthol ($C_{10}H_7OH$) and then a drop or two of concentrated *sulphuric acid* is added. A violet colour confirms the substance as a carbohydrate.

Mollusca The group name for shell fish, e.g. mussels and snails.

molten Said of a *solid* substance

which has been heated until it melts and becomes a *liquid.*

moment of a force The turning effect of a *force,* which is equal to the force multiplied by the distance from the force to the pivot or turning point (the distance is measured at right angles to the line of the force). Moments are called clockwise or anti-clockwise depending on the direction in which the particular force, if it acted on its own, would turn the object about the pivot.

momentum The *mass* of an object multiplied by its *velocity.* See also *conservation of momentum.*

monatomic Having one *atom* in the *molecule* as argon, Ar, or helium, He.

monobasic acid An *acid* with only one *atom* of *hydrogen* in the *molecule* which can be replaced by a *metal* to form a *salt,* e.g. hydrochloric acid, HCl or ethanoic acid (acetic acid), CH_3COOH. Such acids form only one series of salts, chlorides or ethanoates (acetates).

monochromatic light *Light* of one colour, or more accurately, light of one *frequency.*

monoclinic *Crystals* of a particular shape having three unequal axes, two at right angles to each other. One variety (*allotrope*) of sulphur forms monoclinic or prismatic crystals.

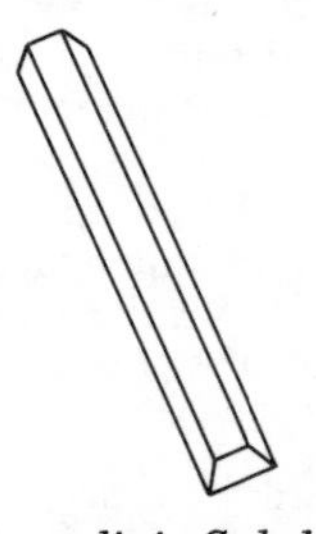

Fig. 113 Monoclinic Sulphur Crystal

monocotyledon A group of flowering plants whose *seeds* have only one *cotyledon,* e.g. corn, lily.

monomer The simple unit from which, by repetition, *polymers* are made, e.g. *ethene* (C_2H_4) in polythene, or *styrene* ($CH_2{=}CH{-}C_6H_5$) in polystyrene.

monovalent Having a *valency* of one – capable of combining with one *atom* of hydrogen or chlorine, e.g. sodium, the ammonium ion, the nitrate ion.

mordant Substances which fix a *dye* to a fabric which cannot be dyed directly.

mortar 1. A mixture of calcium hydroxide, sand and water (sometimes also cement) used in building for holding bricks together. It sets hard on exposure to the air. 2. A strong bowl used with a *pestle* for grinding a solid to a powder.

moss A plant of the group *Bryophyta.* Mosses have a simple structure, no true root *system* and reproduce by *spores* and *gametes.*

motor nerve A *nerve* which carries messages from the *brain* or *spinal cord* to the *muscles.*

motor spirit See *petrol.*

mould A name given to kinds of *fungus* which grow on jam or cheese or similar substances.

moulting 1. The splitting of the *skin* of a young *insect,* making it able to grow. See *ecdysis.* 2. The dropping out of *feathers* or *hairs* so that new ones can grow.

moving coil instrument An *ammeter, galvanometer* or *voltameter* in which a pivoted coil carrying an *electric current* turns in the *magnetic field* of a *permanent magnet* against the action of springs and moves a pointer over a scale. (Fig. 114.)

moving iron instrument An *ammeter* or *voltmeter* in which a piece of iron moves when the current flows through a fixed coil, setting up a *magnetic field.* There are several different types.

mucilage A sticky substance made by some plants, e.g. seaweeds.

mucus The slippery substance produced by some animals, e.g. in the mouth and throat, to help in the swallowing of food.

muffle furnace A small oven heated by gas or electricity and suitably lined so that heat does not easily escape.

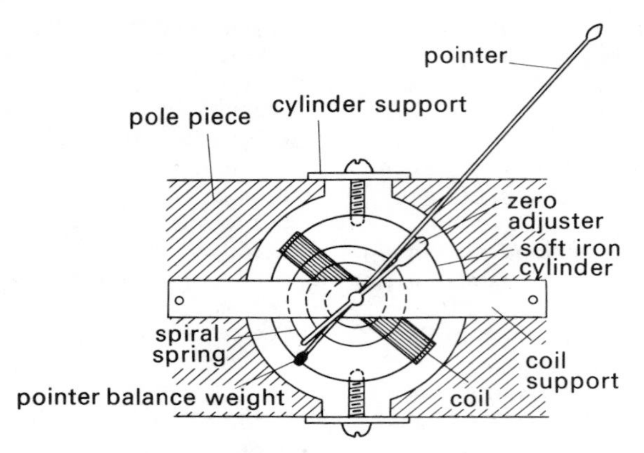

Fig. 114 The Moving Coil Galvanometer

Multiple Proportions, Law of When two *elements,* A & B, join together to form more than one *compound,* the different *masses* of B which unite with a fixed mass of A bear a simple relationship to each other, e.g. in carbon monoxide, 16 g of oxygen join with 12 g of carbon and in carbon dioxide 32 g of oxygen join with 12 g of carbon. The masses, 32 and 16 g, are in the simple ratio of 2:1. This is apparent from the formulae, CO and CO_2.

multiple reflections When two flat (plane) *mirrors* are placed at an angle to each other and an object is put between them, many images can be seen. If the mirrors are at right angles to each other 3 images will be seen, if they are at sixty degrees to each other 5 images, etc. If the mirrors are parallel to each other a very large number (infinite) of images will be seen.

muscle The flesh of the body. Muscles will stretch and pull back, and work usually on *bones,* so that movements of the body take place. The strength of the body and the speed of an animal depend on the muscles.

muscle fibres The thin threads of elastic *cells* which make up a *muscle.*

mutation The sudden change in the chemical structure of a *gene.* The result is the appearance of a new character, e.g. a plant which usually has red flowers produces seeds which grow into plants with blue flowers.

mycelium The very thin threads (*hyphae*) which make the growing part of a *fungus.*

myopia Short sight. A fault in the eye which causes a person to be unable to see distant objects clearly. It is corrected by a *diverging lens* used in spectacles.

Myriapoda The group name for centipedes and millepedes.

myxoedema A *disease* caused by the failure of the *thyroid gland.*

myxomatosis A *virus disease* which kills rabbits.

nano- A prefix which means one thousand millionth or 10^{-9}. It can be joined in front of any *SI unit*, e.g. 1 nanosecond = 10^{-9} second. The symbol is n.

naphthalene $C_{10}H_8$ A solid *hydrocarbon* with a pungent smell; crystallizes as white flakes. Sold as moth balls to put among clothing.

nasal To do with the nose, e.g. nasal canal.

nascent state When gases such as oxygen are first made in a reaction they are probably in the form of single *atoms* and are said to be nascent (newly-born). In such a state the oxygen (for example) is much more *reactive* than when molecular and the atoms combine readily with other substances (e.g. with some dyes which are bleached) which may be present. Failing this, they join together as molecules.

nastic A kind of movement in plants, e.g. the opening and closing of *flowers* in the morning and at night.

natural gas A mixture of *hydrocarbons* (almost entirely *methane*), obtained from the earth or from under the sea.

natural selection One of the ideas of *Darwin* in his Theory of *Evolution*. He said that nature picks out the best kinds of plants and animals and the weaker ones die. In this way the best characters are passed on.

near point The nearest point to the eye where objects can be seen clearly and without difficulty. The near point is about 25 cm from the normal eye.

nectar The sweet liquid made by *flowers* and collected by bees to make *honey*.

negative charge An *electric charge* of the same type as that on the *electron*.

negative electrode The *electrode* at which *metal ions* in *solution* or in a melt gain *electrons* on being released during *electrolysis*. Also called the *cathode*.

negative ion An *atom* or *radical* which has gained one or more *electrons*. Such *ions* are, e.g. Cl^-, OH^-, SO_4^{2-}. They are released at the anode during electrolysis, and are called *anions*.

negative pole See *cathode*.

Nelson cell A cell in which sodium chloride solution is *electrolyzed* to produce chlorine and hydrogen to leave a solution of *sodium hydroxide*.

Nematoda A group name of round worms. Many nematodes, e.g. hookworm, are harmful *parasites* in man. Many other kinds live in the *soil*.

neon Ne At.No. 10 R.A.M. 20 A colourless, invisible gas present in small amounts in the air and used in illuminated discharge signs, producing a red glow. One of the argonons in *Group* O of the *periodic table*, it forms no known compounds.

nephridium A tube used for *excretion*, especially in worms. In the earthworm there are two nephridia in every *segment*.

nerve A string like structure which carries messages (*nerve impulses*) through the body. *Sensory* nerves carry impulses from the *sense organs* to the *central nervous system*. *Motor* nerves carry impulses from the central nervous system to *muscles* and *glands*. A nerve is made up of *nerve fibres* surrounded by a fatty *membrane*.

nerve cell A single *cell* of the *nervous system*. Each cell connects with others to form pathways for *nerve impulses*. See also *neuron*, *nerve fibre*.

nerve fibre One of the many fine threads inside a *nerve*. Each fibre is part of a *neuron*, the *cell body* of which is usually in the *central nervous system*. A nerve fibre carries impulses in one direction only, but both *sensory* and *motor* fibres may be found in a mixed nerve.

nerve impulse The message carried by a *nerve fibre*. An impulse is controlled by *chemical* and *electrical* means.

nervous system The *brain*, *spinal cord* and all the *nerves* of the body.

Nessler's reagent A *reagent* used for the detection of *ammonia* in solution. It is an alkaline solution of potassium iodide and mercury (II) iodide and it produces a yellow or brown colour on exposure to ammonia.

neural Having to do with the *nervous system*.

neural arch The part of a *vertebra* (Fig. 166) around the *spinal cord*.

neuron A *nerve cell*, made up of *cell body*, *axon* and *dendrons*. The whole *nervous system* is built up of neurons.

neutral A *solution* with a *pH* = 7; it is neither *acidic* nor *alkaline*.

neutral equilibrium If a body is in *equilibrium* and then it is moved slightly to one side it may stay in its new position. If it does so it was in neutral equilibrium, e.g. a ball on a level table. See also *stable* and *unstable equilibrium*.

neutral point A point in a *magnetic field* where there are no *lines of force* and therefore where the strength of the field is zero.

neutralization 1. A chemical reaction between an *acid* and an *alkali* (or *base*) until neither is in excess and the solution is neutral. The properties of the original acid and base completely disappear and a new substance, a *salt*, is formed, e.g. sodium hydroxide and hydrochloric acid give a neutral solution of sodium chloride. 2. With strong acids and bases, ie. ones which are fully *ionized*, neutralization may be regarded as a combination of hydrogen ions (H^+) and hydroxyl ions(OH^-) to form water, $H^+ + OH^- \rightarrow H_2O$.

neutron A basic particle found in the nuclei (*nucleus*) of all *atoms* (except normal hydrogen). It has no *electric charge* but a *mass* about the same as that of the *proton*.

newton The *SI unit* of *force*. The symbol is N. If a force of one newton acts on a body of *mass* one *kilogram* it will produce an *acceleration* of one metre per second squared.

Newton's Law of Gravitation The *force* of attraction between two particles is *proportional* to the *mass* of one multiplied by the mass of the other, divided by the square of the distance apart.

Newton's Laws of Motion 1. A body moves steadily along a straight line or stays at rest unless an unbalanced force acts on it. 2. The rate at which the *momentum* of a body changes is *proportional* to the force which acts on the body. The change in momentum is the same direction as the force. This also means that, unless the mass of the object is changing, the force is proportional to the *mass* multiplied by the acceleration. 3. If body A sets up a force on body B then B will set up an equal force on A but in the opposite direction.

nichrome wire A wire made from an *alloy* of nickel and chromium, which is a good conductor of heat and electricity and does not readily burn or oxidize when heated. It is used in electric fires for carrying the heating current.

nickel Ni At. No. 28 R.A.M. 58.7 A silvery-grey *metal* similar to iron with a high melting point. Not readily attacked by air or water. Forms *salts* which are frequently green and give green solutions. Used as a *catalyst* in many reactions involving *hydrogen* (e.g. the manufacture of margarine) and, deposited on other metals, makes them resistant to corrosion. Other uses are in coinage, in food-handling plants and in the form of wire in small computers.

nictitating membrane The third eyelid of a bird, used to keep the eye clean.

nife cell The trade name of a

type of *accumulator* which uses iron and nickel oxide plates in potassium hydroxide solution.

nitrates Salts of *nitric acid*; they contain the NO_3^- ion and are all soluble in water. They can be identified as nitrates by the *brown ring test* (Fig. 23).

nitre See *potassium nitrate*.

nitric acid HNO_3 A *fuming*, *corrosive acid* made in the laboratory by heating sodium nitrate with concentrated sulphuric acid. Colourless when pure, but usually yellow due to dissolved nitrogen dioxide. A powerful oxidant, it attacks most metals. Industrially it is made by passing *ammonia* and air over a heated platinum *catalyst* and dissolving the oxides of nitrogen formed in water (or in dilute acid already prepared). See *flow diagram* (Fig. 75). It is used in the manufacture of nitrates for *fertilizers* and for *explosives*.

nitric oxide See *nitrogen monoxide*.

nitrides *Binary compounds* of *nitrogen* with a *metal*, e.g. magnesium nitride, Mg_3N_2. Usually prepared by passing nitrogen or ammonia over the heated metal.

nitrifying bacteria See *nitrogen bacteria*.

nitrite A compound (*salt*) of *nitrous acid* and a metal or other group, which contains the NO_2^- *radical*, e.g. sodium nitrite, $NaNO_2$.

nitrochalk A quick-acting *fertilizer*.

nitrogen N At. No. 7 R.A.M. 14 An odourless, invisible gas making up about 78% of the atmosphere, It is not very reactive, but does combine with a number of elements under suitable conditions. Present in *proteins* and essential for life. Used in the manufacture of *ammonia* and of compounds used as *fertilizers.*

nitrogen bacteria *Bacteria* which live in the *soil*. They are important in getting the nitrogen out of the remains of plants and animals. Some are *nitrifying*, breaking down *humus*, others are *nitrogen fixing*, using the nitrogen of the air in the soil and passing it on to plant roots, e.g. beans.

nitrogen cycle The circulation of *nitrogen compounds* in nature.

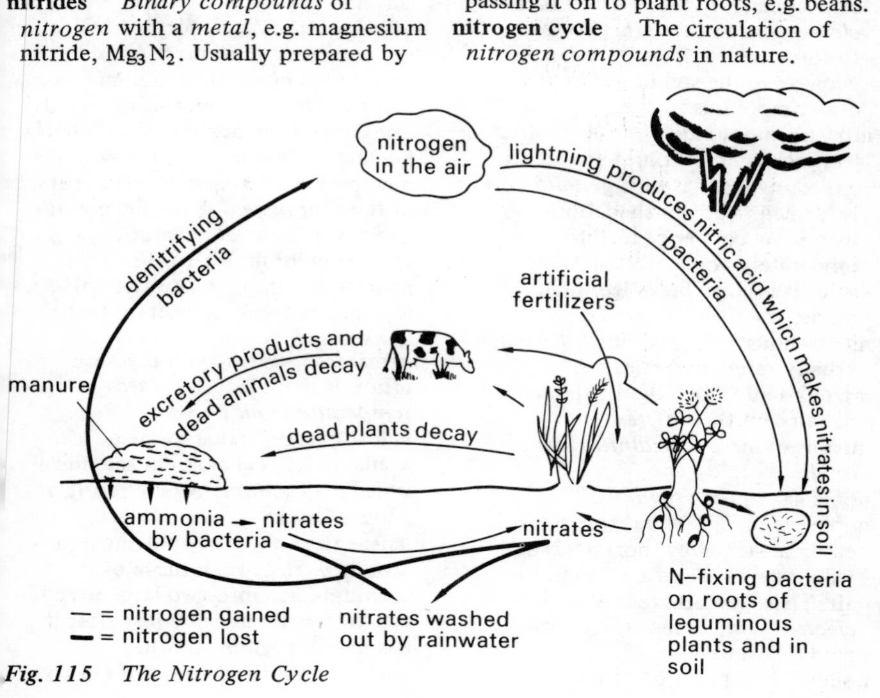

Fig. 115 The Nitrogen Cycle

Inorganic nitrogen compounds in solution in the soil are taken in by plants and changed into *proteins.* These are eaten by animals and reconverted into simpler substances which are returned to the soil in waste products and by decay. *Bacteria* in the soil change these compounds into others more suitable for the plants to absorb.

nitrogen dioxide NO_2 A brown *gas* with a *pungent* smell; formed by heating *nitrates* other than those of the *alkali metals.* It is usually prepared in the laboratory by heating lead nitrate as this is anhydrous. Also formed when *nitrogen monoxide* mixes with air. Dissolves in water to form a mixture of nitrous and nitric acids. On cooling, nitrogen dioxide *molecules associate* in pairs to form *dinitrogen tetroxide* (N_2O_4), a pale yellow liquid. This change is *reversible.*

nitrogen fixation *Combination* of atmospheric *nitrogen* with other *elements* as in the *Haber process* to form ammonia or by means of *bacteria* in the soil to form nitrogenous compounds.

nitrogen monoxide NO A colourless gas which becomes brown on exposure to air as *nitrogen dioxide* is formed. Made in the laboratory by adding copper or lead to moderately concentrated nitric acid. It used to be called nitric oxide.

nitrogenous Containing *nitrogen*; usually refers to *fertilizers.*

nitrous acid HNO_2 Unstable acid; its salts are the *nitrites.*

nitrous oxide See *dinitrogen oxide.*

noble gas See *argonons.*

noble metal A *metal* which is not easily attacked by most *acids* and will not *corrode* or *tarnish* in the air. These metals are low in the *electrochemical series*, e.g. silver, gold, platinum.

node 1. A point of zero *amplitude* in a *stationary wave*. 2. A joint in a *stem*. Leaves grow from nodes. See *twig* (Fig. 162).

nodule A small swelling on a *root*, e.g. on bean roots where the nodules contain *nitrogen fixing bacteria.*

non-conductor A solid, liquid or gas which does not *conduct* electricity (or heat) well, e.g. sulphur, porcelain, alcohol, oxygen. An *insulator.*

non-electrolyte A liquid which will not *conduct* electricity because it has a *covalent* structure, e.g. alcohol, benzene.

non-endospermous A *seed* which does not contain any *endosperm*, e.g. beans.

non-luminous A *flame* which is not white or yellow and does not give out much light, e.g. bnrning ethanol, or a mixture of natural (or coal) gas and air.

normal A line drawn at right angles to a surface, especially at a point where a ray of light strikes that surface.

normal salts A *salt* formed by the complete *neutralization* of an *acid* by a *base*, e.g. sodium chloride, NaCl; copper sulphate, $CuSO_4$; sodium acetate, CH_3COONa; Such salts do not contain any replaceable *hydrogen atoms* as part of the *acid radical*, but they are not necessarily *neutral* in solution, e.g. sodium chloride with a *pH* of 7 is neutral, but copper sulphate (pH 5) is *acidic* and sodium acetate (pH 9) is *alkaline.*

normal temperature and pressure (*abbr.* N.T.P.) See *standard temperature and pressure.*

nuclear energy *Energy* made available from changes in the nuclei (*nucleus*) of *atoms*, which result in a loss of *mass.*

nuclear fission The splitting of an atomic *nucleus* (like those of uranium-235) into two large parts and two or three *neutrons*, after it has been struck by another neutron. A large amount of *energy* is set free at the same time.

nuclear power Power obtained by the release of *nuclear energy*.

nuclear reactor A structure containing a *fissile* substance such as *uranium* and a *moderator*. This is arranged so that the energy is continually released in a controlled manner. A *chain-reaction* is involved as *neutrons* are given out. See also *atomic pile*.

nucleic acid The substance in the *nucleus* which has important work to do in *cell division*. See DNA.

nucleon A particle found in the *nucleus* of an *atom*, i.e. a *proton* or a *neutron*.

nucleus 1. The centre part of the *atom*. It is made up of *protons* and *neutrons*, except in the case of normal hydrogen which has no neutrons. Thus the nucleus has a positive *electric charge*. Nearly all the *mass* of the atom is found in the nucleus as the other part, the *electrons* which move in *orbits* round the nucleus are of relatively very low mass. However the nucleus has a diameter only one ten thousandth of that of the atom.
2. The most important part of a *cell*. The nucleus controls all the work in the cell, including *cell division*.

nuclide Any particular type of *atom* whose *nucleus* contains a stated number of *protons* and a stated number of *neutrons*.

nutrition Getting and dealing with food. It includes *ingestion, digestion, absorption* and *assimilation* in animals and *photosynthesis* in plants.

nylon An artificial *fibre* made from organic materials such as *adipic chloride* or *sebacoyl chloride* with diaminohexane. It is melted and forced through fine jets to make *filaments*. Used as a dress material and for ropes.

nymph A young *insect* with the same shape as the grown up insect, e.g. dragonfly in its *aquatic* stage.

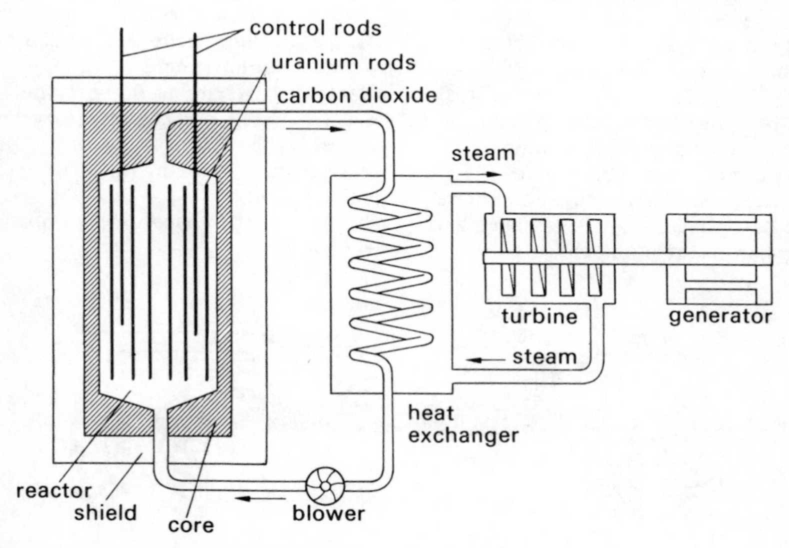

Fig. 116 Nuclear Power Station Layout

objective The *lens,* or group of lenses, in a *microscope* or *telescope* (etc.) which is nearest to the object.

ocellus A part of the *compound eye* of an *insect.*

octahedral Having eight surfaces; a *crystal* of this shape, e.g. rhombic sulphur, *diamond.*

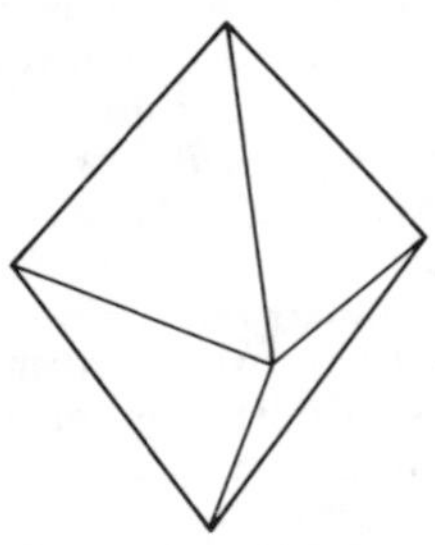

Fig. 117 Octahedral Crystal of Diamond

octane number A number connected with the ability of a *motor spirit* (petrol) to 'explode' at the correct part of the engine's cycle. Heptane, C_7H_{16}, is very poor in this respect and is given a rating of 'O' while iso-octane, C_8H_{18}, is very good and = 100. An octane number of 90 means that the fuel is similar as far as this property is concerned to a mixture of 90% octane and 10% heptane.

oesophagus The first part of the *alimentary canal* (Fig. 5). Food enters the oesophagus when swallowed and is taken to the stomach.

oestrogen A *hormone* produced by the *ovary.* It plays an important part in the control of *egg* production. It also has a great deal to do with *secondary sexual characteristics.*

ohm The *SI unit* of electrical *resistance.* The symbol is Ω. It is the resistance of a *conductor* in which a current of one *ampere* flows when a *potential difference* of one *volt* is put across it.

Ohm's Law The *electric current* flowing through a metal *conductor* at *constant temperature* is directly *proportional* to the *potential difference* between its ends.

ohm-meter An instrument for measuring electrical *resistance* directly.

oil 1. General name for various mixtures of hydrocarbons occurring naturally and needing *fractional distillation* and refining. See also petroleum. 2. A liquid *fat* found in animals and seeds.

oildag See *aquadag.*

olefin(e) See *alkene.*

oleum $H_2S_2O_7$ An *acid* made by dissolving *sulphur trioxide* in concentrated *sulphuric acid.* Also called *fuming* sulphuric acid.

olfactory Having to do with the sense of smell, e.g. the olfactory organ is the nose.

omnivorous Feeding on plant and animal food.

opaque Not *transparent* or *translucent* to *light.*

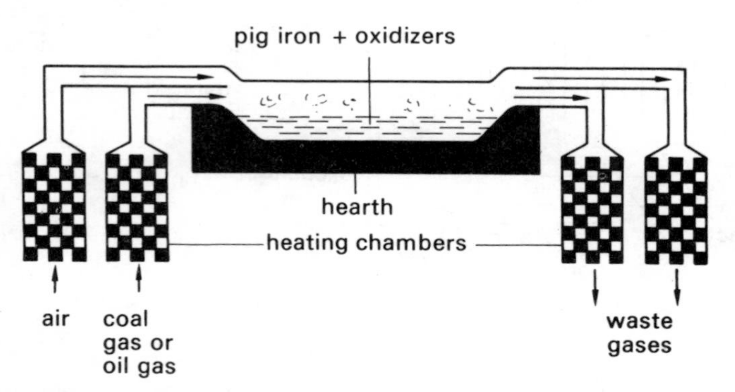

Fig. 118 The Open-hearth Process

open chain See *straight chain.*

open circuit An electrical *circuit* which is not complete, so that no current can flow.

open-hearth process Process used in *steel* manufacture. Pig-iron (*cast iron*) and *scrap* steel along with iron(III) oxide are heated together by producer gas on a hearth in a furnace.

operculum A covering, e.g. the skin over the *gills* of a fish.

optic Having to do with the *eyes,* e.g. the optic nerve carries messages from the eye to the *brain.*

optical centre The point in a *lens* through which a ray of light passes without change in direction. In simple work, and especially for lenses where both surfaces curve equally, it is taken to be on the *principal axis,* midway between the faces.

optical density The more optically dense a material is, the more it will bend light as the light enters the material. See also *refraction, refractive index.*

optics The study of *light.*

orbit 1. The space in the *skull* where the eye is placed. 2. The path of a planet round the sun. 3. The path taken by an *electron* in circulating round the *nucleus* of an *atom.* There are different orbits at various distances from the nucleus.

ore The *mineral* rock or earth from which a metal or non-metal is obtained. Haematite, Fe_2O_3, is the principle ore of iron; galena, PbS, of lead; and calcium phosphate, $Ca_3(PO_4)_2$, of phosphorus.

organ A group of *tissues* making up a large structure in the plant or animal body, e.g. a *leaf,* the *liver.*

organic Having to do with life, e.g. organic evolution.

organic compounds Compounds of carbon other than its oxides, the carbonates and carbonic acid. They frequently contain a large number of *atoms,* do not usually form *ions* (being mainly *covalent*) and often show the property of *isomerism.* Organic compounds form the basis of living matter.

organism A living plant or animal.

origin The part of a *muscle* which is attached to a *bone* which does not move when the muscle shortens. See diagram of *joint.*

origin of species The name given to the idea that new kinds of plants or animals can develop from ones already there, as a result of *natural selection* over a long time. See also *Darwin.*

orlon A synthetic *fibre,* widely used as an imitation fur and in carpets.

orthorhombic A *crystal* system in which magnesium sulphate (Epsom salt) and sulphur occur.

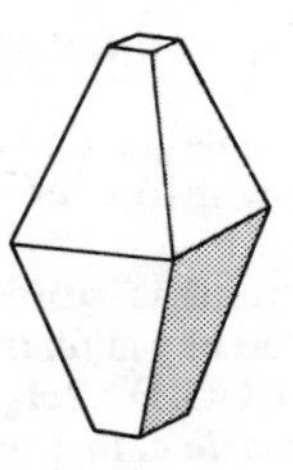

Fig. 119 Orthorhombic Sulphur Crystal

oscillation A complete to and fro movement of any repeating process. See *amplitude* (Fig. 8).

oscilloscope See *cathode-ray oscilloscope.*

osmo-regulation The ability of an animal to control the amount of water lost by *excretion.*

osmosis The way in which the movement of water into and out of *cells* is controlled. If more *sugar* is in the water inside the cell than there is outside, water will pass into the cell through the *semipermeable membrane.*

osmotic pressure The *pressure* developed when water passes through a *semipermeable membrane* into a liquid containing other substances such as salt or sugar, as in *osmosis.*

ossicle Any very small bone as in the *ear.*

oval window The small piece of thin

skin which separates the *middle ear* from the *inner ear.* See *ear* (Fig. 59).

ovary The place where *eggs* in animals and *ovules* in plants are produced. See *reproductive organs* (Fig. 141i).

overtone A *frequency* given out by a source of sound at the same time as the *fundamental* frequency. It is a small whole number times the fundamental frequency and usually much smaller in *amplitude.*

oviduct The tube through which an *egg* passes when it is laid.

oviparous Any animal which lays *eggs.*

ovipositor The *egg* laying tube of an *insect.*

ovulation The breaking off of the *egg* from the *ovary.*

ovule The female part of a plant which will grow into a *seed* after *fertilization.*

ovum The scientific word for an *egg.* Used before the egg is *fertilized.*

oxalic acid $(COOH)_2.2H_2O$ A white, *crystalline, dibasic acid* present in rhubarb leaves and in beetroot and spinach. On heating with concentrated sulphuric acid both carbon monoxide and carbon dioxide are formed. The salts are called oxalates and are used in dyeing, ink manufacture and bleaching.

oxidant A substance which is capable of *oxidizing* another, e.g. manganese (IV) oxide, MnO_2, oxidizes concentrated hydrochloric acid to chlorine. Air and steam are oxidants when iron or magnesium burn in them, the metal oxide being formed.

oxidase An *enzyme* used in the chemical part of *respiration.*

oxidation 1. The addition of oxygen or another *element* or *group* which forms negative ions to a substance, e.g. magnesium → magnesium oxide; iron(II) chloride → iron(III) chloride ($FeCl_2 \rightarrow FeCl_3$). 2. An increase in the *valency* of a metal element, e.g. copper(I) oxide (Cu_2O) → copper(II) oxide (CuO). 3. A loss of *electrons* by an element or group, e.g. $Zn - 2e^- \rightarrow Zn^{2+}$. 4. Removal of hydrogen or another element or group which forms positive ions, e.g. ammonia (NH_3) → nitrogen (N_2).

oxidation number A positive or negative number indicating the *valency* of an *element* in a particular *compound.* The oxidation numbers of all the elements in a compound add up to zero. With simple *ions* such as Cl^- or Sn^{2+} the oxidation number = the charge on the ion, here −1 and +2.

oxide *Binary compound* formed between *oxygen* and another *element,* e.g. calcium oxide, CaO, or sulphur dioxide, SO_2.

oxidize To cause *oxidation* or to be subject to it.

oxidizing agent See *oxidant.*

oxy-acetylene blowpipe A burner in which a controlled *mixture* of *oxygen* and acetylene (*ethyne*) is burned to give a very hot *flame* (up to 3000 °C).

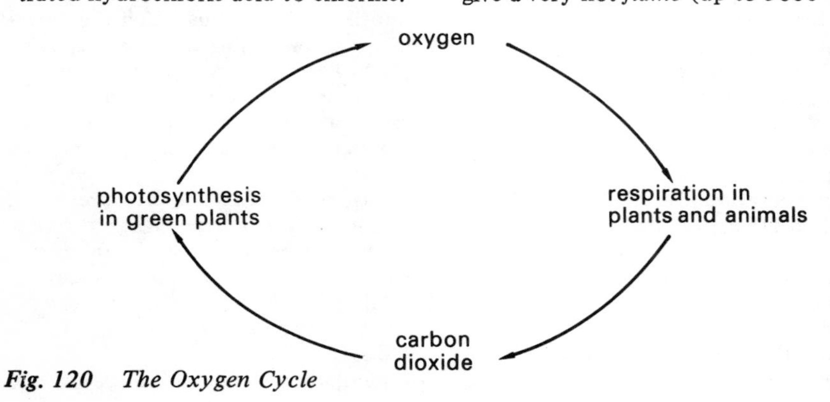

Fig. 120 The Oxygen Cycle

Used in cutting and *welding* metals.

oxygen O At. No. 8 R.A.M. 16 A colourless, odourless gas forming 20% of the air. The most abundant of all *elements* (mainly in the combined state). An active element which joins up with most other elements under suitable conditions, forming *oxides.* Essential to life where, in *respiration,* it breaks down *sugar* to set free *energy,* and necessary for *combustion.* Made from liquid air by *fractional distillation* or–in the laboratory–by heating certain compounds of oxygen. Used in *welding,* cutting, in the manufacture of *steel* and as a liquid in rocket *fuels.*

oxyhaemoglobin The bright red substance in *blood* which carries the *oxygen* to the *cells.* It is made in the *lungs* when oxygen joins with *haemoglobin.* It becomes haemoglobin again when the oxygen is given up to the cells.

oxy-hydrogen blowpipe A burner in which a controlled *mixture* of *oxygen* and *hydrogen* is burnt to give a very hot *flame* (up to 2400 °C). Used in cutting and *welding.*

ozone O_3 An allotrope (*polymorph*) of *oxygen.* A powerful *oxidant.* Used in improving the air in underground railways.

paint A coloured *pigment suspended* in a suitable *liquid* such as linseed oil with turpentine and a *drying agent* added.

palate The roof of the mouth. It separates the *buccal cavity* from the *nasal cavity*. It is often rough, to help in *mastication*.

palea Part of a grass *flower* (Fig. 76 ii). It is left behind when the grain is cleaned before being used for grinding.

palisade layer The *cells* under the upper skin of a *leaf* (Fig. 101 ii). Most of the *chlorophyll* is found here.

palp A feeler near the mouth of an insect or spider. It is sensitive to touch or taste.

pancreas A digestive organ joined to the *alimentary canal* (Fig. 5). It produces *enzymes* which mix with the food in the *intestine*. It also makes the *hormone insulin*.

pancreatic juice The liquid made by the *pancreas*.

paper A web of vegetable fibres such as *cellulose*, which is obtained from wood pulp after the lignin and other unneccessary material have been removed.

papilla A small outgrowth as on the *tongue* or on the inside of the *intestine* (*villus*).

pappus A ring of hairs which help to scatter the *fruits* of plants, e.g. dandelion.

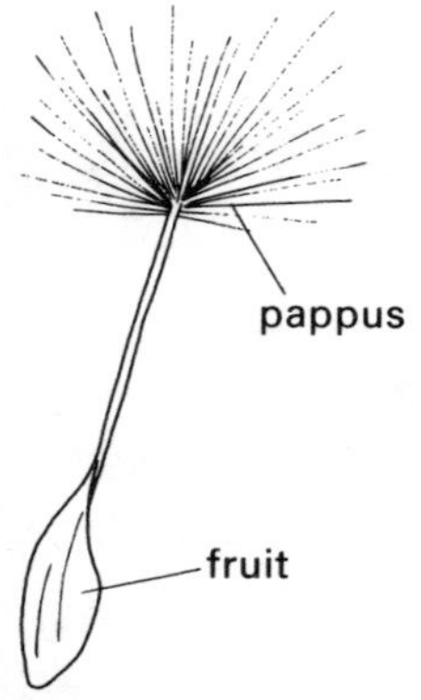

Fig. 121 Pappus

paraboloidal mirror A curved mirror whose cross-section is a parabola. If a source of light is placed at the focus a true parallel beam is formed.

paraffin oil See *kerosine*.

paraffin wax White, *translucent* solid, a mixture of *alkane hydrocarbons* used in the manufacture of waxed paper, candles and polishes. It melts at 50–60°C.

paraffins See *alkanes*.

parallax When viewed from different places an object appears to be in different positions relative to the background. This change in position is called parallax. To prevent mistakes due to parallax when reading instruments, the line from the eye to the pointer (etc.) should be at right angles to the scale. If there is no parallax between two objects then they must be in the same place. This test is often used in *optics*. See *error of parallax* (Fig. 66).

parallel connection of resistors To join *resistors* (etc.) in parallel the first connections of each are joined together and then the second connections are joined. The total *resistance* of all the resistors, of separate resistances R_1, R_2, R_3 etc., is R, where

$$\frac{1}{R} = \frac{1}{R_1} + \frac{1}{R_2} + \frac{1}{R_3} + \ldots.$$

Note that the current flowing in the main circuit is the total of the current flowing in each resistor.

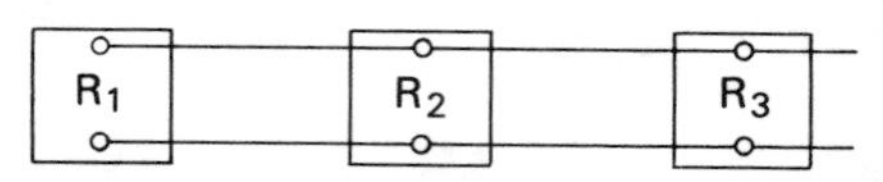

Fig. 122 Parallel Connection of Resistors

parallelogram of forces If two *forces* act on an object then the *resultant* can be found by a scale drawing. Draw a parallelogram with one pair of sides *proportional* in length to the size of the first force and in the same direction as that force, and with the other pair of sides proportional in length to the size of the second force and in the same direction as that force. The diagonal of the parallelogram gives the size and direction of the resultant.

parasite A plant or animal which lives inside or on the outside of another plant or animal (the *host*). The parasite gets all its food from the host and usually causes a *disease*, e.g. tapeworm, potato blight.

parenchyma The soft part of a plant stem or root.

parotid gland One of the glands which make *saliva*. One is found on each side of the lower jaw. When diseased it causes mumps–a swelling of the face and neck.

parthenogenesis A kind of *reproduction* where the eggs from the female develop without *fertilization* by the male, e.g. greenfly, worker bees.

pascal The *SI unit* of *pressure*. The symbol is Pa. The pascal is the pressure set up when a *force* of one *newton* acts, at right angles, to an *area* of one square metre.

Pasteur, Louis A Frenchman (1822–1895) who discovered that *bacteria* cause disease, and that they can be killed by boiling. He made many other important discoveries. See *pasteurization*.

pasteurization The way in which milk is made safe for drinking. The milk is heated and then cooled, killing the harmful *tuberculosis germs*.

pasture Land which grows grass on which animals feed.

patella Part of the *skeleton* (Fig. 148). The kneecap bone.

pathogen Any plant or animal which causes *disease*, e.g. *malaria germs* in the *blood*.

Pavlov, Ivan A Russian scientist (1849–1936) who discovered that *reflex actions* could be brought on by a *stimulus* other than the proper one. If a bell is rung every time a hungry dog is given meat, after a time the dog will make *saliva* when the bell is rung but there is no meat. This is called a conditioned reflex.

peak value The highest value an *alternating current* (Fig. 6) or voltage reaches in either direction.

peat A very early stage in the conversion of wood and vegetation to *coal*. It is found in cold, *temperate regions* and is an accumulation of partly decomposed vegetable material. It is used in many parts of the world as a *fuel* for heating and even in power stations.

pectoral To do with the chest, e.g. the pectoral *girdle* which joins the chest bones to the backbone. Also pectoral *fins* (Fig. 71).

pedicel The stalk of a single *flower* (Fig. 76 i).

pedungle The stalk of a group of *flowers* (Fig. 76 i).

pelagic Living in the upper layers of the sea, e.g. jellyfish.

pellagra Disease caused by a lack of one of the *Vitamin* B group. It affects the *skin* and the *nerves*.

pelvis Part of the *skeleton* (Fig. 148). The bones which support the lower end of the body (pelvic girdle). The leg bones are joined to the pelvis. Also pelvic *fins* (Fig. 71).

pendulum A heavy weight (the bob) on the end of a light string or rod, which is hung so that it can swing regularly to and fro. For a simple pendulum, where the *weight* of the string is taken to be zero, where the angle through which the cord moves is small, and where the *mass* of the bob can be taken to be at one point, the time for one complete swing

$$= 2\pi\sqrt{\frac{\text{length of pendulum}}{\textit{acceleration due to gravity}}}$$

penicillin A substance produced by a *fungus*. It kills *bacteria* and is used by doctors to cure many *diseases*. See *antibiotic*.

penis One of the male *reproductive organs* (Fig. 141 ii). The male *organ* through which the *sperm* is passed into a female *mammal*. It is also used to pass the *urine* out of the body.

penta- Prefix meaning five, e.g. phosphorous pentachloride is PCl_5.

pentadactyl Having five *digits* on the hand or foot.

pentavalent Having a valency of five, e.g. phosphorus. Also called quinquevalent.

penumbra The grey part of a *shadow* where only part of the *light* from the source is stopped by the object. It gets lighter towards the outside edge of the shadow. See also *umbra*.

pepsin One of the main *enzymes* made in the *stomach* and used to start the *digestion* of *proteins*.

peptone A substance formed during the *digestion* of *proteins*.

percentage composition The *masses* of the different *elements* in a *compound* which would be present in 100 g of the compound. These figures are obtained from the expression: % of $\hat{X}$ in compound =

$$\frac{\text{mass of } X \text{ in compound}}{\text{total mass of compound}} \times 100$$

percolate To run through. Said of water which finds its way through *porous* rocks, gravel or sand.

perennate (plants) Being able to survive through the winter.

perennial A plant which can *perennate*. *Herbaceous* perennials are those plants whose roots only stay alive; woody perennials are trees and shrubs.

perfect gas An imaginary gas which obeys *Boyle's law* and *Charles's law* exactly. Its behaviour is shown by the equation

$$\frac{pressure_1 \times volume_1}{absolute\ temperature_1} = \frac{pressure_2 \times volume_2}{absolute\ temperature_2}$$

Real gases come near this behaviour at lower pressures and temperatures well above that at which they become a liquid.

perianth The parts of a *flower* outside the *stamens*, usually the *petals* and *sepals*.

pericardium The very thin skin which is around the *heart*.

pericarp The part of a *fruit* which is around the *seed*(s).

pericycle The layer inside a *stem* or root from which the branches grow.

perigynous A kind of *flower* (Fig. 76 ii) where the *stamens* are at the same level as the *sepals*.

period See *periodic time*.

periodic law The statement that the *properties* of an *element* are related to its *atomic number* and so to its position in the *periodic table*.

periodic table An arrangement of *elements* in the order of their *atomic numbers*. In any vertical column the elements resemble each other closely in chemical *properties* as they all have similar *electronic structures* with the same number of *electrons* in the outer *shell*, e.g. the alkali metals in *Group* I have one such electron and the halogens in Group VII have seven. Each row (period) ends with Group 0 where the outermost shell has its full number of eight electrons and the elements, the argonons or noble gases, are very inactive. A new shell is then added as the next period begins. The number of elements in the first two periods is eight in each, but this increases to eighteen in the third and fourth periods and then to thirty-two as room is made for including a number of very similar elements such as the *transition elements*. The table is still incomplete; new and heavier elements are being made. From the position of an element in the table a good idea of its properties can be obtained.

I	II											III	IV	V	VI	VII	
					1 H Hydrogen 1												4 He Helium 2
7 Li Lithium 3	9 Be Beryllium 4											11 B Boron 5	12 C Carbon 6	14 N Nitrogen 7	16 O Oxygen 8	19 F Fluorine 9	20 Ne Neon 10
23 Na Sodium 11	24 Mg Magnesium 12											27 Al Aluminium 13	28 Si Silicon 14	31 P Phosphorus 15	32 S Sulphur 16	35.5 Cl Chlorine 17	40 Ar Argon 18
39 K Potassium 19	40 Ca Calcium 20	45 Sc Scandium 21	48 Ti Titanium 22	51 V Vanadium 23	52 Cr Chromium 24	55 Mn Manganese 25	56 Fe Iron 26	59 Co Cobalt 27	59 Ni Nickel 28	64 Cu Copper 29	65 Zn Zinc 30	70 Ga Gallium 31	73 Ge Germanium 32	75 As Arsenic 33	79 Se Selenium 34	80 Br Bromine 35	84 Kr Krypton 36
85.5 Rb Rubidium 37	88 Sr Strontium 38	89 Y Yttrium 39	91 Zr Zirconium 40	93 Nb Niobium 41	96 Mo Molybdenum 42	98 Tc Technetium 43	101 Ru Ruthenium 44	103 Rh Rhodium 45	106 Pd Palladium 46	108 Ag Silver 47	112 Cd Cadmium 48	115 In Indium 49	119 Sn Tin 50	122 Sb Antimony 51	128 Te Tellurium 52	127 I Iodine 53	131 Xe Xenon 54
133 Cs Caesium 55	137 Ba Barium 56	139 La Lanthanum 57	178.5 Hf Hafnium 72	181 Ta Tantalum 73	184 W Tungsten 74	186 Re Rhenium 75	190 Os Osmium 76	192 Ir Iridium 77	195 Pt Platinum 78	197 Au Gold 79	201 Hg Mercury 80	204 Tl Thallium 81	207 Pb Lead 82	209 Bi Bismuth 83	210 Po Polonium 84	210 At Astatine 85	222 Rn Radon 86
223 Fr Francium 87	226 Ra Radium 88	227 Ac Actinium 89	Ku Kurchatovium 104														

Relative Atomic Mass Symbol Name Atomic Number														
139 La Lanthanum 57	140 Ce Cerium 58	141 Pr Praseo – dymium 59	144 Nd Neodymium 60	147 Pm Promethium 61	150 Sm Samarium 62	152 Eu Europium 63	157 Gd Gadolinium 64	159 Tb Terbium 65	162.5 Dy Dysprosium 66	165 Ho Holmium 67	167 Er Erbium 68	169 Tm Thulium 69	173 Yb Ytterbium 70	175 Lu Lutetium 71
227 Ac Actinium 89	232 Th Thorium 90	231 Pa Protactinium 91	238 U Uranium 92	237 Np Neptunium 93	242 Pu Plutonium 94	243 Am Americium 95	247 Cm Curium 96	247 Bk Berkelium 97	251 Cf Californium 98	254 Es Einsteinium 99	253 Fm Fermium 100	256 Md Mendelevium 101	254 No Nobelium 102	257 Lr Lawrencium 103

Fig. 123 Periodic Table of the Elements

periodic time The time for one complete *cycle* of events in any regularly repeating process.
periosteum The thin skin around the *bone*. It carries the *blood* to the bone.
periscope An instrument with two *mirrors* or reflecting *prisms* at opposite ends of a long tube which allows one to see over objects, or people, which are in the way. In the case of a submarine the periscope allows the captain to see what is happening above the sea without the whole boat coming to the surface.

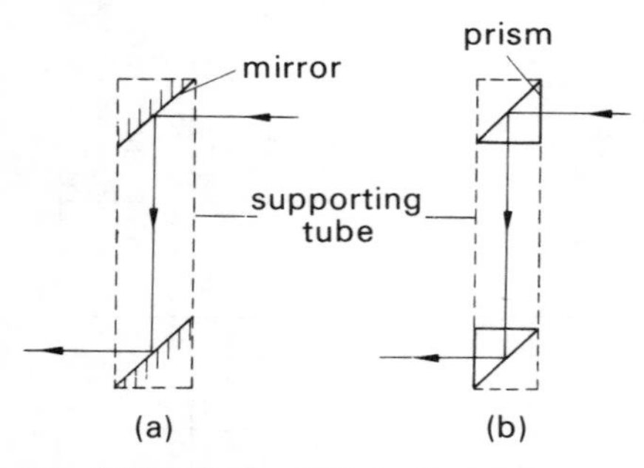

Fig. 124 Path of Light through a Periscope

peristalsis *Muscle* movements which cause the food to be passed through the *alimentary canal.*
peritoneum The thin skin folded around the *organs* in the *abdomen*. When diseased it causes peritonitis.
permanent gas A *gas* which cannot be liquefied by *pressure* alone; the *temperature* must be lowered as the pressure is increased, e.g. oxygen, hydrogen.
permanent hardness *Hardness* in water caused by the presence of dissolved *calcium sulphate* or magnesium sulphate. Permanent hardness cannot be removed by boiling, but is destroyed by adding sodium carbonate or by passing the water through *zeolite* or *permutit.*
permanent magnet A magnet which keeps its magnetism.
permanent teeth The second set of teeth which grow when the first, or milk teeth drop out. They stay during the lifetime unless they become diseased.
permeable membrane A skin which will allow anything in *solution* to pass through.
permutit *Synthetic* sodium *zeolite* which is capable of *softening hard water* by replacing the *calcium ions* in the water by *sodium* ions. It is used in many water softeners and is gradually changed into calcium zeolite and then the softening stops. The sodium zeolite is re-formed by adding *sodium chloride* to the softener and washing the salt through very thoroughly.
peroxide An *oxide* of a *metal* which yields *hydrogen peroxide* on addition of a dilute acid, e.g. sodium peroxide, Na_2O_2. The name, peroxide is often used for hydrogen peroxide itself.
perspex A trade name for polymethyl methacrylate, a *thermosetting plastic* which is usually clear and can replace glass, being less liable to break and much less dense (*density*). It can also be made in various colours and is then *translucent* rather than *transparent*. On heating it is broken down into smaller *molecules* of methyl methacrylate, a liquid.
perspiration *Sweat* produced in the *skin* when the temperature of the body is greater than that of the air. It helps to cool the body.
perversion Another name for *lateral inversion*.
pestle A metal or porcelain club-shaped implement used for grinding solids to powder in a *mortar*.
petal A leaf-like structure, usually coloured, in a *flower* (Fig. 76 ii). The colour attracts insects.
petiole The *stalk* of a *leaf*.
petri dish A shallow circular dish with a lid, usually made of glass or some suitable plastic. It is used for growing cultures of *bacteria* on jelly.
petrify To turn into 'stone'. Objects

exposed for a long time to water containing dissolved calcium hydrogencarbonate, $Ca(HCO_3)_2$, (*temporary hardness*) gradually get coated with a deposit of *insoluble calcium carbonate*, caused by the breaking down of the *unstable* hydrogencarbonate.

petrol A complex *mixture* of *hydrocarbons* with 5–8 *carbon atoms* in the chain. It is used as a *fuel* in internal combustion (motor car) engines. It is also called motor spirit, gasoline or 'gas'.

petroleum A naturally-occurring *mixture* of *hydrocarbons* and other organic compounds. It is this mixture which, on distillation, provides the fractions *kerosine*, *diesel oil*, *paraffin wax*, etc. which are put to various different uses. It is also called *mineral* oil. See *refine* (Fig. 135).

pH A term used to express the degree of *acidity* of a *solution*, i.e. the *hydrogen ion concentration*. A pH of 7 is *neutral*, one of less than 7 shows acidity, increasing as the number gets less, while a value greater than 7 indicates *alkalinity*. See also *universal indicator*.

phagocyte A white *blood cell* which feeds on *bacteria*, so keeping the blood free of *disease*.

phalanges (*sing.* phalanx,) Part of the *skeleton* (Fig. 148). The bones of the fingers or toes.

pharynx The tube leading from the mouth through the throat. It divides into the *oesophagus* and the *trachea*.

phase 1. A state of matter such as *solid*, *liquid*, *vapour* or *gas*. 2. If two *waves* travel so that the crest of one comes with the crest of the other they are said to be in phase. If the crest of one comes with the trough of the other they are said to be completely out of phase.

phenol C_6H_5OH A substituted (*substitution reaction*) *benzene compound* with slightly *acidic* and *disinfecting* properties in solution. When pure it is a white *crystalline* solid with a *characteristic* smell and is used in the preparation of some plastics and some dyes. When impure it is pink. It is corrosive and poisonous.

phenolphthalein An *indicator* which is colourless in *acid* or *neutral solution* and a bright red in *alkaline* solution. It consists of colourless crystals dissolved in alcohol.

phenomenon A happening: something that can be seen to occur.

phenotype A word used in *genetics* to mean the outward appearance of an *organism*. See also *genotype*.

philosopher A wise or learned man; formerly a student of natural or moral science. A lover of wisdom.

phloem The part of a *stem* or *root* used to carry *synthesized* food from one part to another. E.g. from the *leaf* (Fig. 101 ii) where the food is made to the root where it is stored.

phosphates Salts of (ortho) *phosphoric acid*, H_3PO_4, containing the phosphate ion. They are used as fertilizers and *mineral calcium phosphate* is the main source of *phosphorus*. It is also the chief compound present in bones. Phosphates are used in the mineral water industry, in medicines, in dyeing and for fireproofing wood.

phosphor bronze An *alloy* of copper, tin and phosphorus used in making bearings in cars, gears, etc. It is hard and resists *corrosion*.

phosphorescence Property of shining after exposure to light.

phosphoric acid H_3PO_4 There are several phosphoric acids of which this is the most important. It can be made by dissolving *phosphorus pentoxide* in hot water. It is a thick syrupy liquid and forms *salts* called phosphates. Also called orthophosphoric acid.

phosphorus P At. No. 15 R.A.M. 31 A non-metal *element* in Group V of the *periodic table*. It exists in the form of *polymorphs* (allotropes). White (or yellow) phosphorus is a hard waxy solid kept under water because of its low *ignition temperature* and great activity. It is poisonous. The red variety is *amorphous*, non-poisonous and does not take fire when exposed to the air.

phosphorus pentoxide P_2O_5 A white powder formed by heating *phosphorus* in air. It is very *deliquescent*, dissolving in cold water to form metaphosphoric acid, HPO_3, and in hot water to give orthophosphoric acid, H_3PO_4.

photoelectric effect The giving out of *electrons* when a metal surface is hit by *radiation* of a sufficiently high *frequency*. The *energy* of the *light* or *ultra-violet photon* is given to an *atom* which can then send out an electron.

photometry The study of the strengths of light sources and the *illumination* they cause.

photon The basic amount or *quantum* of light (or other electromagnetic *radiation*) energy which behaves in many ways like a particle. The amount of energy carried by a photon is equal to the *frequency* of the radiation multiplied by Planck's *constant*, 6.63×10^{-34} *joule* second.

photosynthesis The making of *sugar* from *carbon dioxide* and water in the cells of a plant which contain green *chlorophyll*. *Light* is needed to give the *energy* for the work to be done. *Oxygen* is also produced. *Equation:* carbon dioxide + water = sugar (with energy) + oxygen.

phototropism The way in which a plant stem or leaf grows in relation to the direction of light, e.g. a plant stem grows towards the light–it is positively phototropic. Also called heliotropism.

phylum A main group of animals (and plants), e.g. all animals with a *spinal cord* belong to the *phylum Chordata.*

physical changes Changes in which no new substance is formed; usually a change in *phase* only, e.g. from solid to liquid. Such changes are usually easily *reversible*.

physiology The study of how a plant or animal performs its different *functions*.

phytoplankton The group of very small plants which float about in water.

pico- A prefix which means one million millionth or 10^{-12}. It can be joined in front of any *SI unit*, e.g. 1 picofarad = 10^{-12} farad. The symbol is p.

pig-iron See *cast-iron*.

pigment 1. A coloured substance, insoluble in water, used to colour a surface. It differs from a *dye* as it does not penetrate the surface and is usually removable by mechanical means such as scraping. 2. Colouring matter in the hair, skin, eyes, etc. 3. When two pigments are mixed the colour formed is the colour or colours that both reflect (*reflection*).

piliferous layer *Cells* which grow out into hairs, e.g. the outside layer of a young *root*.

pinch-cock A clip placed on a piece of rubber tubing connecting a jet to a *burette* to enable the flow of liquid to be controlled. Used in junior work in schools.

Fig. 125 Pinch-cock

pinking See *knocking*.

pinna 1. A *feather*. 2. The part of the *ear* (Fig. 59) which can be seen.

Fig. 126 Pipette

pinnate With parts arranged like a *feather*, e.g. pinnate *leaf* (Fig. 101 i).

pipette A glass tube with an enlarged bulb designed so that a definite volume of a liquid can be drawn into the tube up to a mark on the neck and then released where required. Some pipettes do not have the bulb, but have an enlarged stem which is graduated to deliver different volumes up to, say, 10 cm^3.

pistil A name given to the female part of the *flower* (Fig. 76 iii) the *gynaecium*.

pitch The black residue left in the still after the *destructive distillation* of wood or *coal tar*, or the *fractional distillation* of the latter. It is thick and viscous with a boiling point above 350°C. It is used in waterproofing joints and roofs.

pitchblende A *radioactive* material, mainly uranium oxide, U_3O_8, from which radium was first obtained.

pituitary gland A *ductless gland* under the brain. It makes *hormones* which control the other ductless glands and so is very important in the growth and development of the body. If it is not working properly, it can cause trouble in the development of *sex* and in the formation of *bones*, as well as many other diseases. See *central nervous system* (Fig. 34 i).

pituitrin The older name given to a group of *hormones* of the *pituitary gland*. They help to control the movements during childbirth.

placenta The part of the *womb* which joints the *foetus* (Fig. 77) to its mother. It contains much *blood* and helps to pass food and *oxygen* into the baby and to carry the waste substances away.

plane Flat or level, e.g. a *crystal* has plane surfaces.

plane of cleavage *Crystals* are formed by gradual 'growth' from a *saturated solution* and *planes*, often parallel to the outside surfaces, are generally present.

plankton The floating plants and animals. They are usually carried around by the movements of the water.

plasma The liquid part of *blood*. It is mostly water, with many different substances in it. The red and white blood *cells* are carried in the plasma.

plasmolysis The loss of water from a *cell* causing it to get smaller. If too much water is lost the cell can die.

Plaster of Paris $(CaSO_4)_2.H_2O$ A *hydrated* form of calcium sulphate

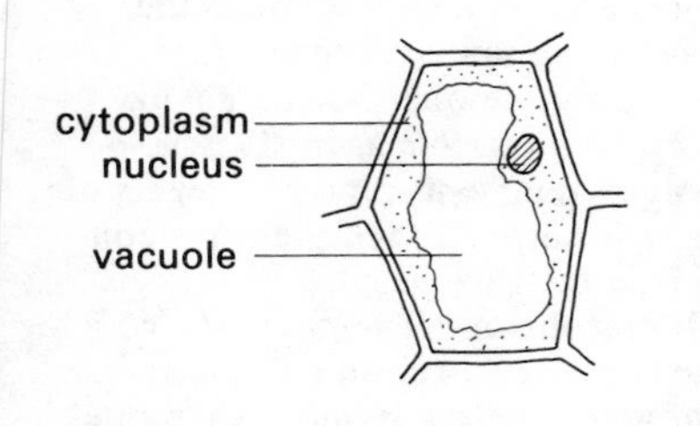

Fig. 127 Plasmolysis

obtained by heating *gypsum*, $CaSO_4 . 2H_2O$, to 120–130°C, allowing to cool and adding water. The mixture evolves heat and slightly expands as interlocking *crystals* of gypsum are re-formed. It is used in making casts and for preventing the movement of broken limbs.

plastic Capable of being moulded by *pressure* and, or heat, e.g. clay.

plastic sulphur A form of sulphur obtained by pouring *molten* sulphur into cold water. It is elastic when first formed, but soon becomes hard as it changes to small rhombic (octahedral) *crystals*.

plastics *Organic* materials, mostly *polymers*, which have certain properties (e.g. good *insulators*) and at some stage in their manufacture have been shaped by heat and, or pressure; Bakelite, ebonite, melamine. They are of two kinds, *thermoplastic* and *thermosetting* and are usually very stable in normal use.

plastid A very small body in a cell, e.g. *chloroplast*.

platelet Very small *cells* in the *blood* which help to harden (clot) the blood on the outside of a wound.

plating Depositing a metal coating by *electrolysis* on a metal forming the *cathode* in a suitable solution.

platinized asbestos Asbestos which has been soaked in a solution of platinum chloride and then heated to leave a black deposit of *platinum* on the *fibres*. Used as a *catalyst* in place of platinum itself.

platinum Pt At. No. 78 R.A.M. 195 A hard silvery *metal* resistant to *corrosion*, heat and acids. Used as a *catalyst*, for some scientific apparatus and for jewellery.

pleural cavity The space in which the *lungs* (Fig. 109 i) are found. The skin around the lungs is the pleural *membrane*.

plexus A network of *nerves*, e.g. *solar plexus*.

plumule The part of a *seed* which will grow into the stem and leaves.

pneumatic trough A large vessel used in the collection of gases over water.

polarization When a *cell* supplies an *electric current* the *potential difference* gradually falls because the cell becomes polarized. Chemical action in the cell sets up a *back e.m.f.* Polarization can be reduced by having extra chemicals (depolarizers) in the cell, e.g. the manganese dioxide in a *Leclanché cell*.

pole 1. (Magnet) See *magnetic pole*. 2. (Mirror) The centre of the surface of a *spherical mirror*.

pollen The fine dust, usually yellow, made by the *stamens* of a *flower* (Fig. 76 iii). It carries the male *cells*.

pollination The carrying of *pollen*, usually by wind or insects, from the *stamens* of a flower to the female *stigma* of the same flower (*self pollination*), or another flower of the same sort (*cross pollination*).

poly- Prefix meaning 'many' or 'several'.

polymer A *carbon compound*, usually of high *molecular mass*, which contains a small group repeared many times, e.g. polyethene (*polythene*), $—CH_2—CH_2—CH_2—CH_2$

polymerization 1. Conversion of an *unsaturated compound* with a small *molecule* to one with a repeating group, e.g. ethene ($CH_2 = CH_2$) to polyethene (*polythene*), ($—CH_2—CH_2—CH_2—$). 2. Combination of several molecules to form a more *complex* molecule having the same *empirical formula* as the simple one. Such a change may be *reversible* as in styrene→polystyrene→styrene or irreversible as when ethyne (acetylene) is converted to benzene.

polymorphism The *property* which certain *elements* and *compounds* show of existing in more than one different *crystalline* state; sulphur as rhombic or monoclinic, calcium carbonate as calcite or arragonite.

See *allotropy*. Each different form is called a polymorph and the substances are said to be polymorphic.

polyp A kind of simple animal, shaped like a tube and fixed at one end, e.g. hydra.

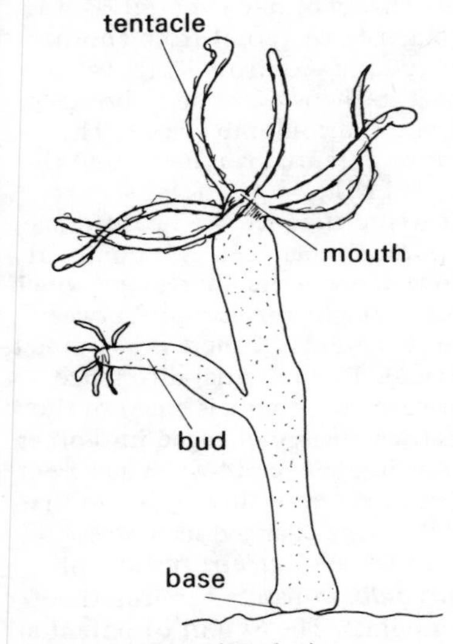

Fig. 128 Polyp (Hydra)

polystyrene A *polymer* formed by heating *styrene*, $CH_2 = CH—C_6H_5$, with a suitable catalyst to form a chain $—CH_2—CH(C_6H_5)—CH_2—CH(C_6H_5)—$. It is a *transparent* material shaped by heat. It may be expanded to many times its original volume by air or carbon dioxide, becoming *opaque*. It is used as an *insulator* against heat or noise and as a packaging material. In this form it is very *brittle*.

polythene A trade name for polyethylene. A *thermoplastic polymer* prepared from *ethene*, C_2H_4. It is usually clear and *transparent*, but may be coloured. Used in sheet form for packaging. It is a good *insulator* and is much used for covering electrical wires and apparatus.

polyvinyl chloride (*abbr.* PVC) A *polymer* (and *plastic*) made from the *monomer*, vinyl chloride, $CH_2 = CHCl$. In the rigid form it is used for pipework and in the plastic form for sheets (impervious to water) in mouldings and as an insulator for electric cables.

porcelain A smoothly-glazed pottery made first by heating strongly a mixture of kaolin (china-clay) and other materials containing silica. It is used for making such laboratory ware as evaporating dishes and crucibles.

porous A material which allows water (etc.) to pass slowly through it.

porous pot An unglazed earthenware useful for absorbing (*absorption*) liquids or for *cracking* purposes at a high temperature.

portal vein A *vein* which carries *blood* back towards the heart but does not go directly to the heart, e.g. the *hepatic* portal vein which carries blood from the *intestine* to the *liver*. See diagram of *blood circulation*.

positive charge An *electric charge* of the same type as that on the *proton*.

positive ion An *atom* which has lost one or more *electrons*. It is called a *cation*.

positive rays Positively charged (*electric charge*) particles formed in a *discharge tube* by *ionization*. If the *cathode* has a hole in it, the positive rays can be seen (with difficulty) as a stream on the far side of the cathode to the *anode*.

posterior The back end of an animal. The back legs are posterior to the front legs.

potassium K At. No. 19 R.A.M. 39 A light, soft element belonging to the *alkali metals* of Group I of the periodic table. It rapidly attacks water, *oxidizes* in the air and burns with a violet or lilac *flame* (this appears red if looked at through blue glass). It is essential for *respiration* in plants and animals and its salts find use as fertilizers or in explosives. One *isotope* of potassium is *radioactive*.

potassium-40 A naturally-occurring *radioactive* form of *potassium* of relative atomic mass 40.

potassium chlorate $KClO_3$ A white *crystalline compound* which yields oxygen on heating. It is used in explosives, in the making of matches and for the relief of sore throats.

potassium dichromate $K_2Cr_2O_7$ A red crystalline solid soluble in water. Used as an *oxidant*: its solution is changed to green chromium sulphate in the presence of dilute sulphuric acid and a suitable *reductant*. Addition of concentrated sulphuric acid to the solid produces a solution which is a good cleanser for laboratory apparatus.

potassium hydroxide KOH A white, waxy, *deliquescent solid* readily soluble in water to give a strongly *alkaline solution* which will absorb carbon dioxide. It used to be called caustic potash.

potassium iodide KI A white *crystalline solid* readily soluble in water. Its solution is used as a solvent for *iodine*. Used in medicine and in photography.

potassium nitrate KNO_3 A white *crystalline solid* obtained industrially from sodium nitrate (Chile saltpetre). It is one of the *constituents* of *gunpowder*, yields oxygen on heating and is very soluble in hot water. It is used as a *fertilizer* and in glass manufacture. It used to be called saltpetre or nitre.

potassium permanganate $KMnO_4$ A purple *crystalline solid* very soluble in water. It is a strong *oxidant*. Its solution, acidified with dilute sulphuric acid, is turned colourless by many *reductants*. It is used as a *disinfectant*, as a stain and in many chemical reactions.

potential The property of an *electric field* which (rather like temperature and pressure) governs the direction of flow of (positive) *electric charge*. Positive electric charge will always flow from high positive potential to low positive potential or from positive potential to negative potential. Negative electric charge flows in the opposite direction. The earth is taken to be at zero potential. The potential at a point is equal to the energy gained by a charge of one *coulomb* as it is brought to the point from another a very long way from it. It is also equal to the *work* done in bringing up the one coulomb charge. The *SI unit* of potential is the *volt* and the symbol is V.

potential difference The difference in *potential* between two points. It shows the electrical 'pressure' which pushes an *electric current* between the two points (if there is a complete circuit). The potential difference between two points is equal to the electrical energy changed into other forms in one second when a current of one *ampere* is flowing, or to the total energy changed into other forms when an *electric charge* of one *coulomb* passes between the two points. The *SI unit* of potential difference is the *volt* and the symbol is U or V.

potential energy The *energy* which a body has because of its position. When a body is raised up from the earth its potential energy is equal to its *mass* multiplied by the *acceleration due to gravity* multiplied by the height it was raised. The *SI unit* is the *joule* and the symbol is E_p, V or ϕ.

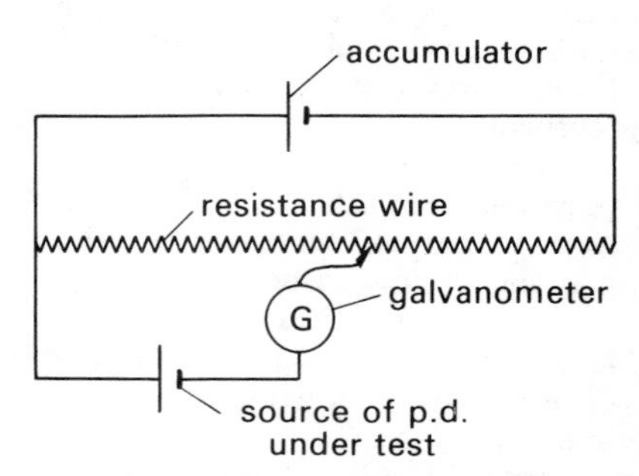

Fig. 129 The Potentiometer

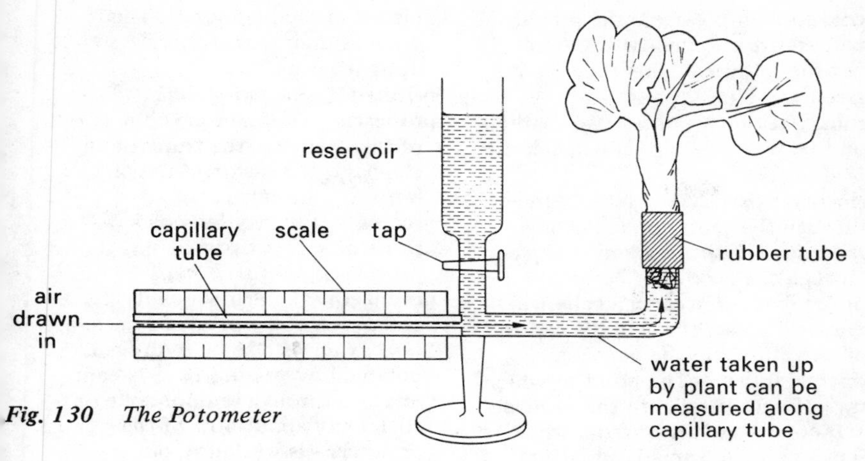

Fig. 130 The Potometer

potentiometer An instrument for measuring *potential difference*. A metre length of *resistance* wire is stretched across a board and the ends are joined to an *accumulator*. The potential difference to be measured is connected between one end and a moving connection on the wire with a *galvanometer* in this part of the circuit as well.

potometer An instrument used to measure the amount of water taken in by a plant under different conditions.

power 1. (Electrical) The *electric current* multiplied by the *potential difference*. The *SI unit* is the *watt*, see 3. 2. (Lens) One divided by the *focal length*. If the focal length is in metres, the power is given in *dioptres*. 3. (Mechanical) The rate of doing work, or the work done divided by the time taken. The *SI unit* is the *watt* and the symbol is *P*.

power pack An *electric power* supply unit.

precession If an object is spinning like a top and sideways forces act on it, then the top of the axis about which it is spinning moves round (precesses) in a circle.

precipitate See *precipitation*.

precipitation The formation of an *insoluble solid* (precipitate) by a chemical reaction between *solutions*, e.g. barium chloride solution mixed with any *aqueous* sulphate will give a precipitate of barium sulphate. This is an example of *double decomposition*.

predator A hunting animal, e.g. a lion.

pregnancy The state of a woman or any female *mammal* when she has a baby developing inside the *womb*.

prehensile Able to seize hold of something, e.g. a man's hand or a bird's foot.

premolar One of the teeth between the *canine* and the *molars*. Premolars are used for cutting or grinding food. See *tooth* (Fig. 159 ii).

pressure The *force*, pushing at right angles to a surface, divided by the area of the surface. The *SI unit* is the *pascal* and the symbol is *p* or *P*.

primary cell A *cell* which produces (rather than stores) an electric current by chemical methods. See also *Daniell cell, dry cell, Leclanché cell, secondary cell, simple cell* and *voltaic cell*.

primary coil The coil of an *induction coil* or *transformer* which is connected to the electricity supply and which produces the *magnetic field* which works the instrument.

primary colour A *colour* which cannot be made by mixing other colours. For *pigments* the primary colours are red, yellow and blue, but for coloured lights they are red, green and blue.

primary feathers The large feathers on the *wing* of a bird which help the bird to fly.

principal axis The reference line through the centre of a *lens* or *spherical mirror*. In a lens it passes through the *centres of curvature* of the lens surfaces, in a spherical mirror it passes through the centre of curvature and the *pole*.

principal focus The point to which rays of light parallel to the *principal axis* come together or from which they appear to spread out, after they have passed through a *lens* or have been reflected by a *spherical mirror*.

principle of flotation A floating body pushes aside its own *weight* of the *fluid* in which it is floating. This is a special example of *Archimedes' principle.*

principle of moments When several *forces* act on an object and it stays in *equilibrium*, then the total of the *moments* which act clockwise about any point equals the total of the *moments* which act anti-clockwise about that point.

prism A block of glass (etc.) with a base and a top usually in the shape of a triangle and with sides in the shape of a rectangle. It is often used to produce a *spectrum* by *refraction.*

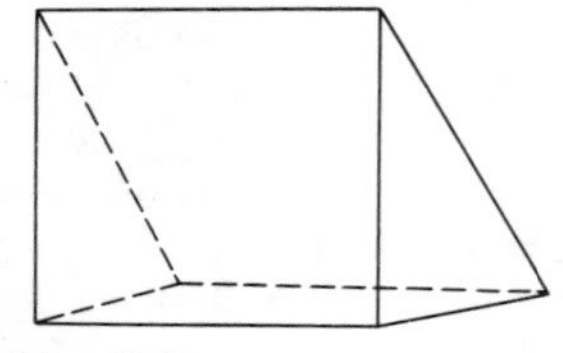

Fig. 131 Prism

prism binoculars Two *telescopes* fixed side by side (one for each eye) but so arranged that the light is totally internally reflected (*total internal reflection*) by two *prisms* placed in each telescope. This reduces their length and gives an upright image.

prismatic See *monoclinic*.

proboscis A long tube coming out of the head, e.g. the trunk of an elephant; the sucking tube of a butterfly, or mosquito.

process The way by which something is performed, e.g. the process of *photosynthesis*.

producer gas A low-quality gas composed of *nitrogen* and *carbon monoxide* with some hydrogen, obtained by passing an air/steam mixture through red-hot coke or coal. Only about 40% of the producer gas will burn, but it is easy and economical method of producing an industrial gas for heating furnaces.

progesterone A *hormone* made in the *ovary* after *ovulation*. It controls the conditions for the proper growth of the unborn baby.

progressive waves *Waves* which continually spread out from the source.

proleg One of the short legs on a *caterpillar*.

pronate The position of the hand when the back is uppermost. The opposite of *supinate*.

propagate To increase in numbers, e.g. most plants propagate by *seeds*.

propane C_3H_8 A colourless *flammable gas* of the *alkane* series of *hydrocarbons*. Used commercially. In liquid form it is known as liquid petroleum gas.

property Some *characteristic* which describes how a substance appears or behaves under certain conditions, e.g. oxygen has the properties of being an invisible, tasteless, odourless gas, slightly soluble in water, not flammable but able to support burning. Sulphur has the properties of being a yellow solid capable of crystallizing in two different forms and having a low melting point.

proportional If two quantities are connected so that if one is doubled then the other doubles or if one is halved the other halves, etc., then they are said to be directly proportional or just proportional, e.g. increase in length and stretching *force* in *Hooke's law*. If the connection is such that if one is doubled the other halves etc., then they are said to be inversely proportional, e.g. pressure and volume in *Boyle's law*.

prostate gland A *gland* connected with the *male reproductive system*. It makes the liquid in which the *sperms* swim.

protease Any *enzyme* which breaks down *proteins* in *digestion*.

proteins Complex *organic compounds* of high *relative molecular mass* (up to 10 000 000) which all contain carbon, hydrogen, oxygen and nitrogen and often phosphorus and sulphur as well. They are present in all living matter and are essential for a balanced diet. Colourless and tasteless, they are *decomposed* by heat and broken down by acids. Meat, cheese and eggs are rich in proteins, which are derived from simple *amino acids*.

prothallus The simple green plant which grows from a fern *spore*. The prothallus produces the *gametes*. After *fertilization*, a new fern plant is formed.

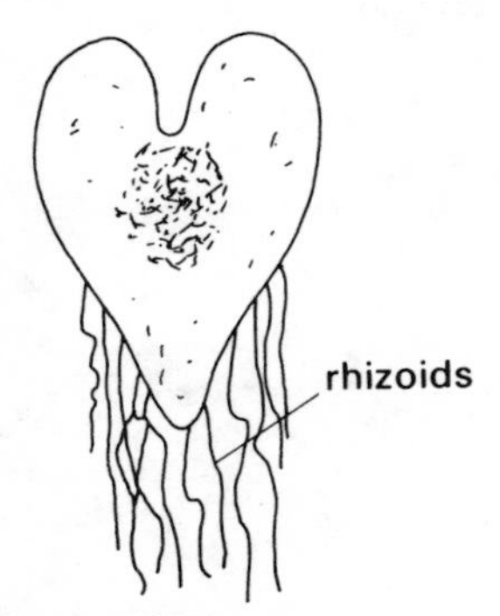

Fig. 132 Prothallus

prothorax The part of an *insect* from which the front legs grow.

proton The *nucleus* of the ordinary hydrogen *atom*. It is a basic particle and is found in the nuclei of all other atoms. The number of protons in the nucleus is equal to the atomic number of the element. It has an *electric charge* equal in size to that of an electron but the charge is positive. Its mass is about the same as that of a *neutron* and about eighteen hundred and forty times that of an electron.

protoplasm The jelly-like substance in the *cells* of all plants and animals. It is made up of many different chemicals mixed up in water. It is the living substance.

protozoans Very small animals made up of only one *cell*, e.g. amoeba.

pruning Cutting back a plant in order to make new growth.

pseudopodium A part of the *protoplasm* of an animal or a *cell* which is pushed out and encloses a food particle.

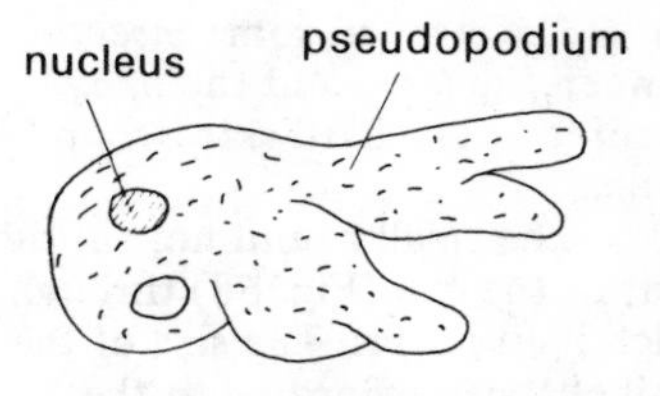

Fig. 133 Pseudopodium of Amoeba

pteridophyte Any plant belonging to the fern group.

ptfe --- $(CF_2)_n$---A *polymer* made from the *monomer*, tetrafluoroethene, C_2F_4, the hydrogen atoms in ethene being replaced by fluorine ones. It is chemically inert and is used as a coating on kitchenware to provide a surface to which milk or fat will not stick. Its full name is polytetrafluoroethene. One trade name is Teflon.

ptyalin An *enzyme* made in the *salivary glands* which changes *starch* to *sugar*.

puberty The stage in a boy's or

girl's life when the *reproductive organs* begin to work.

pubis 1. The lower part of the *abdomen.* 2. A bone which forms part of the *pelvis.*

pulley A wheel with a hollowed edge around which a rope or string can pass, to make a simple *machine.*

pulmonary Having to do with the *lungs.*

pulse A wave of *blood* sent from the *heart* through the *arteries.* By counting the pulse, the rate of the heart beat can be found. Doctors use it as a test of good health.

pulverized Broken down into a fine powder, e.g. pulverized coal is used in some power stations in the furnaces.

pumice A piece of organic lava full of holes caused by escaping gases. It is used as an abrasive and also for absorbing gases.

pungent Powerful, penetrating, strong—used in describing the smell of something such as ammonia or sulphur dioxide.

pupa The stage in some *insects* between the *larva* and the *imago.*

puparium The hard skin around a *pupa.*

pupil The small round hole in the front of the *eye* (Fig. 68) through which light passes. The size of the pupil changes according to the brightness of the light.

pus The yellow liquid which comes out of a wound. It is formed from dead *bacteria* and *blood.*

PVC See *polyvinyl chloride.*

pump An instrument with at least two *valves* for raising a liquid or for forcing a *fluid* to move along a pipe etc. See also *force pump, lift pump.*

pure spectrum A *spectrum* in which the colours can be clearly seen because they do not mix again after having been separated by the *prism,* etc. A *lens,* or lenses, are needed as well as the prism.

pyloric sphincter The *muscle* which surrounds the *pylorus.*

pylorus The opening between the *stomach* and the *duodenum.*

pyramid Part of the *kidney* into which the *urine* goes before being taken to the *bladder* through the *ureter.*

pyrenoid A small particle where *starch* is stored in simple plants, e.g. spirogyra.

pyrene Trade name for a *fire extinguisher* containing *tetrachloromethane.*

pyrites A brass-coloured *mineral* (known as fools' gold) often found in coal. Iron pyrites contains a sulphide of iron, FeS_2, and copper pyrites contains $CuFeS_2$.

pyrometer An instrument for measuring the *temperature* of very hot bodies by studying the *radiation* they give out.

quadravalent See *tetravalent.*

quadriceps muscle The *muscle* at the front of the upper leg which straightens the knee *joint.*

quadruped Any animal with four legs.

qualitative analysis A system of *analysis* in which tests are carried out in order to identify the particular *elements* or *groups* present in a compound without regard for the amount of each one.

quality of sound The same basic note (*fundamental*) played on different instruments does not seem the same. This difference in quality (or timbre) is a result of the presence of various *overtones* in different amounts.

quantitative analysis The *analysis* of a *mixture* or *compound* in order to determine by chemical means the amount of each *element* or *group* present.

quantity of electricity The *electric current* flowing multiplied by the time time for which it flows. The *SI unit* is the *coulomb* and the symbol is *Q.* See also *ampere-hour.*

quantum *Energy* can only be found as a whole number times a very small basic amount or quantum. The size of the quantum depends on the radiation involved. See also *photon.*

quartz A naturally-occurring *crystalline* form of *silica,* SiO_2. It is widely distributed in all forms of rocks and is usually white and *opaque,* but may be clear and colourless. It expands so little when heated that it can be used for such articles as *crucibles* and furnace tubes; these can be *quenched* in cold water when hot without cracking.

queen The *egg*-laying female of *insects* which live in groups, e.g. bee.

quenching Rapid cooling; white-hot steel is quenched in oil or cold water in order to harden it.

quicklime See *calcium oxide.*

quill The stem of a *feather.*

raceme The name given to the grouping of *flowers* where the oldest flower is at the bottom of the flower stem.

rachis The *quill* of a *feather.*

radian The *SI unit* of angle. The symbol is *rad.* If a circle is drawn and a distance marked along the edge equal to the radius of the circle, then the angle made by joining the ends of this marked distance to the centre of the circle is one radian.

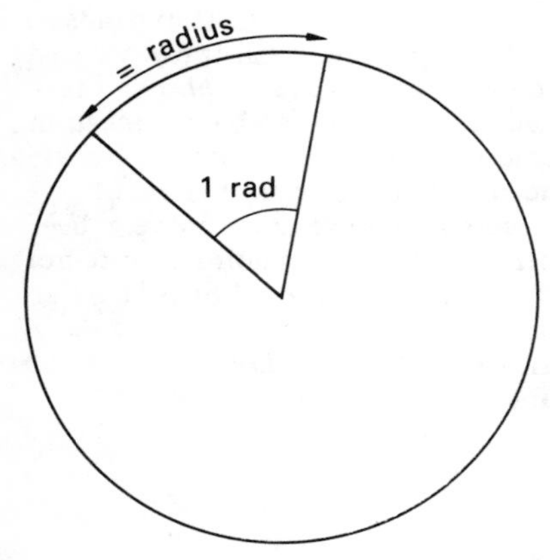

Fig. 134 Radian

radiation 1. The carrying of *energy* by *waves,* particularly by infra-red radiation, which are members of the *electromagnetic spectrum.* The energy is released as the wave is absorbed by a surface. 2. The members of the *electromagnetic spectrum.* 3. *Alpha* and *beta particles* given out by *radioactive* materials.

radical A group of *atoms* present in *compounds* which are normally incapable of a separate existence, but which behave as a single unit in *chemical reactions.* They may be *positive* or *negative ions* or neutral groups, e.g. *sulphate* radical, SO_4^{2-}; *hydroxyl* radical, OH^-; *ammonium* radical, NH_4^+; methyl radical, CH_3-.

radicle The part of a *seed* which will grow into the *root.*

radioactive A radioactive material gives out *alpha* or *beta particles* or *gamma radiation* from the nuclei (*nucleus*) of its *atoms.* The atoms change into different atoms as a result of this natural process.

radioisotope A *radioactive isotope* of an *element.* Some, such as *potassium-40* occur naturally, but many are made artificially by bombarding elements with *neutrons* in an *atomic pile.*

radio waves Radiation which forms the part of the *electromagnetic spectrum* with the longest *wavelengths.* They are used for sending information over long distances as in radio, television and radar.

radium Ra At. No. 88 R.A.M. 226 Rare *Group* II *metal* similar to barium, but *radioactive.* Used in the treatment of cancer.

radius Part of the *skeleton* (Fig. 148). It is one of the two bones in the forearm. The other is the *ulna.*

radius of curvature The radius of the sphere of which a curved surface is part. The term is very often used in connection with *lens* and *spherical mirror* surfaces. See also *centre of curvature.*

radon Rn At. No. 86 R.A.M. 222 A naturally-occurring *radioactive* gas given off by radium on disintegrating. It belongs to the *argonon* family of Group O of the *periodic table* and is used in the treatment of cancer.

radula The rough *tongue* of a snail.

random walk The path of an object which moves a short way in one direction, then a short way in another direction, then a short way in a third direction, etc. A *molecule* of a gas moves in this way, changing direction each time it hits another molecule. If the random walk consists of equal steps in different directions, the average, *resultant* distance gone from the start can be shown to be the square root of the number of steps multiplied by the length of one of them.

rare gases See *argonons.*

rarefaction A space where the *pres-*

sure is reduced, particularly in a sound *wave*.

ratemeter An instrument used with a *Geiger-Muller tube*. It shows the average rate at which the *radiations* are arriving and may also supply the electricity to work the tube.

ray diagram A drawing made to find out the position and size of the *image* formed by a *lens* or *spherical mirror*. Only two rays, whose paths are especially easy to draw, need be used.

ray floret The large outer '*petals*' of the *flowers* of the daisy group. See *compositae* (Fig. 41).

rayon A number of man-made *fibres* based on *cellulose;* formerly known as artificial silk.

react To act on or together. When a substance such as ammonia is mixed with hydrogen chloride the two act on each other (react) to produce a new *compound*, ammonium chloride. We say that a reaction has taken place. The ease with which the reaction occurs depends on the activity (reactivity) of each substance taking part and these are called the reactants. See also *chemical reaction*.

reactive Able to enter fairly readily into *chemical changes*, e.g. ethene, chlorine, sodium.

reactivity series See *electrochemical series*.

reactor See *nuclear reactor*.

reagent A reagent is simply a substance which takes part with another to bring about a *chemical change*.

real and apparent depth When an object is seen through a substance other than air it appears to be nearer than it actually is, due to the *refraction* of the light. It can be shown that real depth divided by apparent depth equals the *refractive index* of the sub-substance.

real image A type of *image* produced by a *lens* or *spherical mirror* where the light actually passes through the image position. A real image can therefore be seen on a screen if one is placed at the image position. See also *virtual image*.

receiver A vessel in which the distillate from a *distillation* is collected.

receptacle The top of the flower stalk from which grow the parts of the *flower* (Fig. 76 ii).

receptor A general name given to those parts of the body which are sensitive to a *stimulus*, e.g. *eye, skin* receptors.

recessive Used to describe a *character* of a plant or animal which, although present, does not show if the opposite *dominant* character is also present, e.g. the blue colour of eyes which does not show if the brown colour is also present.

rectification The process of turning an *alternating current* into one which flows in one direction only. A *diode* is often used for this purpose.

rectified spirit See *ethanol*.

rectilinear propagation The usual path of light is along straight lines, unless *diffraction* happens.

rectum The last part of the *alimentary canal* (Fig. 5). It opens to the outside at the *anus*.

rectus A *muscle* which causes the *eye* to move around in its *socket*.

red blood cells The most numerous of the *blood cells*. They are red because they contain *haemoglobin*. They are important because they carry *oxygen* around the body. See *cell* (Fig. 33ii).

red lead See *dilead(II) lead(IV) oxide*.

reductant A substance capable of bringing about *reduction*. Some important reductants are hydrogen, carbon, powdered magnesium, carbon monoxide, sulphur dioxide, hydrogen sulphide.

reduction A process involving (*a*) The removal of *oxygen* or a group which forms *negative ions;* (*b*) The gain of *hydrogen* or a group which forms *positive ions;* (*c*) A decrease in the *valency* of the electropositive constituent; (*d*) The addition of an *electron*(s) to an *atom* or *ion*, i.e. a gain of electrons.

refine To remove impurities during the process of purification. (Fig. 135.)

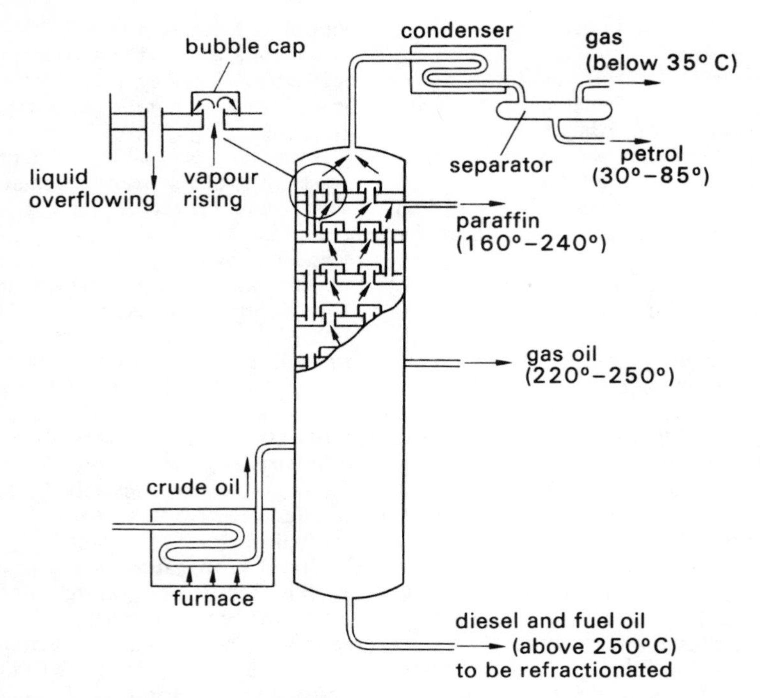

Fig. 135 The Refining of Crude Petroleum by Fractional Distillation

reflecting telescope A *telescope* in which the *objective* is a *concave* mirror rather than a *converging lens.*

reflection The sending back of light (etc.) when it falls on to a *mirror* surface. The reflected ray is the ray which leaves the mirror. The angle of reflection is the angle between this ray and the *normal* to the mirror at the point where the incident ray (*incidence*) struck it.

plane mirror
angle of incidence
angle of reflection
incident ray
reflected ray
normal

Fig. 136 Terms Used in Reflection

Reflection, Laws of 1. The angle of *reflection* is equal to the angle of *incidence.* 2. The incident ray, the *normal* (at the point of incidence) and the reflected ray are all in the same plane (flat surface).

reflex action An action in the body which is not under the control of the will power, e.g. sneezing, *knee jerk,* secretion of saliva. (Fig. 137.)

reflex arc The pathway of a *nerve impulse* which has to do with a *reflex action.*

reflux To boil liquids or a liquid and solid in a flask to which is attached a vertical *condenser.* The vapour *condenses* in this and flows back, thus preventing *evaporation.* (Fig. 138.)

refraction The bending of light (etc.) as it passes from one substance to another. The refracted ray is the one which travels in the second substance. The angle of refraction is the angle this ray makes to the *normal* to the

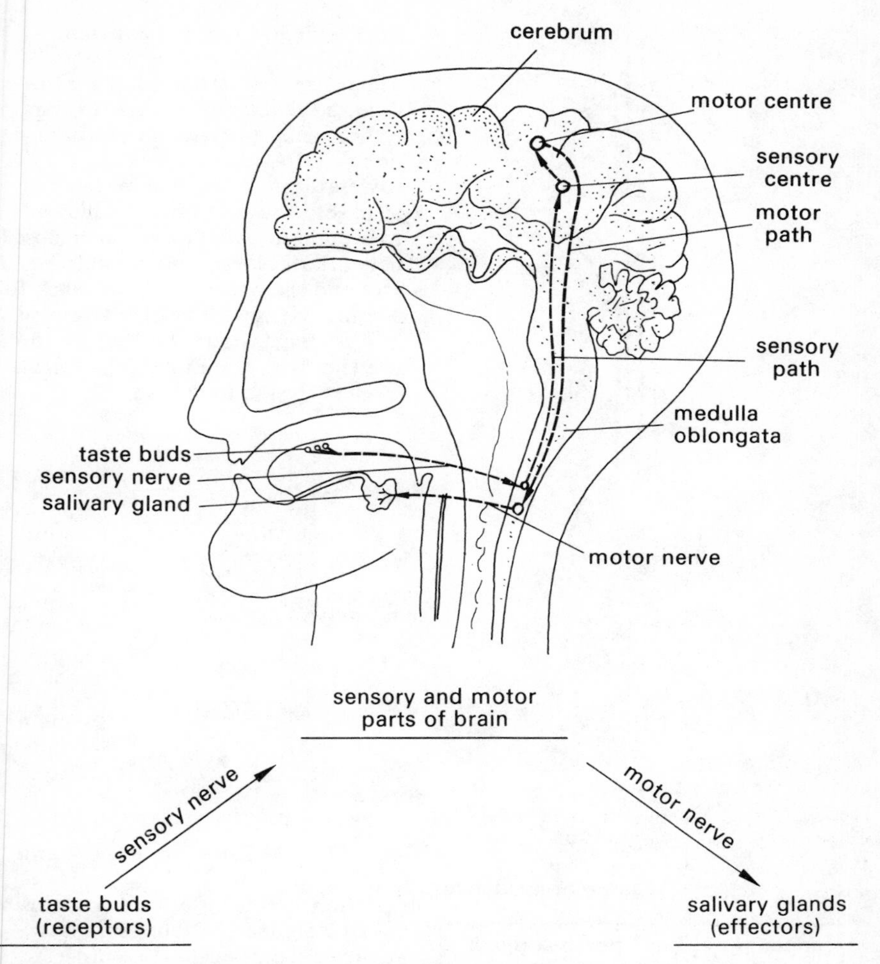

Fig. 137 The Secretion of Saliva – a reflex action.

surface where the incident ray (*incidence*) struck it. The emergent ray is the one which passes out of the second substance. (Fig. 139.)

Refraction, Laws of 1. The sine of the angle of *incidence* divided by the sine of the angle of *refraction* is a *constant* for all light of a given *colour* or *frequency* going from one particular substance to another. (Snell's law). 2. The incident ray, the *normal* (at the point of incidence) and the refracted ray are all in the same plane (flat surface).

refractive index The *constant* which shows the amount of bending or *refraction* when light goes from one particular substance to another. It is equal to the sine of the angle of *incidence* divided by the sine of the angle of *refraction* and is equal to the *velocity* of light in the first substance divided by the velocity of light in the second substance. Standard values are given when light is going from a *vacuum* into a named substance. It has no *units* (as it is a ratio) but the symbol is *n*. See also *refraction, laws*

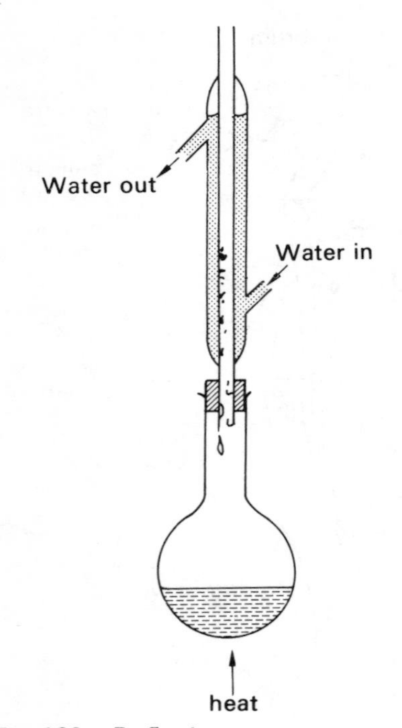

Fig. 138 Refluxing

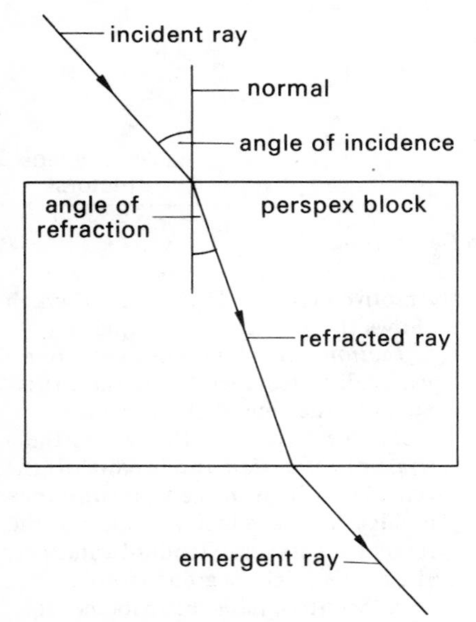

Fig. 139 Refraction through a Rectangular Perspex Block

of, Snell's law, real and apparent depth.

refractories Materials such as *silica,* lime and *fireclay* which are resistant to heat and can be used as furnace linings.

refrigerator A machine for keeping food (etc.) cold. A liquid is allowed to evaporate (*evaporation*) in a closed network of pipes inside the refrigerator and the *vapour* is turned back to a liquid outside the refrigerator case. The cooling produced by evaporation together with good insulation (*insulator*) keeps the food cold.

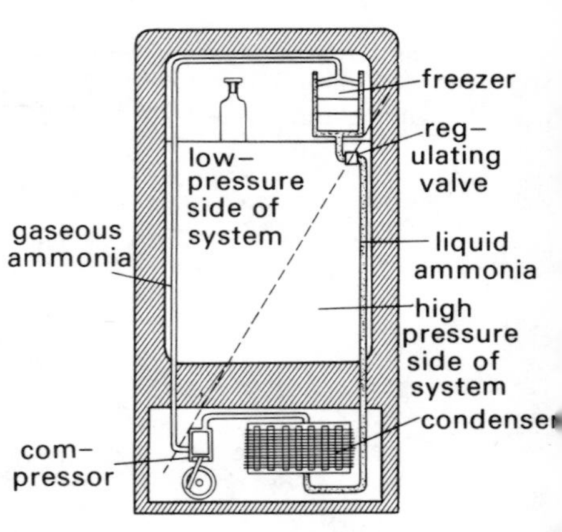

Fig. 140 The Domestic Refrigerator

regelation If ice is put under *pressure* it can melt and if the pressure is removed freeze again. This process, called regelation, will allow a wire carrying a heavy weight to pass slowly through a block of ice without the ice becoming two separate blocks.

regeneration 1. To make again, e.g. in the *water-softener* the sodium *zeolite* is gradually changed into calcium zeolite. The softening action then ceases. The sodium zeolite can be regenerated by flushing excess sodium chloride through the softener. 2. The ability of a plant or animal to grow a new part if damaged, e.g. a lizard can grow a new tail.

relative atomic mass This used to be called atomic weight. 1. Originally defined as the ratio of the *mass* of one *atom* of an *element* compared with that of one atom of *hydrogen*, but now it is compared with the mass of one atom of *carbon* on a scale where the atom of carbon is reckoned as twelve units. 2. The smallest amount of an element present in one *molecule* of any of its *compounds*. 3. The number of *grams* of an element which contain *Avogadro's constant* of atoms (6×10^{23}). If expressed in grams the relative atomic mass is called the *gram-atomic mass* (gram-atomic weight) or *mole*.

relative molecular mass This used to be called molecular weight. 1. The *mass* of one *mole* of *molecules* of an *element* or *compound* compared with that of one mole of *hydrogen atoms* (more accurately with one mole of *carbon* atoms on the scale where carbon = 12). 2. The mass of one molecule of an element of compound compared with that of one atom of carbon on the scale where carbon = 12. 3. The sum of all the *relative atomic masses* in one molecule of an element or compound, e.g. $H_2 = (2 \times 1) = 2$: $CO_2 = 12 + (2 \times 16) = 44$: $C_2H_5OH = (2 \times 12) + 6 + 16 = 46$.

relative density The *density* of a substance divided by the density of water. It has no *units* as it is a ratio. The older name for this ratio is *specific gravity*.

relative vapour density The number of times a definite *volume* of a *gas* is heavier than the same volume of *hydrogen* under the same conditions of *temperature* and *pressure*. Numerically, the r.v.d. equals half the *molecular mass* (weight) of the gas.

relay A magnetic switch. A current flowing in a coil sets up a *magnetic field* which moves a lever. The lever can move one or several sets of connections to switch other *circuits* on or off.

renal Having to do with the *kidney*, e.g. renal artery.

rennin The substance in the *stomach* which causes *milk* to become solid. Sold as rennet.

reproductive organs Those parts of the body of a plant or animal which have to do with *reproduction*, e.g. the *ovary*, the *flower*. (Fig. 141.)

reproduction The ability of a plant or animal to increase in number. It may be *sexual*, *asexual* or *vegetative*.

reptile A *vertebrate* animal whose body is covered with dry scales. The shelled *eggs* are laid on or in the ground, e.g. snake, crocodile, turtle.

residue The *solid* remaining in a flask or other vessel after *evaporation* or on a *filter paper* after *filtering*.

resins Hard *amorphous solids* obtained from the sap of trees and certain plants. When these are cut down the sap comes out and, on *evaporation*, leaves a hard glassy solid, the resin. Resins contain *long-chain molecules* and are *insoluble* in water, but generally *soluble* in *organic solvents* (in this they differ from *gums*). They are used in making *varnishes*, paper, rubber substitutes and adhesives. Some *plastics* are considered artificial resins.

resistance The resistance of a *resistor* is equal to the *potential difference* across the resistor divided by the *current* which flows through it. The *SI unit* is the *ohm* and the symbol is *R*. See also *parallel* and *series connection of resistors*.

resistivity The *resistance* of a wire is equal to its length divided by its cross-sectional *area* and then multiplied by a *constant* which depends on the metal from which the wire is made. This constant is called the resistivity (specific resistance) of the metal. The *SI unit* is ohm-metres and the symbol is ρ.

resistor A *component* in an electrical *circuit* whose main property is that it possesses *resistance*.

resolution The ability to see two objects, which are close together, as two separate things rather than as one. This is helped by an instrument like a *microscope* or a *telescope*.

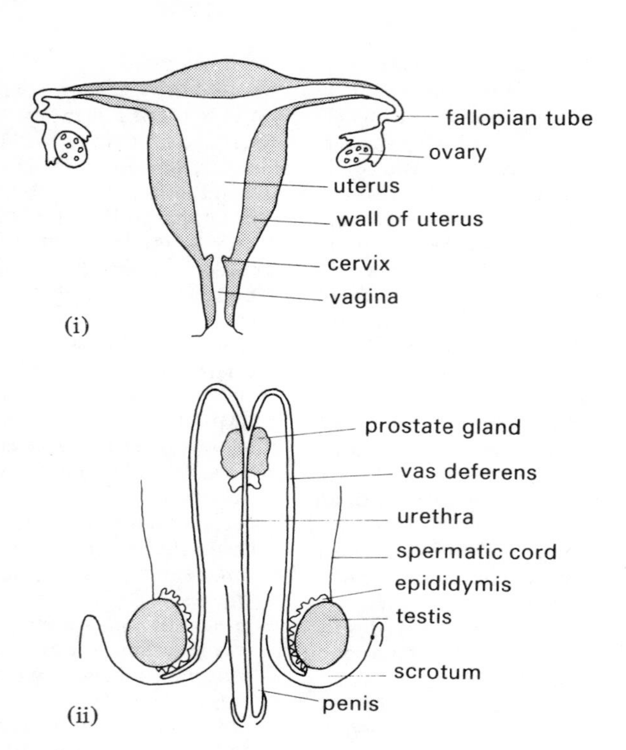

Fig. 141*(i) Female Reproductive System (human) (ii) Male Reproductive System (human) (simplified)*

resonance If a body is being made to vibrate by an outside source and the *frequency* of the source happens to be the same as the frequency of natural vibration of the body then the amplitude of the vibration is larger than for all other frequencies. This is called resonance and can become so violent as to damage the resonating body. See also *Barton's pendula, free vibration, forced vibration.*

respiration The *process* by which *energy* is released in living *organisms.* External respiration is the term often used for the movement of *gases* in and out of the organism. Internal respiration is the term used for the chemical processes which result in the release of energy in the *cell.*

response The action which follows a *stimulus,* e.g. when the *eye* is threatened, the response is to close the eyelid.

resultant The *vector* which is formed by the adding together of two or more other vectors. It has the same effect as the other vectors have when they act together. See also *parallelogram* and *triangle of forces.*

retardation The rate at which the *velocity* of an object decreases with time. See also *acceleration, deceleration.*

retina The back of the *eye* (Fig. 68) which is sensitive to light. In the retina the light is changed to *nerve impulses* which are then carried to the *brain* through the *optic nerve.*

retort 1. A piece of glass apparatus

with a large bulb and long neck in which certain preparations and *distillations* can be carried out. 2. A closed container lined with *refractory* bricks, in which coal is heated in the production of coal gas.

reversible A term applied to a *physical* or *chemical change* in which the original substances in their original states can be re-formed by altering the conditions under which the change has been brought about.

reversible reaction A *chemical reaction* which can occur in both directions by altering the conditions, i.e. *temperature* and/or *pressure* as a rule. It is usually incomplete unless one of the products is removed as it is formed. Otherwise, it proceeds to *equilibrium,* when the forward and backward reactions go on at the same rate, e.g. if ethanol and ethanoic (acetic) acid are heated together they form ethyl ethanoate and water which then react together to re-form the original substances. In the chemical equation for the reaction, reversed half-arrows are used: ethanol + ethanoic acid $\rightleftharpoons$ ethyl ethanoate + water; iron + steam $\rightleftharpoons$ iron(II) iron(III) oxide + hydrogen. See *dissociation.*

rheostat A *resistor* whose *resistance* can be changed, often by a sliding connection, so as to control the electric current flowing in a circuit.

rhesus factor A substance found in the *blood* of some humans. It is handed on to children from the parents. Rh plus means that the substance is present, Rh minus means that is is absent. It is possible that trouble may come if the mother is Rh minus and the father is Rh plus.

rhizoid A fine root-like structure found in some plants, e.g. *mosses, liverwort.*

rhizome A stem which grows under the ground and produces new plants as it grows, e.g. iris, mint. See *vegetative reproduction* (Fig. 164).

rhombic See *orthorhombic.*

rhombohedral A *crystal* system in which calcite, a form of calcium carbonate, occurs.

rib Part of the *skeleton.* Twelve pairs of ribs protect the *lungs.*

riboflavine A *vitamin* of the B group found in *milk, yeast, liver* and *egg.* If it is absent from food, *skin diseases* are likely.

rickets A disease due to lack of Vitamin D. The signs are soft bones in children, causing bent legs.

rider A small piece of wire which can be moved along the *beam* of certain types of analytical *balances* so as to add a very small mass to that already on the pan.

ringworm A skin disease caused by a *fungus.* It can be passed from an animal to man.

ripple tank An instrument for studying the behaviour of water *waves.* It often consists of a shallow container of water with a lamp fixed above it so that the shadows of the waves are seen on a white surface placed below the glass, or perspex, base of the container.

rochelle salt $COOK(CHOH)_2COONa.4H_2O$ Sodium potassium tartrate. A white *crystalline* salt which has a variety of uses, (*a*) as a *constituent* of *Fehling's solution;* (*b*) in the preparation of *baking powder;* (*c*) in *microphones* and *loud-speakers* where the distortion of the crystals by changes in pressure produces a kind of electricity (piezo-electricity).

rock salt A hard, natural, crystalline form of *sodium chloride,* often coloured by other minerals. The form in which sodium chloride is mined.

'Rocksil' wool The trade name for a *fibrous mineral* used in place of *asbestos wool* for insulation or fireproofing. It is used in chemical reactions where an *inert* absorbent (*absorption*) material is needed.

rod A *cell* in the *retina* of the *eye.* It works best in weak light. See *cone.*

rodent An animal which gnaws its food with its front teeth, e.g. rat, squirrel.

room temperature and pressure (abbr. rtp) Conditions, usually taken as 20 °C and 760 mm Hg, under which the *molar volume* of a gas is 24 000 cm^3 (24 litres).

root The part of the plant which grows underground from the *radicle* of the *seed*. It keeps the plant firm in the ground and takes in water and food from the soil. It has different shapes.

root cap The very end of a *root* (Fig. 142ii). The *cells* are small and protect the young root's growth.

root hair The long narrow *cells* which grow out from young *roots*. The thin walls allow soil liquids to be taken into the root.

root mean square value The root mean square (r.m.s.) value of an *alternating current* (Fig. 6) is the effective value of that current. This is the value of the steady current which will cause the same heating in a *resistor* as the alternating current does. It is equal to the peak value divided by the square root of two, 1.414. The name comes from the mathematical process by which this value can also be calculated. All the values of the current are squared, the average of the squares found and then the square root of this average is the required value.

root pressure The *pressure* caused by the taking in of water in young *roots*. It may cause the water to rise up in the root to supply the growing plant when it is in the young stage.

Ross, Sir Ronald The English scientist (1857–1932) who discovered the cause of *malaria*.

rostrum A beak, e.g. as in a bird or beetle.

rotation of crops The growing of different crops each year in the same ground. The different crops use different food substances and so the *soil* is put to the best use, e.g. corn may be followed by beans and then by root crops.

round window A thin spot in the wall between the *middle ear* and the *inner ear*.

royal jelly The mixture of *pollen* and *honey* which is fed to young queen bees.

ruminant An animal which swallows

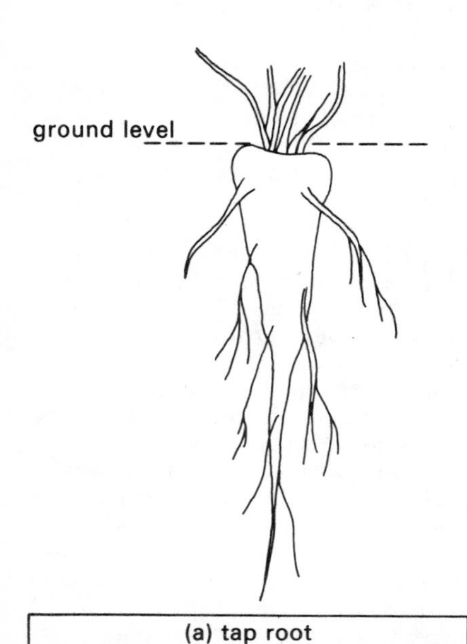

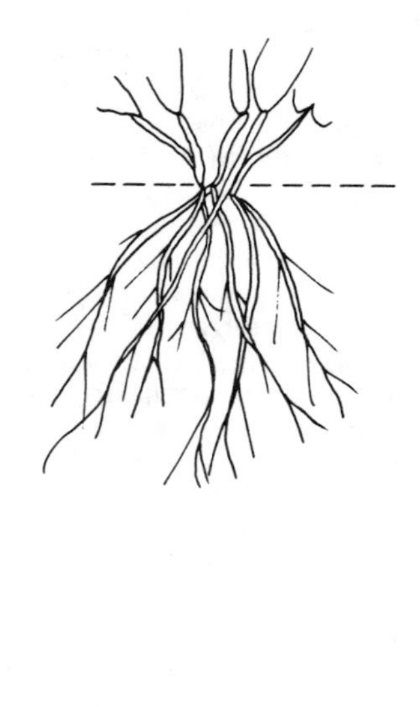

Fig. 142(i) *Root Formations*

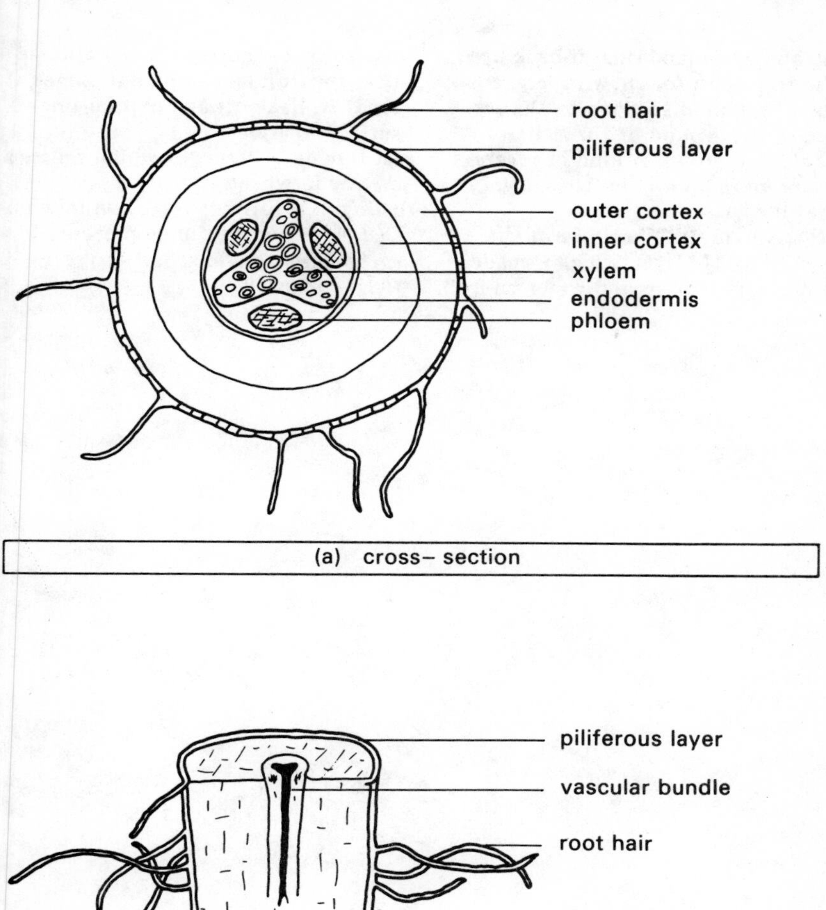

(a) cross– section

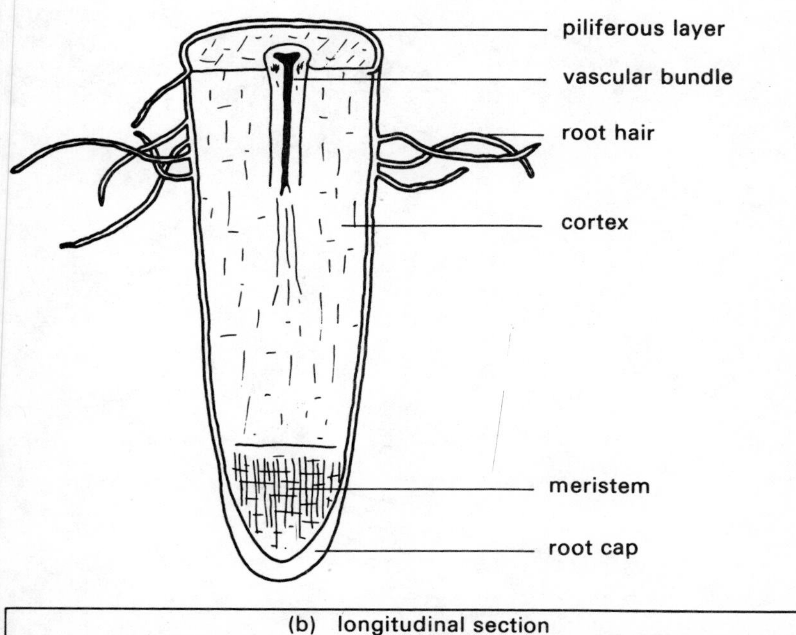

(b) longitudinal section

'ig. 142 (ii) Root structure

its food and then brings it back up into its mouth for chewing, e.g. cow.

runner A *stem* growing on the surface of the ground. It forms new plants as it grows. A kind of *vegetative reproduction* (Fig. 164), e.g. strawberry.

rust A hydrated form of *iron(III) oxide,* $Fe_2O_3.xH_2O$, which is made by the action of air and water on iron. As the rust occupies a larger volume than the iron from which it comes, it tends to flake off and expose a new surface to attack.

rust fungus A *fungus* which causes a *disease* in wheat.

rusting A form of *corrosion* in which *rust* is formed. It can be prevented by *cathodic protection.* See also *sacrificial corrosion.*

sacculus Part of the liquid-filled space of the inner *ear* (Fig. 59).

sacrificial corrosion In order to keep *iron* or *steel* in contact with air and water from *rusting* a block of a *metal* which is higher in the *electrochemical* series than iron is attached to it. This metal, being more reactive, is attacked in place of the iron. Usually, magnesium *alloyed* with some aluminium and zinc is used and in water it becomes the *anode* of an electrolytic *cell* with the iron or steel as the *cathode*. The anode undergoes sacrificial corrosion.

sacrum Part of the *pelvis* which is joined to the *vertebral column*. See *skeleton* (Fig. 148).

sal ammoniac See *ammonium chloride*.

sal volatile Commercially impure ammonium carbonate used as a stimulant to overcome a feeling of fainting.

saliva The watery liquid made in the *salivary glands*. It is used partly to start the *digestion* of *starch* and partly to make it easier for food to be swallowed. It contains *amylase* and *mucus*.

salt 1. A *compound* containing a metal or *positive radical* (other than hydrogen) along with a non-metal or *negative* radical. 2. A substance which, along with water, is produced by the action of an *acid* and a *base*. 3. An *ionic* compound which is neither an acid nor a base. Sodium nitrate, barium chloride, calcium hydrogencarbonate, ammonium sulphate are all salts. Salts may be produced by (*a*) reacting an acid with a base; (*b*) reacting an acid with a metal; (*c*) reacting an acid with a metal *carbonate;* (*d*) mixing two salts already in solution to produce an *insoluble one;* (*e*) the direct union of *elements* or compounds. 4. The common name for sodium chloride.

salt gas See *hydrogen chloride*.

salting-out The removal of an *organic compound* from its *solution* in water by addition of *brine* in which it is *insoluble*. The term is applied to the separation of *soap* from *glycerol*. The mixture is poured into a concentrated solution of sodium chloride. The soap is precipitated (*precipitation*) and can be removed and dried.

saltpetre See *potassium nitrate.*

sand 1. Hard granular form of *silica,* mixed with other compounds such as iron(III) oxide (which colours it brown). 2. The larger particles of the *soil.* A sandy soil will allow water to pass through quickly.

sandbath A metal plate containing sand on which glass vessels may be heated.

sap The watery liquid which passes through a plant. It contains food substances.

saponin A substance obtained from chestnuts by boiling them with water. It acts as a frothing agent in producing a *lather* as in a *foam fire extinguisher.*

saprophyte A plant without *chlorophyll* which gets its food from other dead or rotting plants or from the waste of animals, e.g. most fungi (*fungus*).

saturated compound An *organic compound* is said to be saturated when all the carbon atoms are linked by single *covalent bonds,* e.g. ethane C_2H_6 or H_3C-CH_3. Such compounds do not form *addition compounds.*

saturated solution A solution which cannot *dissolve* any more of a particular solid at a given *temperature.* Some of the solid must also be present as such to ensure that the solution is completely saturated.

saturated vapour A *vapour* which has has as much substance in the vapour form as possible, at a particular *temperature.* It is often found in a closed space when there is some of the liquid present. Note that the liquid continues to evaporate, but as many *molecules* return to the liquid state as leave it.

saturation vapour pressure (*abbr.* S.V.P.) The *pressure* of a *saturated vapour.* It changes with the *temperature* but not with the volume of the vapour if it remains saturated.

scalar A quantity which has no special direction connected with it. **See** also *vector.*

scale 1. A *deposit* of *calcium carbonate* (or other calcium or magnesium compounds) left in kettles and boilers in which *hard water* is boiled or in pipes through which this water (hot) has been passed. It is also called fur. 2. One of the flat plates on the outside of many animals, e.g. fish. Also on the wings of butterflies and moths. 3. A small leaf on the outside of a bud.

scaler An instrument used with a *Geiger-Muller tube.* It counts the total number of *radiations* which have arrived and may also supply the electricity to work the tube.

scapula The shoulder blade bone of the *skeleton* (Fig. 148).

scar The mark left when a wound heals or when a leaf falls.

scavenger An animal which gets its food from the dead remains of other animals, e.g. vulture.

sciatic Having to do with the leg, e.g. sciatic *nerve.*

scintillation Tiny flashes produced by *alpha* and *beta particles* or *x-rays* striking a screen coated with zinc sulphide or certain chemicals known as phosphors.

scion The part of a plant which is *grafted* on to a *stock.*

sclerotic Very firm, giving support, e.g. the sclerotic layer of the *eye* (Fig. 68). Sclerosis is the hardening of tissues, e.g. arterio sclerosis is the hardening of the walls of *arteries.*

scrap metal *Iron* and *steel* which has been used for some purpose no longer required and is now available for mixing with fresh pig-iron (*cast-iron*) and used in the production of steel by the *open-hearth* process.

screw jack A simple *machine* where the *effort* turns a screw thread and so raises a heavy *load.*

scrotum The skin bag which holds the *testes* in a man. See *reproductive organs* (Fig. 141ii).

scum The whitish deposit of calcium or magnesium stearate formed in *hard water* when *soap* is added.

scurvy A skin disease caused by a lack of Vitamin C.

sea water The oceans contain approximately 2.8% sodium chloride, 0.4% magnesium chloride, 0.2% magnesium sulphate, 0.1% calcium sulphate and 0.1% potassium chloride along with many other dissolved minerals. These are the average amounts of the above obtained on evaporation. The composition differs considerably from that of water from inland lakes.

sebaceous glands *Glands* which occur in the *skin* (Fig. 149). They make an oily substance which keeps the hairs soft and waterproof.

sebacoyl chloride $ClOC(CH_2)_6COCl$ A *reagent* used in the preparation of one form of *nylon.*

second The *SI unit* of time. It is a basic SI unit and the symbol is s. The second is 9 192 631 770 multiplied by the time for a Caesium-133 atom to vibrate in a certain way.

secondary cell A *cell* which stores, rather than produces, electric current by chemical methods. See also *accumulator, Nife cell, primary cell.*

secondary coil The coil of an *induction coil* or *transformer* in which the wanted *potential difference* or *current* is induced.

secondary feathers The smaller *feathers* on the *wings* of a bird.

secondary sexual characters The outward signs of the different *sexes,* e.g. the beard of man, the colours of some birds.

secondary thickening The extra wood added to a tree every year.

secretin The *hormone* made when food goes into the *duodenum.* Secretin goes in the *blood* to the *pancreas* and 'tells' the pancreas to make its *enzymes.* It also 'tells' the *liver* to make *bile.*

secretion A substance made in the *cells* of a gland, e.g. the pancreas secretes *enzymes;* the *liver* secretes *bile.*

sedative A *drug,* such as potassium bromide, which produces quietness, calm and possibly sleep.

sediment Solid matter falling to the bottom of a liquid; a precipitate (*precipitation*).

sedimentary rocks Rocks formed in layers, usually by plant or animal remains settling on other rocks, e.g. *coal.*

seed The reproductive (*reproduction*) part of a *flower* (Fig. 76iv), formed after *pollination* and *fertilization.* The seed has an outer skin (*testa*) inside which is the young plant *embryo* with a store of food. After *germination,* the seed first grows into a *seedling* and then into a plant.

segment A part, e.g. one of the divisions of an earthworm or one of the parts of an insect's leg.

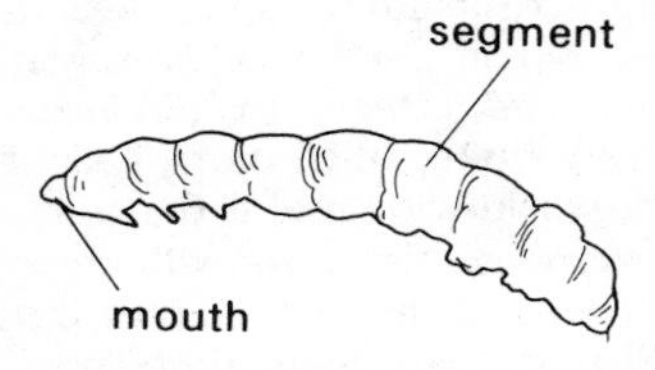

Fig. 143 Segmentation in the Caterpillar

selective discharge When a *current* is passed through an *electrolyte* containing several *ions,* the *cation* lowest in the *electrochemical series* is the one which is normally liberated first. Thus copper ions before hydrogen and hydrogen ions before calcium. Similarly hydroxyl ions before chloride or sulphate ions. This order of discharge may be altered by other conditions such as *concentration* or the nature of the *electrodes.*

selective membrane The protoplasmic (*protoplasm*) layer of a *cell* which allows some substances to pass through but not others, e.g. common salt with small *molecules* will pass through but *proteins* with large molecules will not.

self pollination When the *pollen* of a *flower* lands on the *stigma* of the same flower.

semen The liquid (seminal fluid) which contains the male *sperms.* It is passed out through the *penis* into the female body in a *mammal* during *sexual reproduction.*

semicircular canals Parts of the inner ear (Fig. 57) which are concerned with balance.

semiconductors A group of materials (e.g. germanium and silicon) which come between *insulators* and *conductors* in their ability to carry an electric current. They form the basis of *transistors* and some (non-*thermionic*) *diodes.*

semipermeable membrane See *selective membrane.*

sense organ A part of the body which is able to receive stimuli (*stimulus*), e.g. *eye, skin.*

sensitivity The ability of a plant or animal to 'know' when changes are taking place around it.

sensory nerve Another name for *afferent*—carrying messages to the *brain* or *spinal cord.*

sepals The outer parts of a *flower,* usually green. They are used to protect the flower *bud.*

separating funnel See *dropping funnel.*

sepsis Poisoning, usually by *bacteria.*

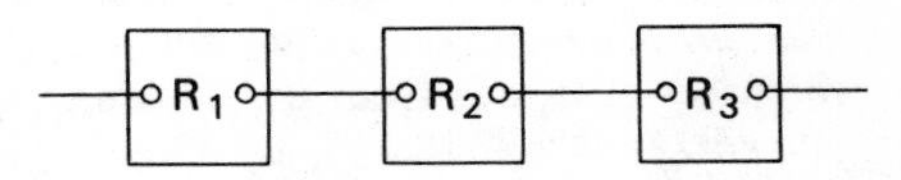

Fig. 144 Series Connection of Resistors

series connection of resistors To join *resistors* (etc.) in series the second connection on the first resistor is joined to the first connection on the second resistor. Then the second connection on the second resistor is joined to the first connection on the third resistor, etc. The total *resistance* of all the resistors (of resistances R_1,

R_2, R_3, R_4 etc., is R where $R = R_1 + R_2 + R_3 + R_4$. Note that the same current flows through each resistor.

serum The clear yellow liquid part of *blood.* It is sometimes used to fight against a *disease* because it may have the right *antibodies.*

sex Male or female. It is decided by the *chromosomes.* See also *X-chromosome.*

sex linkage When a character is carried on the same *chromosome* as the sex, e.g. the bright colours of many male birds.

sexual reproduction The way by which the young of plants or animals are made by the joining together (*fertilization*) of a male *cell* and a female cell. Most plants and animals use this way. See *asexual* and *vegetative reproduction* and *zygote.*

shadow The area of darkness caused by an object coming between a lamp and a screen. See also *penumbra, umbra.*

shaft The middle part of a *feather.* See *barb* (Fig. 14).

shell The *electrons* in the *atom* are arranged in a series of shells or *orbits* at different distances from the *nucleus,* each shell containing a fixed maximum number of electrons. These are numbered outwards from the nucleus, 1, 2, 3 etc. or lettered K, L, M etc. and the number of electrons in the outer shell determines the *valency* of the *element.* The *energy* of the electrons increases outwards from the nucleus. See *energy levels.*

shellac A *thermoplastic resin* formed on some trees and used in varnishes.

sherardizing A process of *rust*-proofing small *iron* or *steel* objects such as screws by coating them with *zinc.* The articles are enclosed in a container of powdered zinc and zinc oxide and heated to about 300 °C.

shivering *Muscles* under the *skin* moving when cold. It is an example of *involuntary action* and helps to warm the *blood* in the *skin.*

shoot The *stem* and *leaves* of a plant.

short circuit When a *battery* (etc.) has its connections joined together by a thick wire, etc., so that too much current flows, it is said to be short circuited.

short sight See *myopia.*

shunt A low value *resistor* connected across an *ammeter* so that most of the current in the circuit flows through the shunt and only enough to work the ammeter actually flows through it. The value of the shunt controls the range of currents that the ammeter can measure.

SI units The only set of *units* that should be used in science today. All quantities are measured in the basic SI units, the *metre,* the *kilogram,* the *second,* the *ampere,* the *kelvin,* the *candela* and the *mole* or in the necessary "mixtures" of them. Some of these mixtures have special names, e.g. the *newton* for the metre kilogram per second squared. (See also *joule, pascal, watt, coulomb, volt, ohm, farad, lumen, lux* and *hertz.*) The advantage of SI units, apart from their world-wide use, is that the answer to a calculation will always be in SI units if the information is put in SI units. The decimal prefixes (*atto-, femto-, pico-, nano-, micro-, milli-, centi-, deci-, deca-, kilo-, mega-, giga-,* and tera-) can be joined in front of any SI unit. (SI comes from the French 'Système International d'Unités'.)

side chain An *aliphatic radical* or series of groups attached to a *benzene* ring, e.g. the methyl group in methyl benzene, (toluene), and the $CH_2 = CH-$ group in *styrene.*

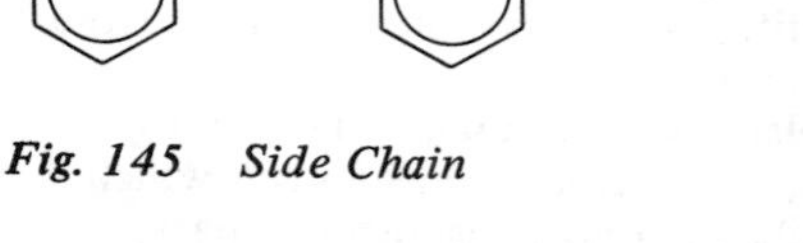

CH_3 $CH = CH_2$

Fig. 145 Side Chain

sieve tubes The fine tubes in a plant through which food substances in water travel to other parts, e.g. the *sugar* made in the leaves passes down

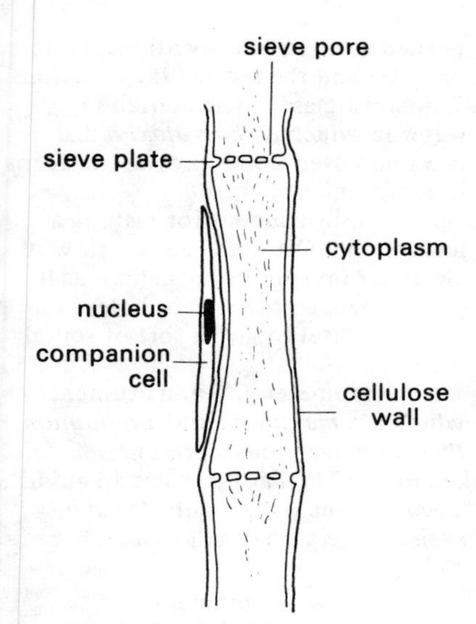

Fig. 146 *Sieve Tube*

to the roots through the sieve tubes. See *phloem*.

sign convention An agreement about the meaning of plus and minus signs when using the *lens formula,* e.g. that all distances measured to *real* objects, *images* and *principal foci* are counted positive and that all distances measured to *virtual* objects, *images* and *principal foci* are counted negative.

silage Cattle food made from plants cut when they are green and then fermented (*fermentation*).

silica SiO_2 Silicon dioxide; hard, insoluble, white or colourless. Usually found impure as *sand* or as *quartz* or flint. Because of its high resistance to heat it expands very little on heating and is used for lining furnaces and retorts and for making *crucibles* and other laboratory apparatus. It is also used in the production of *glass.*

silica gel A hard, *amorphous,* granular solid made by heating silicic acid, H_2SiO_3, very strongly to leave *silica.* Because it is *hygroscopic* it is used very considerably as a *drying agent.*

silicates Salts of various silicic acids. Present in many rocks as complex silicates of aluminium, magnesium, calcium, etc. They are often coloured, glassy in appearance and insoluble in water although sodium silicate does dissolve in hot water to form a solution known as water-glass. Some natural silicates, the *zeolites,* have the property of *softening hard water.*

silicon Si At. No. 14 R.A.M. 28 A non-metallic *element* placed in Group IV of the periodic table. It is similar to carbon in many of its properties and occurs mainly as *silica,* SiO_2. It is used in some *alloys* and is the second most abundant element found in nature.

silicones Substances containing *aliphatic* groups along with *chains* of *silicon* and *oxygen atoms.* The ones with low *relative atomic masses* are used as oils, lubricants and polishes and for waterproofing garments. Others are plastic-like substances useful as electrical *insulators.*

silt The very small particles of *soil* washed down by a river.

silver Ag At. No. 47 R.A.M. 108 A comparatively unreactive *metal* element not affected by air or water and used in coinage and in jewellery. Its *compounds* are used in photography and the metal is the best *conductor* of electricity.

silver nitrate $AgNO_3$ A white, *crystalline compound,* the commonest silver *salt.* It is used in medicine, in chemical *analysis,* in marking inks and for removing small growths such as warts from the skin. It used to be called lunar caustic.

simple cell A *cell* consisting of two different plates, e.g. copper and zinc, in a liquid, e.g. sulphuric acid. It suffers from *local action* and *polarization.* It is also called a *voltaic cell.*

sinus A hollow space inside a *bone.*

siphon 1. A curved pipe or tube (as shown) used for transferring a liquid from a level A to a lower one B by first raising it to a still higher level C. The tube is first filled with the liquid and as the *pressure* at C is lower than that at A, this causes the liquid to be

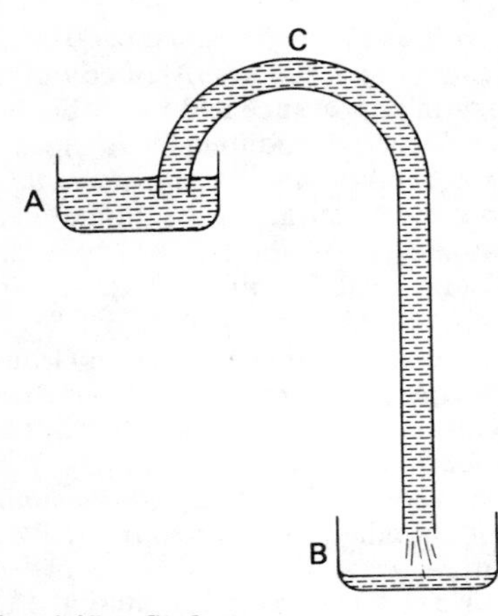

Fig. 147 Siphon

pushed from A to C by atmospheric pressure and then down the tube to B. 2. A stout glass bottle containing water in which *carbon dioxide* has been dissolved under pressure to form *soda-water.*

siren An instrument for making a loud sound. Often the steady flow of air under pressure is broken up as it passes through the slits or holes in a turning wheel to give a sort of sound *wave.*

six's thermometer An instrument which is a *maximum* and a *minimum thermometer* joined into one.

skeleton The hard parts of an animal. They may be on the outside (*exo-skeleton*), e.g. the shell of a crab, or

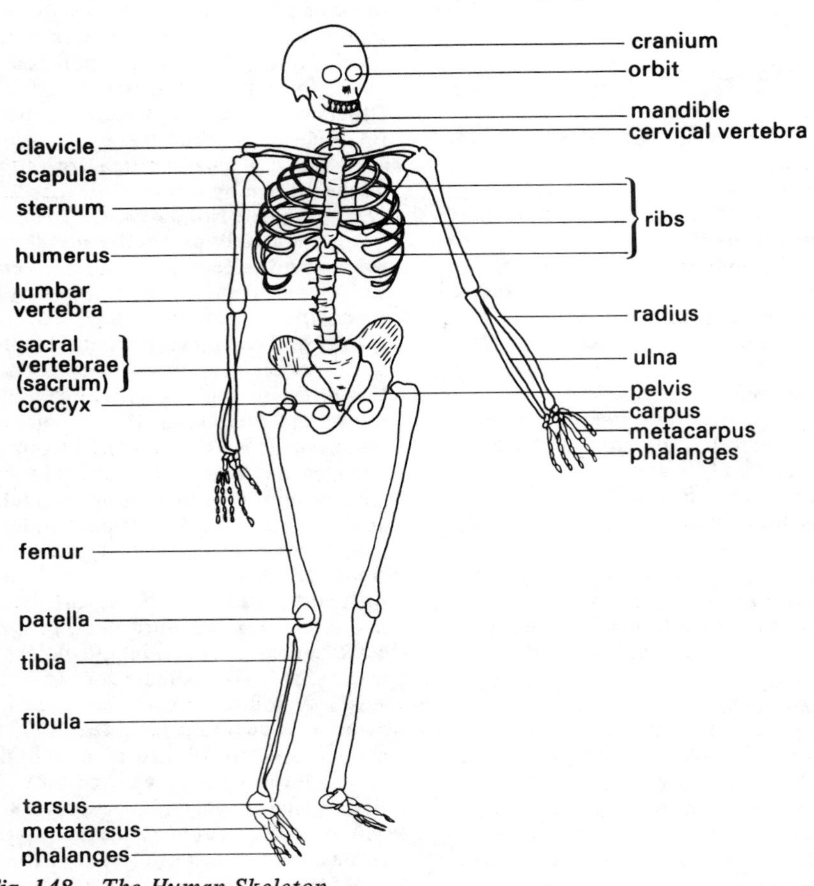

Fig. 148 The Human Skeleton

they may be inside (*endoskeleton*), e.g. the bones of a man.

skin The outer covering of the body, except where there is an *exoskeleton.* The skin is made up of different layers and has many uses, e.g. protection, *sensitivity.* (Fig. 149.)

skull Part of the *skeleton.* It is made up of the *cranium* which contains the *brain,* the upper *jaw* and the bones which surround the *sense organs* in the head.

slag 1. Waste solid, non-metallic material formed during the extraction of some elements from their ores which are mixed with *sand* and possibly *limestone* before being heated. Iron formed in the blast furnace is covered with a layer of molten slag (calcium silicate) which prevents it re-oxidizing. Phosphorus is obtained from calcium phosphate and this with sand again forms a slag of calcium silicate. 2. Waste material from coal mines left after the coal has been removed.

slaked lime See *calcium hydroxide.*

sleeping sickness A *disease* of humans and cattle carried by the tsetse fly. The germ is a protozoan (*protozoa*) called trypanosoma. It affects the *nervous system.*

slip ring A metal ring on the axle of a *generator* or *dynamo* which has a sprung connection, or brush, pressing on to it. If two slip rings are used, one joined to each end of the coil, *alternating current* will flow in the outside *circuit.* See also *commutator.*

slurry A mass of fine particles suspended in water or just mixed with water, e.g. cement and water.

smallpox A *skin disease* caused by a *virus.* See *vaccination.*

smelling salts Impure ammonium carbonate used for reviving a person about to faint.

smelting Heating a *metal ore* with *limestone* and *coke* to produce the metal along with a *slag* on top of the molten metal.

smog A *mixture* of *smoke* particles with *fog* or *mist.*

smoke A *colloidal suspension* of very small *solid* particles, usually of *carbon,* in the air.

Snell's Law See *refraction, laws of.*

soaps *Salts* of the *alkali metals* and *aliphatic acids* such as stearic (heptadecanoic), oleic or palmitic acids. Made by the action of *sodium hydroxide* on a *fat* followed by *hydrolysis.* Soaps produce a good *lather* in *soft water* and a *scum* with *hard water.* They are used for washing purposes and for removing grease. Sodium salts give hard soaps and potassium ones soft soaps.

socket The space on one *bone* into which another fits.

soda A term applied to various sodium compounds such as the hydroxide (*caustic soda*), the carbonate (*washing soda* or *soda ash*) and the hydrogencarbonate (*baking soda*).

soda ash Anhydrous sodium carbonate, Na_2CO_3.

soda-lime A solid obtained by *slaking calcium oxide* (lime) with a solution of *sodium hydroxide* (caustic soda) instead of with water and then drying by heat. Soda-lime can be obtained in small lumps and is used as a good absorbent for carbon dioxide, both in the laboratory and for purifying the air when men are working in a confined space. It is often coloured by a dye.

soda-water Water containing dissolved *carbon dioxide* forced in under *pressure.* It is usually kept in a glass bottle (a *siphon*) and can be forced out as a jet by releasing the pressure and so setting free the gas.

sodium Na At. No. 11 R.A.M. 23 A very soft grey *metal* which rapidly *tarnishes* in air and reacts exothermically with water to form *sodium hydroxide* solution with the liberation of *hydrogen:* it may take fire. It is very high in the *electrochemical series* and belongs to the *alkali metals* of *Group I* of the *periodic table.* It occurs naturally as the chloride (rock salt) and is obtained by the *electrolysis* of the *fused* salt. It forms many

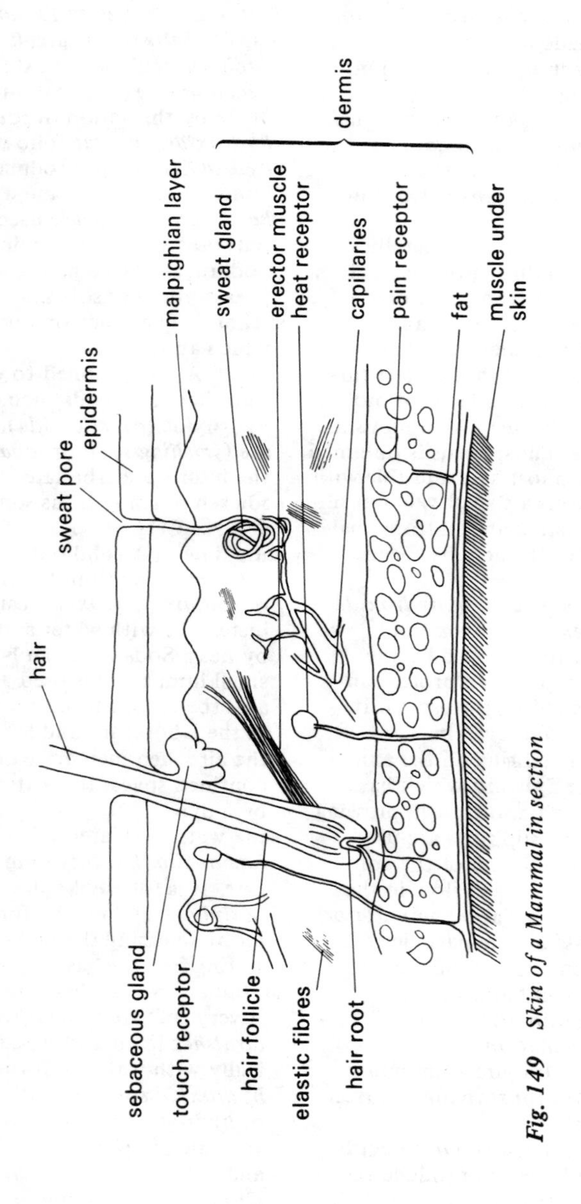

Fig. 149 Skin of a Mammal in section

widely-used compounds and they and the metal give an orange-yellow colour to a Bunsen flame.

sodium bicarbonate See *sodium hydrogencarbonate.*

sodium carbonate Na_2CO_3 The product obtained by passing carbon dioxide into aqueous sodium hydroxide, evaporating the solution and leaving to crystallize. It is made industrially by the *ammonia-soda* or Solvay *process.* It is very soluble in water to give an alkaline solution and crystals obtained from this have the formula, $Na_2CO_3.10H_2O$, being known as washing soda, or soda ash. In the anhydrous form, soda ash is used in softening hard water. One of the main uses of sodium carbonate is in the manufacture of glass.

sodium chloride NaCl The most widely-occuring sodium compound. Sodium is obtained by *electrolysis* of the *fused salt* while electrolysis of the solution gives sodium hydroxide, chlorine and hydrogen. It is found in all the liquids of the animal body and is necessary for many processes in the body. Also known as common salt.

sodium hydrogencarbonate $NaHCO_3$ If carbon dioxide is passed into a *saturated solution* of sodium carbonate, sodium hydrogencarbonate eventually precipitates (*precipitation*) out because of its low solubility. Industrially it is obtained by the *ammonia-soda process.* It is a white powder which readily decomposes on warming to give sodium carbonate, carbon dioxide and steam. Known also as baking soda it is a constituent of baking-powder, is used as a mild antiseptic for treating burns and is of use in the relief of indigestion. It is also called sodium bicarbonate.

sodium hydroxide NaOH A compound formed in solution when sodium acts on water and is obtained industrially by the *electrolysis* of

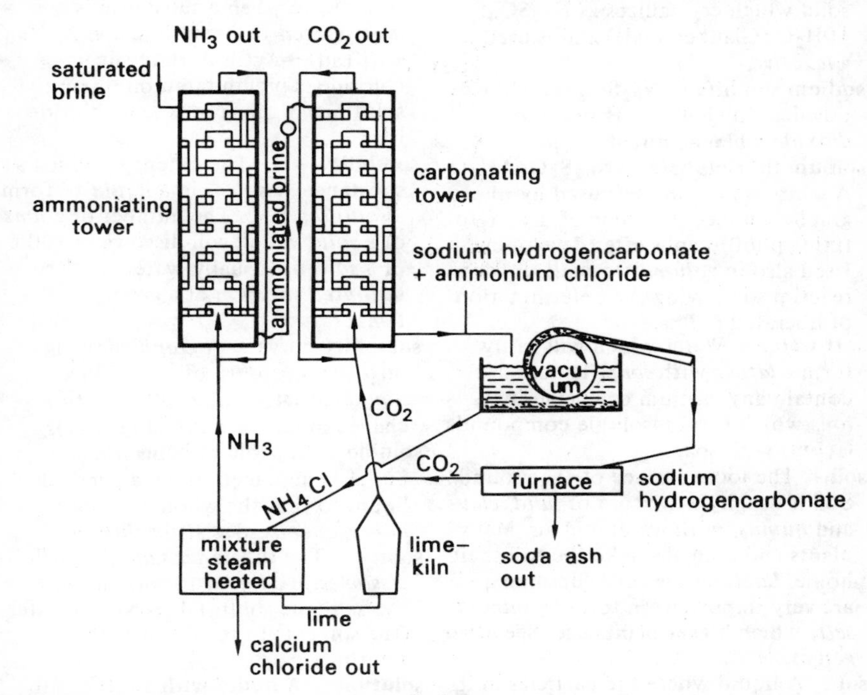

Fig. 150 Solvay Process for Sodium Carbonate

brine. It is a white waxy solid, very *deliquescent* and very soluble in water to form a strongly *alkaline* solution. It is used in the manufacture of soap and rayon. It is also called caustic soda.

sodium nitrate $NaNO_3$ A compound of sodium which occurs in large deposits in northern Chile. Much used as a *fertilizer*, it is a white *crystalline*, *deliquescent* solid. Also called *Chile* saltpetre.

sodium potassium tartrate A white crystalline solid used in the preparation of *Fehling's Solution* and, mixed with sodium hydrogencarbonate, as a *laxative* under the name of Seidlitz powders. It is also called Rochelle salt.

sodium silicate Na_2SiO_3 White crystalline solid soluble in hot water to form 'water-glass', used for preserving eggs (it seals up the shells) and as an adhesive.

sodium sulphate Na_2SO_4 A white solid which crystallizes as $Na_2SO_4.10H_2O$ (Glauber's salt) and is used as a *laxative*.

sodium sulphite Na_2SO_3 A white powder which gives off *sulphur dioxide* when acidified.

sodium thiosulphate $Na_2S_2O_3.5H_2O$ A white *crystalline salt* used by photographers under the name of 'hypo' for fixing photographs after developing. Used also in *volumetric analysis* for reactions involving the determination of liberated *iodine*.

soft water Water which will easily form a *lather* with *soap*. It does not contain any calcium or magnesium *ions* which form insoluble compounds (*scum*) with soap.

soil The loose surface of the ground. Soil is usually a mixture of *sand*, *clay* and *humus*, with water and air. Many plants and animals make the soil their home. *Bacteria* are very common and are very important in forming *mineral salts* which bigger plants use. See *nitrogen cycle*.

sol A liquid where the particles in it are not properly dissolved, but which are so small that they will not settle down.

solar plexus A network of *nerves* near the backbone, level with the *diaphragm*. The nerves control some *involuntary actions* of the body, e.g. *digestion*.

solder An *alloy* of tin with lead, copper and zinc, used for joining *metals* together. It has a low *melting point* and is used in the molten state. Soft solder contains tin and lead and brazing solder is tin and zinc.

solenoid A coil carrying an electric current and used for its magnetic effects.

solid That *state* of matter in which the particles (*atoms, molecules, ions*) have no movement other than a very limited vibration about a fixed position. A solid has a definite shape and also a definite *volume* for a fixed *mass* at a constant *temperature*. The symbol is s and sometimes c, e.g. $H_2O(s)$ stands for ice. It is also used to indicate when a substance is *precipitated* in a reaction, e.g. $Ag^+(aq) + Cl^-(aq) \rightarrow AgCl(s)$. Here silver chloride is precipitated on mixing solutions of silver *ions* and chloride ions.

solubility 1. The extent to which a substance *dissolves* in a *liquid* to form a *solution*. 2. The number of *grams* of a *solid* which will dissolve in 100 g of a *solvent* (usually water) to form a *saturated solution* at a given *temperature*.

solubility curve A *graph* showing how the *solubility* of a *solid* in a *liquid* (usually water) alters with a change in *temperature* (Fig. 151).

soluble Capable of being *dissolved*, i.e. of being taken up by a particular liquid so that the whole becomes *homogeneous*. See also *solution*.

solute The substance (*solid*) which *dissolves* in a *liquid* to form a *solution*, e.g. salt (the solute) dissolves in water (the solvent) to form brine (the solution).

solution A *liquid* with another substance *dissolved* in it. A mixture of

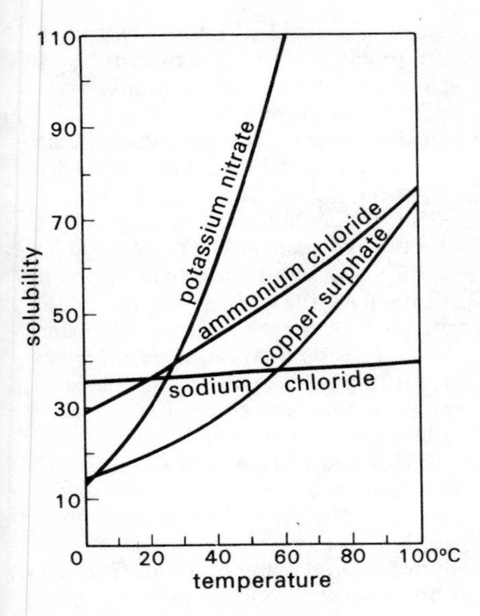

Fig. 151 Solubility Curves

two or more different substances which have become *homogeneous.* The term solution usually applies to the dissolving of a solid in a liquid, but it could be a gas dissolving in the liquid or even of a solid dissolving in another solid as with *alloys.*

Solvay process See *ammonia-soda process, sodium carbonate* (Fig. 150).

solvent A substance, usually a *liquid,* which is able to *dissolve* other substances. The most common solvent is water, but solids and gases may also form solutions, e.g. in alloys one metal may be considered as dissolving the other. A solvent is in the same *physical state* as the solution formed.

sonometer An instrument consisting of a metal wire stretched over two supports on a wooden base. It is used to study how the *tension* in the wire, and its *mass* and length change the note given out when it is set vibrating.

soot Impure *carbon* obtained when *fuels* are burnt in a limited amount of air. Soot obtained from *coal* contains *nitrogen compounds.*

sound waves Movements of the air which affect the sensitive parts of the *ear.*

sparklet bulb *Liquid carbon dioxide* stored under *pressure* in a steel bulb. When the bulb is screwed on to a *siphon* containing water the gas is released and dissolves in the water to make *soda water.*

spatula A long thin implement used for removing or transferring *solids.* Usually made of *stainless steel, nickel,* or *plastic.*

spawn The *eggs* of fish or frogs.

specialization The word used to show how plants and animals or parts of them can be changed in order to carry out particular tasks, e.g. the hand is specialized for holding things, a fish is specialized for breathing in water.

species A group of plants or animals which are like each other and which can *reproduce* together, e.g. all dogs belong to the same species.

specific An adjective which now means 'per unit *mass*' or divided by mass as in *specific charge* or *specific heat capacity* etc.

specific charge The *electric charge* on a particle divided by its *mass.*

specific gravity The older name for *relative density.*

specific heat capacity The amount of heat *energy* needed to raise the *temperature* of one *kilogram* of a substance through one *kelvin.* The *SI unit* is *joules* per kilogram kelvin and the symbol is *c.*

specific latent heat 1. (Fusion) The amount of *energy* needed to change one *kilogram* of a solid to liquid without a change in *temperature.*
2. (Vaporization) The amount of energy needed to change one kilogram of a liquid to vapour without a change in temperature. In both cases the *SI unit* is *joules* per kilogram and the symbol is *l.*

specific resistance See *resistivity.*

spectator ions The *ions* remaining in *solution* during a *double decomposition reaction* and which can be replaced by others without altering the precipitate (*precipitation*) produced. Barium sulphate can be precipitated by mixing solutions of sodium sulphate and barium chloride or of potassium sulphate and barium nitrate. In the first case the sodium and chloride ions take no part in the reaction and are spectator ions; in the second case the spectator ions are the potassium and nitrate ions. Essentially each reaction is $Ba^{2+}(aq) + SO_4^{2-}(aq) \rightarrow BaSO_4(s)$.

spectrum The pattern formed when the *frequencies,* and therefore *colours,* in *light* are separated, by a *prism* or *diffraction grating,* and shone onto a screen. The same idea is also applied to other *radiations,* as in the *electromagnetic spectrum.* See also *absorption spectrum, continuous spectrum, line spectrum.*

speed The rate at which an object is travelling, equal to the distance covered divided by the time taken. The *SI unit* is metres per second and the symbol is *u, v* or *w.* See also *velocity.*

sperm The male *cell* of an animal. Also called a spermatozoon.

sperm duct The tube which carries *sperms.*

sperm sac The palce where *sperms* are stored.

spherical mirror A *mirror* whose surface is part of a sphere (ball). See also *concave, convex.*

sphincter muscle A *muscle* which can close a passage, e.g. the *pyloric sphincter muscle,* and the anal sphincter.

spinal column Another name for the *vertebral column.*

spinal cord The main *nerve* stem leading from the *brain* and passing backwards through the *vertebral column.* It is connected by nerves to all parts of the body.

spinal nerve A *nerve* which is attached to the *spinal cord.*

spinal reflex An action which is controlled by the *spinal cord,* e.g. the *knee jerk reflex.*

spindle A group of fine threads formed in the *cell* when a *nucleus* is starting to divide. The *chromosomes* are pulled apart by the threads.

spinneret The part of the spider which makes the web.

spinthariscope A simple instrument for watching the *scintillations* caused by some *radioactive* material. The scintillations are seen, through a lens, by the eye when it has been made more sensitive to weak light, by being in the dark for a few minutes.

spiracle A breathing hole in the side of an insect's body. Also used for breathing holes in a fish and whale.

spleen An *organ* of the body near the *stomach.* It has to do with the making and storing of *blood cells.*

spongy layer The *cells* of the lower half of a *leaf.* The cells are loosely packed to allow for plenty of air to move around inside the leaf. (Fig. 153.)

spontaneously Occurring suddenly without any apparent external treatment. A pile of oily rags may suddenly burst into *flame* because slow *oxidation* of the oil (*exothermic*) causes a gradual rise in *temperature* and this may eventually exceed the *ignition temperature* required for a flame to be produced. Hydrogen and chlorine mixed and exposed to bright sunlight may suddenly combine explosively.

sporangium The container of the *spores* in a moss, fern or fungus. See *hypha* (Fig. 94).

spore A *reproductive cell* formed when a spore mother cell divides. No *fertilization* takes place. The spore can grow directly into a new plant. See *asexual reproduction, hypha* (Fig. 94).

sporophyte A plant which makes only *spores,* e.g. a fern. See *alternation of generations.*

spout The breathing hole of a *tadpole.*

spring balance An instrument for measuring the *weight* of an object. When the object is hung from the *balance* a spring is stretched (see *Hooke's law*) and moves a pointer over a scale.

stable compound Not easily changed or decomposed. The term is used for *compounds* such as sodium chloride. On the other hand, hydrogen peroxide may be described as *unstable* because it gradually changes on standing in the light to water and oxygen.

stable equilibrium If a body is in *equilibrium* and then it is moved slightly to one side it may return to its first position. If it does so it was in stable equilibrium, e.g. a ball inside a *concave* surface. See also *neutral* and *unstable equilibrium.*

stable nuclide A *nuclide* which does not change as time passes. Such a nuclide is not *radioactive.*

stainless steel A *steel alloy* containing nickel and chromium in certain proportions so that it is not affected by water, air or dilute acids.

stalactites and stalagmites When rain water containing dissolved *carbon dioxide* flows through the ground it takes in more of the gas and is able to dissolve any *calcium carbonate* present in the rocks and eventually caves are hollowed out. The water seeping through the roof contains *calcium hydrogencarbonate* and as it evaporates it leaves behind a deposit of *limestone* on the roof. This gradually builds up to produce a stalactite. Similar evaporation from drops falling on the floor of the cave produces stalagmites and these may eventually join up to form pillars of rock.

stalk A thin stem, e.g. *leaf* stalk, *flower* stalk.

stamen The part of a *flower* (Fig. 76iii) which makes the *pollen.*

standard cell A *cell* which has an accurately known *e.m.f.* and which can be used as a standard with a *potentiometer.*

standard solution A *solution* containing a definite *mass* of *solute* in a definite mass or volume of the solution. Most frequently the term refers to one litre (dm^3) of an *aqueous* solution containing one *mole* (or some fraction of a mole) of solute, e.g. 0.1 M sodium hydroxide contains $0.1 \times 40 = 4$ g of the solid in 1000 cm^3 of solution.

standard temperature and pressure (*abbr.* S.T.P. or stp). A temperature of 273.15 K (0 °C) and a pressure equal to that of a column of mercury 760 mm high. It is also called *normal temperature and pressure.* In comparing the volumes of gases these are adjusted to what they would be at standard temperature and pressure.

standing wave See *stationary wave.*

stannic The former name for *tin compounds* where the metal has a *valency* of four, e.g. $SnCl_4$, stannic chloride.

stannous Former name for *tin compounds* in which the metal has a *valency* of two, e.g. stannous chloride, $SnCl_2$. These salts are good *reductants.*

stapes The *stirrup* bone of the *middle ear.* See *ear* (Fig. 59).

starch The *carbohydrate* made by most plants in *photosynthesis.* It is usually stored in *seeds* or *roots* and is the flour of wheat and other grain.

state The form of a substance under definite conditions of *temperature* and *pressure,* i.e. *solid, liquid* or *gas.* The vibration of the particles, atoms, molecules, ions, is greatest in the gaseous state and least in the solid state. See also *kinetic theory.*

static electricity The study of *electric charges* at rest, also called *electrostatics.*

stationary wave The pattern formed when a *wave* and its *reflection* cross. There are parts with no *amplitude* at any time (*nodes*) and others where the amplitude reaches the largest value (*anti-nodes*). The distance between the nodes is equal to half the *wavelength* of the first wave.

steam H_2O Water which has been heated to its *boiling point* so that it is in the *gaseous state.* The *mist* produced when water boils is often incorrectly called steam—it really consists of droplets of water formed by *condensation* of some of the steam.

steam oven An oven, heated by steam, which is used for drying laboratory articles.

steel An *alloy* of *iron* with 0.1–1.5%

of *carbon* and small amounts of other elements such as silicon, phosphorus and manganese. It is prepared by the *open-hearth* and *Bessemer* processes or the L–D method. See *stainless steel, mild steel.*

steel wool A thick tangled mass of iron or steel *filaments* used for abrasive and cleaning purposes. It is also called iron wool.

steelyard An instrument for weighing heavy objects. It is a pivoted *lever* and the object is hung from a hook near the pivot. A weight is then moved along a *calibrated* arm until it balances.

stem The part of a plant from which the *leaves* grow. It is usually upright but may grow on the surface or even under the ground.

stem tuber A swollen *stem* under the

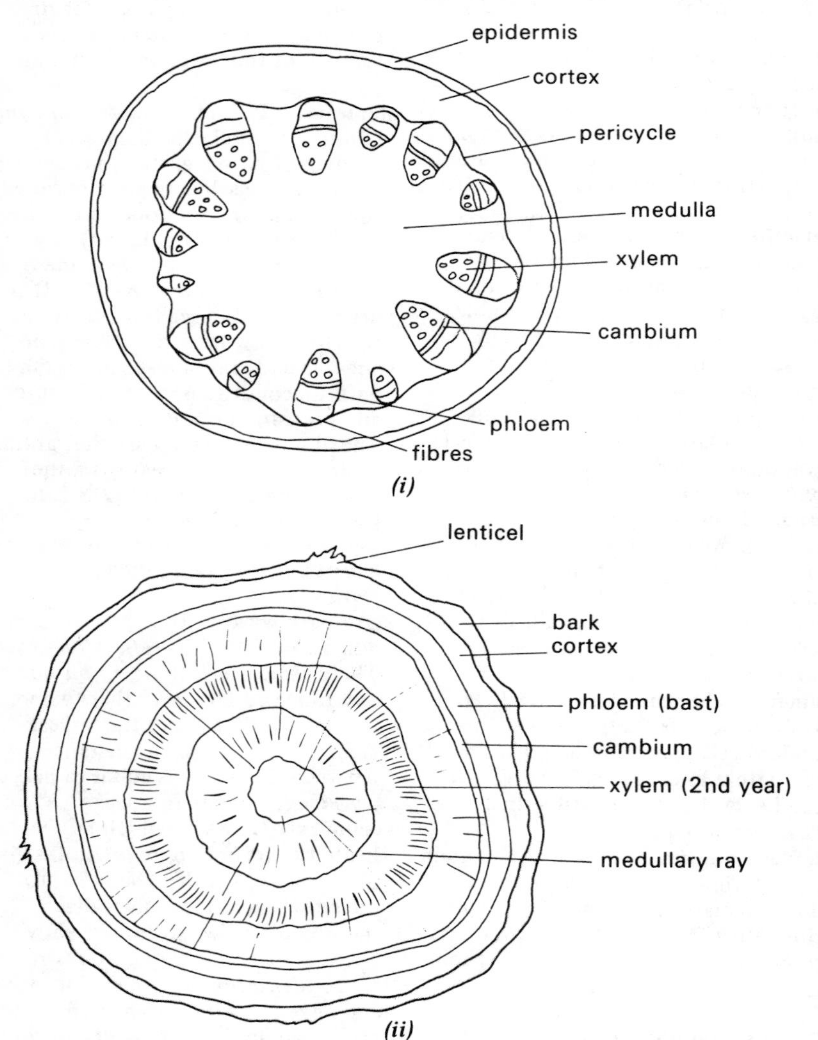

Fig. 152(i) *A Young Stem in cross-section* *(ii)* *Three Year Old Stem in cross-section*

ground, e.g. potato.

steradian The *SI unit* of solid angle. The symbol is sr. If a circle of area r^2 is drawn on the surface of a sphere (ball) of radius r, then the solid angle made by joining the edge of the circle to the centre of the sphere is one steradian.

stereoscopic vision Seeing things with both eyes, so that the object seems solid. Each *eye* gets a slightly different view of the object.

sterilization 1. A method of killing *bacteria.* 2. When animals are treated so that they are not able to reproduce (*reproduction*).

sternum Part of the *skeleton* (Fig. 148). The breastbone. The ribs are joined in front to the sternum.

stigma The part of the *flower* (Fig. 76iii) on which the *pollen* lands.

still A piece of apparatus used for producing a pure liquid by *distillation.* Most laboratory stills are used for producing *distilled water.* The term is also used in connexion with alcoholic drinks such as whisky which are produced by distillation.

stimulus A happening in or around a plant or animal which causes the plant or animal to respond (*response*), e.g. light is a stimulus which causes plant stems to grow toward it.

stipule A small leaf-like growth at the base of a leaf *stalk.*

stirrup The *stapes.*

stock A plant *stem* on to which a *scion* is grown, e.g. in fruit trees.

stoma A small hole in a *leaf,* through which gases pass. There are many stomata (stomates) on each leaf, usually on the underside. The size of the hole is controlled by the *guard cells.*

stomach The part of the *alimentary canal* (Fig. 5) which opens out from the *oesophagus.* It is used to store and to *digest* food. The liquid made by the stomach is the *gastric juice.*

straight chain An *organic compound* containing several carbon *atoms* linked in such a way that not more than two carbon atoms are joined to each other, e.g. butane, $CH_3CH_2CH_2CH_3$. See also *branched chain.*

stretch receptor The part of a *muscle* which sends a *nerve impulse* to the *brain* when the muscle is stretched.

stroboscope An instrument which helps one to see a moving wave etc., by making it appear to be still. It is based either on a flashing light or on a turning wheel with several slits in it, arranged so that the moving object is always in the same position when it is lit up or seen by the eye.

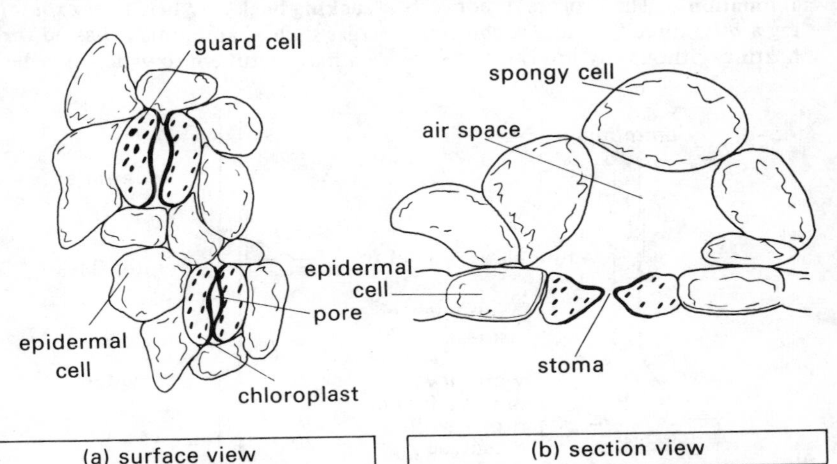

Fig. 153 Stoma

strong acid An *acid,* such as hydrochloric acid or sulphuric acid, which is almost completely *ionized* in *solution* at any concentration.

structural formula The formula of a compound set out in such a way as to show how the atoms in the molecule are arranged, e.g. ethanol (C_2H_6O),

$$CH_3CH_2OH \text{ or } \begin{array}{ccccccc} & & H & & H & & \\ & & | & & | & & \\ H & - & C & - & C & - & O - H; \\ & & | & & | & & \\ & & H & & H & & \end{array}$$

2-methyl propane
(C_4H_{10}), $CH_3CH(CH_3)CH_3$ or

$$(CH_3)CH_3 \text{ or } \begin{array}{ccccccccc} & & H & & H & & H & & \\ & & | & & | & & | & & \\ H & - & C & - & C & - & C & - & H; \\ & & | & & | & & | & & \\ & & H & & | & & H & & \\ & & & H - & C & - H & & & \\ & & & & | & & & & \\ & & & & H & & & & \end{array}$$

sulphurous acid (H_2SO_3),
$$\begin{array}{l} H-O \diagdown \\ \qquad\quad S=O \\ H-O \diagup \end{array}$$

style The *stalk* of the *stigma.* See *flower* (Fig. 76iii).

styrene $C_6H_5CH{=}CH_2$ A colourless liquid which is used in making *polystyrene* and some artificial rubbers. It is also called phenyl ethene.

subcutaneous Below the *skin.*

sublimation The process of converting a *solid* directly into a *vapour* on heating without any intermediate *liquid* state.

sublime When certain substances, e.g. iodine or ammonium chloride, are heated they change directly from the *solid* state to a *vapour,* often with *decomposition.* These substances are said to sublime and, on cooling, the original solid is re-formed as a sublimate on the cooler parts of the vessel in which it is being heated. E.g.
$I_2(s) \rightleftharpoons 2I(g)$;
$NH_4Cl(s) \rightleftharpoons NH_3(g) + HCl(g)$.

sublingual Below the *tongue,* e.g. sublingual *salivary gland.*

submaxillary Below the jaw, e.g. submaxillary salivary gland.

substitution reaction A reaction in which one *atom* of a *compound* or a *radical* is replaced by another atom or group of the same *valency,* e.g. methane, CH_4, may become chloromethane, CH_3Cl or methanol, CH_3OH; sulphuric acid, $SO_2(OH)_2$, may become sulphuryl chloride, SO_2Cl_2.

succulent fruit A juicy *fruit,* e.g. plum. Birds eat the flesh and leave the *seeds.*

succus entericus The liquid secreted (*secretion*) by the *duodenum.* It contains *enzymes* which help in *digestion.*

sucker A *stem* growing from under the ground to form a new plant. See *vegetative reproduction* (Fig. 164).

sucking back When a very *soluble* gas such as ammonia is passed through a narrow tube into water in order to

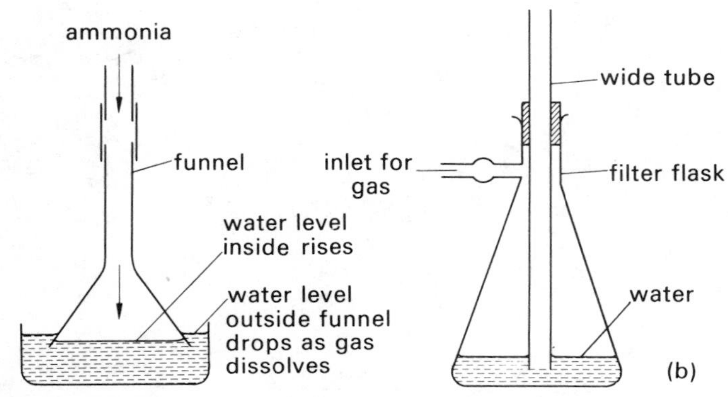

Fig. 154 "Sucking Back" Action of a Very Soluble Gas

make a *solution,* the water tends to rise rapidly up the tube and may enter the vessel in which the gas is being prepared. In order to prevent this some device, such as those shown here, should be fitted to the delivery tube. A *Bunsen valve* may also be used.

sucrose $C_{12}H_{22}O_{11}$ A *double sugar* (Cane sugar or beet sugar); *fermentation* by *enzymes* breaks it down, firstly into *glucose* and then into *ethanol* and carbon dioxide.

suction pressure The ability of a living *cell* to draw water into itself. See *osmosis.*

sugar A sweet-tasting *carbohydrate* with the *formula* $C_6H_{12}O_6$ (simple sugar such as *glucose*) or $C_{12}H_{22}O_{11}$ (a double sugar such as *sucrose*).

sulphates *Salts* of sulphuric acid, H_2SO_4, containing the $SO_4{}^{2-}$ *ion.* All common metal sulphates excepting those of lead and barium, are soluble in water and a solution of a sulphate always produces a white precipitate (*precipitation*) on addition of aqueous barium chloride. This precipitate differs from that produced with a *sulphite* in that it is insoluble in dilute hydrochloric acid.

sulphides Salts of hydrogen sulphide, H_2S. Only the sulphides of those metals near the top of the *electrochemical series* are soluble in water and most metal sulphides may be precipitated (*precipitation*) by passing hydrogen sulphide gas into a solution of another salt of the metal such as copper sulphate. These are often coloured and help to identify the metal present. Many sulphides can be formed by direct combination of an element with sulphur. The sulphide *ion* is S^{2-}. Most sulphides release hydrogen sulphide when acidified and can be recognized by this action.

sulphites Salts of *sulphurous acid,* H_2SO_3, containing the $SO_3{}^{2-}$ *ion.* On addition of a dilute acid and warming *sulphur dioxide* is given off.

sulphur S At. No. 16 R.A.M. 32 A yellow, solid non-metallic *element* obtained from underground deposits by the *Frasch process* (Fig. 82). It occurs in *polymorphic* (allotropic) forms, rhombic and monoclinic, both crystalline. Sulphur belongs to Group VI of the *periodic table* and is used mainly in the manufacture of sulphuric acid and for *vulcanizing* rubber. It burns readily with a blue flame to form sulphur dioxide. Insoluble in water, it dissolves in *carbon disulphide, benzene,* toluene and xylene. It is used in the making of many *proteins.*

sulphur dioxide SO_2 A colourless pungent gas obtained by burning sulphur or adding a dilute acid to a *sulphite.* It dissolves in water to form a solution of *sulphurous acid,* H_2SO_3.

sulphuric acid H_2SO_4 A colourless, corrosive oily liquid obtained from sulphur dioxide by the *contact process.* It is a *strong, dibasic acid* which forms salts called *sulphates* and *hydrogensulphates.* It is a very important industrial chemical used as the *electrolyte* in car batteries, in the refining of oil and in the manufacture of *fertilizers* such as ammonium sulphate. It used to be called oil of vitriol.

sulphurous acid H_2SO_3 A *strong dibasic acid* known only in *solution.* It is prepared by dissolving *sulphur dioxide* in water and forms *salts* called *sulphites* and *hydrogensulphites.*

sulphur trioxide SO_3 A white crystalline solid obtained from *sulphur dioxide* and air in the *Contact Process* for the manufacture of sulphuric acid. It dissolves in water to form sulphuric acid and in sulphuric acid itself to produce *oleum* or fuming sulphuric acid, $H_2S_2O_7$.

sunlight The main form of *energy* which enables green plants to make their own food. The energy of sunlight becomes stored in the plant and can be used for growth. Animals get their energy from plants by eating them.

supercooling The cooling of a liquid below its *freezing point* without any solid being formed. If a small *crystal* of the solid is added or the liquid is shaken it often turns quickly to the solid.

superheated steam Prepared by (*a*) heating water under a *pressure* greater than atmospheric so that it boils above 100 °C or (*b*) passing steam through a strongly-heated tube so that it is at a high temperature. It is completely dry.

superior ovary Where the *ovary* of the *flower* grows above the *petals.*

superphosphate A phosphatic *fertilizer* made by treating *calcium phosphate* with 70% *sulphuric acid;* it is a mixture of calcium sulphate and calcium tetrahydrogendiphosphate.

supinate The hand held so that the back is downwards. The opposite to *pronate.*

supporter of combustion A gas which will allow other substances to burn in it, but which does not itself burn, e.g. many substances will burn in air, oxygen or chlorine; these gases are supporting combustion.

surface tension The extra *forces* which are found in the surface of a liquid. They try to make the surface as small as possible. This is why small drops of liquid are spheres (balls) as this shape has the smallest surface area for a given volume. The forces will also support an object like a needle placed carefully on the surface.

suspension A precipitate (*precipitation*) in which the particles are of such a size that they do not settle down, e.g. if a dilute acid is added to a solution of sodium thiosulphate a suspension of sulphur is produced. Mist is an example of a suspension of a liquid (water) in a gas (air).

suspensory ligament A part of the *eye* which helps to control the shape of the *lens.*

sweat The liquid produced by the *skin* in hot conditions. When the sweat dries, the skin is cooled.

sweat gland The part of the *skin* (Fig. 149) which makes sweat by taking out water from the *blood* in the skin.

sweat pores The small holes in the *skin* through which the *sweat* comes out.

swim bladder The *air bladder* of a fish.

symbiosis The living together of two *organisms.* Each is useful to the other, e.g. *bacteria* live in the roots of *legume* bearing plants. The bacteria get *sugar* from the plant, which gets *nitrogen compounds* from the bacteria.

symbol 1. A method of representing the *atoms* of *elements* by means of certain letters, e.g. Ar for argon, Na for sodium (Latin name is *Natrium*), Cl for chlorine, etc. 2. A letter or group of letters used to represent a quantity or a *unit.* The symbol for a quantity is printed in *italics* but the symbol for a unit is in upright type, e.g. *m* stands for mass and kg for kilogram.

sympathetic nervous system The part of the nervous system which is self-acting and does not depend on the *brain.* It is made of groups of *nerve cells* around the *spinal cord.* It comes into action when danger threatens. See also *solar plexus*

synapse The place where two *neurons* join end to end. A *nerve impulse* can pass through a synapse.

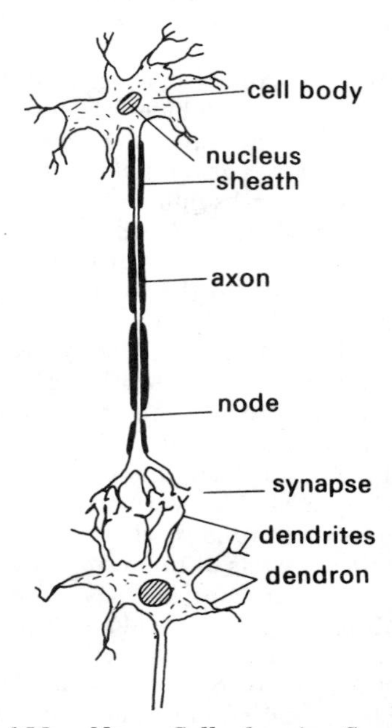

Fig. 155 *Nerve Cells showing Synapse*

synovial fluid The liquid in a *joint* (Fig. 97i). It keeps the bones oily so that they do not wear away.

synthesis 1. The formation of a *compound* from its *elements.* 2. The building up of a compound with large *molecules* from simpler units, e.g. ammonium chloride (NH_4Cl) from from ammonia (NH_3) and hydrogen chloride (HCl).

synthesize To prepare by *synthesis.*

synthetic Not naturally-occurring; man-made, artificial.

syringe A glass or plastic instrument (see diagram) in which gases or liquids may be collected and the volume measured.

systemic Having to do with the whole body, e.g. *systemic circulation, systemic insecticide.*

systole The state of the *ventricles* of the *heart* when the *muscles* tighten. *Blood* is forced out into the *arteries.*

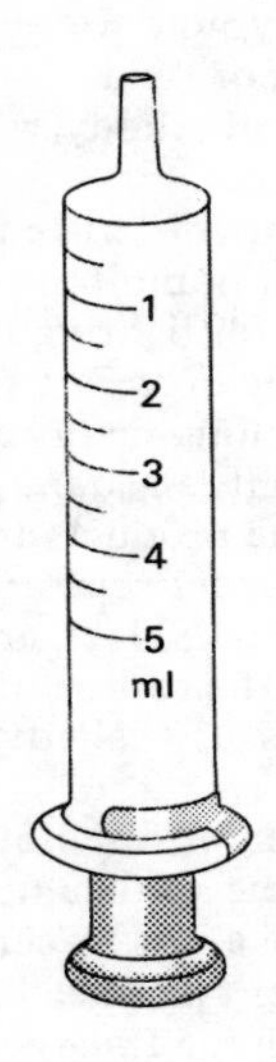

Fig. 156 Syringe

tactile Sensitive to touch.
tadpole The young water living stage of a frog or toad.
tail The part of a body which is behind the *anus.*
taper A 'string' of twisted cotton threads dipped in molten *paraffin wax* and used for lighting purposes.
tap funnel See *dropping funnel.*
tap root The long main *root* (Fig. 142 i) of a plant, e.g. *dandelion.*
tar One of the products formed during the *destructive distillation* of *coal.* It is thick, black and viscous and is the source of many chemicals. It is also called gas tar or coal tar.
tarnish A stain or discolouration produced on the surface of a *metal* exposed to the air or to certain gases, e.g. silver tarnishes in contact with air containing traces of hydrogen sulphide as black silver sulphide is formed.
tarsus Part of the *skeleton.* The *bones* of the back part of the foot. (Fig. 148).
tartaric acid $C_4H_6O_6$
COOH—CH(OH)—CH(OH)—COOH
Dihydroxysuccininc acid. A white crystalline *organic acid* found in grapes and used in *dyeing*, in making *baking-powder* and in '*health salts*'.
taste bud A group of *cells* mainly on the *tongue*. The cells are sensitive to different *chemicals.* A *nerve impulse* is sent to the *brain* which recognizes different tastes.
tear gland The part at the top of the *eye* which makes tears. These keep the eyes clean.
teat The tube through which the *milk* comes out in a *mammal.*
teat pipette A glass or plastic tube fitted with a rubber or *PVC* teat and used for transferring small amounts of liquids or gases from one vessel to another.
teflon See *PTFE.*
telephone An instrument containing a *microphone* and an *earpiece.* The microphone changes sound *waves* into electrical currents. These are carried by wires over long distances and then, in the earpiece of another telephone, are changed back to sound waves.
telescope An instrument for looking at distant objects. The *objective* is a *converging lens* (or *mirror,* see *reflecting telescope*) which forms an image which is then seen through another converging lens which acts like a *magnifying glass.* This is an astronomical telescope and the final image is upside down, which does not matter in astronomy. If the first image is seen through a *diverging lens*, then the final image is the right way up. This is a Gallilean telescope.
temperate regions Those parts of the earth between about $23°-67°$ latitude.
temperature The measurement of how hot or cold an object is. The *SI unit* is the *kelvin* and the symbol is T for *absolute* temperature and t or θ for *celsius* or *centigrade temperature.*
temporary hardness *Hardness* in water which can be destroyed by *boiling*; it is caused by the presence of soluble calcium or magnesium hydrogencarbonate which decomposes on heating to form the insoluble carbonate.
tendon The tough, strong string which joins a *muscle* to a *bone*. See *Achilles tendon, joint* (Fig. 97 ii).
tendril Part of a *stem* or *leaf* which is used to twist around other stems for support, e.g. pea.
tensile strength A measure of the ability of a substance (usually a *metal*) to withstand stretching.
tension A *force* which is a pull.
ter- Prefix meaning 'three'.
tergum The covering on the back of an *insect.*
terminal The part of an electrical component, often in the form of a screw, to which wires can be joined.
terminal bud The main *bud* on the end of a *twig* (Fig.162).

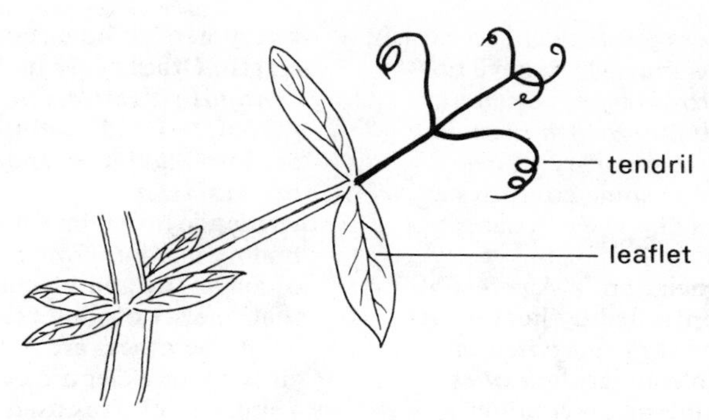

Fig. 157 Tendrils

terminal velocity The *constant velocity* reached in time by an object as it falls through a substance. It happens when the force causing *acceleration* is balanced by the force of resistance set up by the substance.

terrestial magnetism See *earth's magnetism.*

tervalent See *trivalent.*

terylene A strong artificial *fibre*—a *polymer* made from terephthalic acid and ethylene glycol (dihydroxy-ethene). It is used in the manufacture of clothing materials and *textiles*.

testa The outer skin of a *seed.*

testis The male *reproductive organ* (Fig. 141 ii) which produces the *sperm.*

test tube A glass tube, closed at one end, in which *chemical reactions* are carried out.

tetanus A condition of *muscles* where they are permanently tight-ened. It may be caused by *bacteria* and then is called lockjaw.

tetra- Prefix meaning 'four'.

tetrachloromethane CCl_4 A dense, colourless, sweet-smelling liquid, a *substitution* product of *methane.* It is used for the removal of grease and oil stains, as an industrial *solvent* and in some *fire extinguishers.* It used to be called carbon tetra-chloride.

tetrahedron A *solid* figure, bounded by four triangular *faces.*

tetravalent Having a valency of four. Also called quadravalent.

textiles Woven *fibres* or cloths using *filaments* of animal, vegetable, mineral or synthetic origin, e.g. cotton, wool, asbestos, nylon, tery-lene.

thalamus Part of the *brain. Nerve impulses* passs from the body through the thalamus to the *cerebrum.*

thallus A simple plant body without stems and leaves, e.g. seaweed, *liver-wort* (Fig. 107).

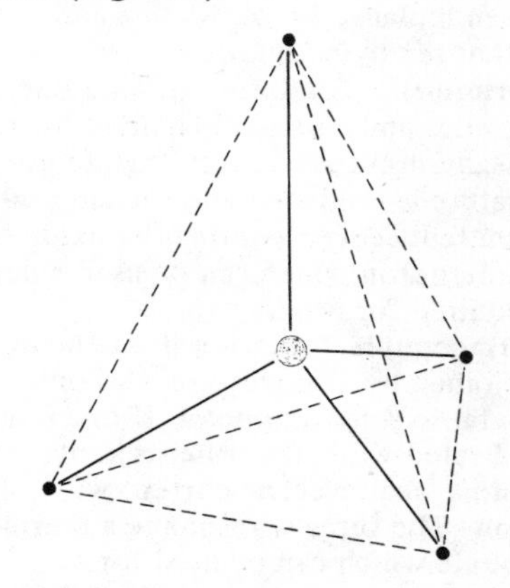

Fig. 158 Tetrahedral Arrangement of Valency Bonds in Methane (the dotted lines form a tetrahedron)

theory Ideas put forward to explain certain facts, but which have not yet been proved by *experiment.*

therm 100 000 *British thermal units* (btu). A quantity of *heat energy* used in some countries as the basis for the sale of coal gas or natural gas.

thermal dissociation A *reversible reaction* controlled by heat. In it a *compound* is *decomposed* on heating into simpler *molecules* which recombine on cooling; such compounds are *dinitrogen tetroxide* and *ammonium chloride.*

thermionic emission The giving out of *electrons* from a very hot wire or tube, as from the *cathode* of a *thermionic valve* or *cathode-ray tube.*

thermionic valve A closed glass tube containing a heated *cathode*, an *anode* and often other *electrodes.* Most valves have a very good *vacuum* inside them. A valve with just the anode and cathode is called a *diode* and is used for rectification. A *triode* has one extra electrode, a tetrode two, a pentode three, etc. (see *grid*). Many valves have now been replaced by *transistors* and other *semiconductor components.*

thermit(e) A mixture of aluminium powder and iron oxide ignited by a magnesium fuse. A very high temperature is produced and the aluminium reduces (*reduction*) the oxide to molten iron which can be used where required for *welding.*

thermocouple A wire of one metal is joined by two wires of a second metal to a *galvanometer*. If one joint is heated while the other is kept cold a small electric current will flow. The three wires make a thermocouple which can be used for measuring the temperature of the hot joint. Different metals are used for different temperature ranges.

thermometer An instrument for measuring *temperature.* The simplest form has a liquid, like mercury, in a bulb at one end of a closed tube with a very narrow hole (bore) along its length. Other types include the *constant volume air thermometer, pyrometers* and *thermocouples.* See also *maximum* and *minimum thermometer.*

thermopile An instrument containing many *thermocouples*, often of antimony and bismuth, so fixed that one set of joints can be warmed while the others are not heated. This gives a larger electric current than a single joint. It is used to detect *infra-red radiation.*

thermoplastic A *plastic* material which can be melted and then cooled again repeatedly without any change in its properties.

thermos flask A container used to keep hot liquids hot or cold liquids cold, because it greatly reduces the rate at which heat energy leaves or enters the liquid. It has a double glass wall, silver-coated on the inside (to reduce energy movement by *radiation*) and with a *vacuum* between the inner glass surfaces to prevent energy movement by convection. The glass, and the cork used to close the flask, are bad conductors of heat. It is also called a vacuum flask.

thermosetting plastic A plastic material which can be moulded when warm and then allowed to set, but this operation can be carried out once only. Strong reheating causes charring and *decomposition*, not softening.

thermostat An instrument for keeping the *temperature* of some object or container *constant.* The expansion and contraction of some part of the thermostat controls the flow of fuel (e.g. gas) or switches the electricity supply off and on.

thiamine *Vitamin B,* found in *yeast,* meat and unpolished grain. Lack of thiamine causes the disease *beri-beri.*

thistle funnel A long tube with an enlarged head through which *liquids* can be easily poured into a flask.

thorax The chest. Inside is the thoracic cavity which contains the *heart* and *lungs*. There are twelve thoracic *vertebrae* and twelve pairs of *ribs* supporting the thorax, with the *sternum* in front. The floor of the thorax is the *diaphragm*.

thrust A *force* which is a push.

thymus A *gland* just in front of the *heart*. It controls *growth* in the young animal and gets smaller as the animal gets older.

thyroid A *gland* in the front of the neck. It makes the *hormone thyroxin* which influences growth and activity of the body. If it gets too big it causes *goitre*.

thyroxin The *hormone* produced by the *thyroid gland*. It contains *iodine*. If the body makes too much, it causes over-activity; if not enough is made, it causes *cretinism*.

tibia Part of the *skeleton* (Fig. 148). The larger of the two bones in the lower part of the leg.

ticker timer An instrument, worked by an *alternating current*, which makes fifty or one hundred dots per second on a moving strip of paper. It is used to study the *velocity* and *acceleration* of objects in experiments.

tidal air The air which flows in and out of the *lungs* during breathing.

timbre See *quality of sound*.

tin Sn At. No. 50 R.A.M. 118.7 A silvery *metal* which melts at a fairly low temperature when heated and so becomes *oxidized*. It is not affected by air or water at room temperature. Used in many *alloys* such as solder and for making tin plate.

tin plate Thin sheet iron coated with a layer of tin deposited electrolytically or by dipping. It is used for making cans for food.

tincture A solution of a drug or medicine in *ethanol*, e.g. tincture of iodine.

tissue A group of *cells* which work together for a special purpose, e.g. the *brain* is made of nervous tissue.

tissue culture Growing *cells* in special liquids away from the body, in order to study how they work.

tissue respiration The chemical part of *respiration* in the *cells*, resulting in *energy* being produced.

titration The basis of *volumetric analysis*. The experimental determination of the *concentration* of a *solution* (A) by running into a measured volume of it a second solution (B) from a burette until all of (A) is used up and the reaction between them is complete. The reaction may be one of *neutralization*, *oxidation*, *reduction*, etc., and the end point is noted by means of an *indicator*.

toluene C_7H_8 An *arene* (aromatic) *hydrocarbon*, the second member of the series. It is a colourless *flammable liquid* similar to *benzene*. It occurs in coal tar and is used as a solvent for sulphur and in the preparation of dyes, drugs and explosives. It is also called methyl benzene.

tongue The movable *muscle* in the *buccal cavity*. It helps to mix food with *saliva* and bears the *taste buds*.

tonsils Two *glands* near the back of the *tongue*. They help to protect the throat from *disease*.

tooth A hard bony (*bone*) growth from the skin over the jaw. Teeth differ in the work they have to do in different animals, and in different parts of the mouth. Each tooth in the *mammal* is fixed into the jawbone. There are usually two sets of teeth—the *milk teeth* and the *permanent teeth*. The four kinds of teeth are *incisors*, *canines*, *premolars* and *molars*. (Fig. 159.)

top-pan balance A single-pan balance on which the object to be weighed is placed: its *mass* is recorded directly on a scale without the use of external masses (weights). See diagram. (Fig. 160.)

torricellian vacuum The *vacuum* formed above the mercury in a *barometer*.

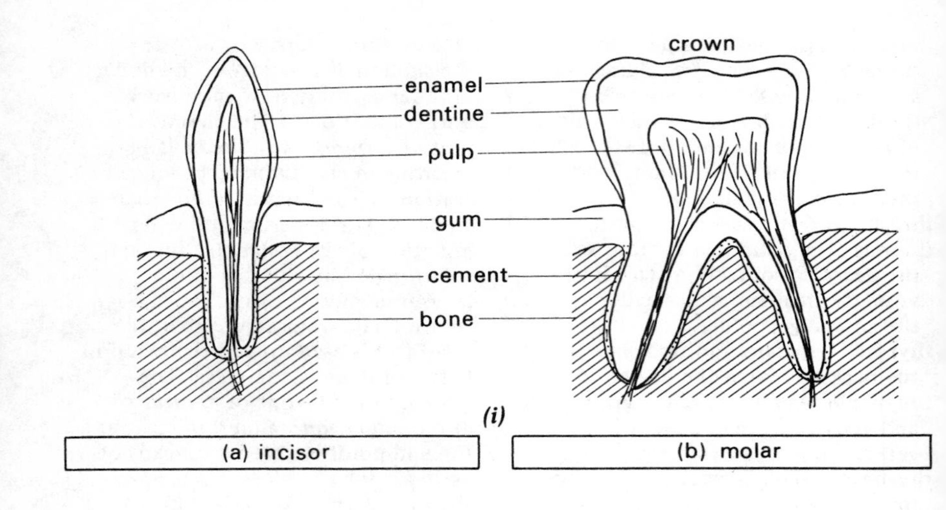

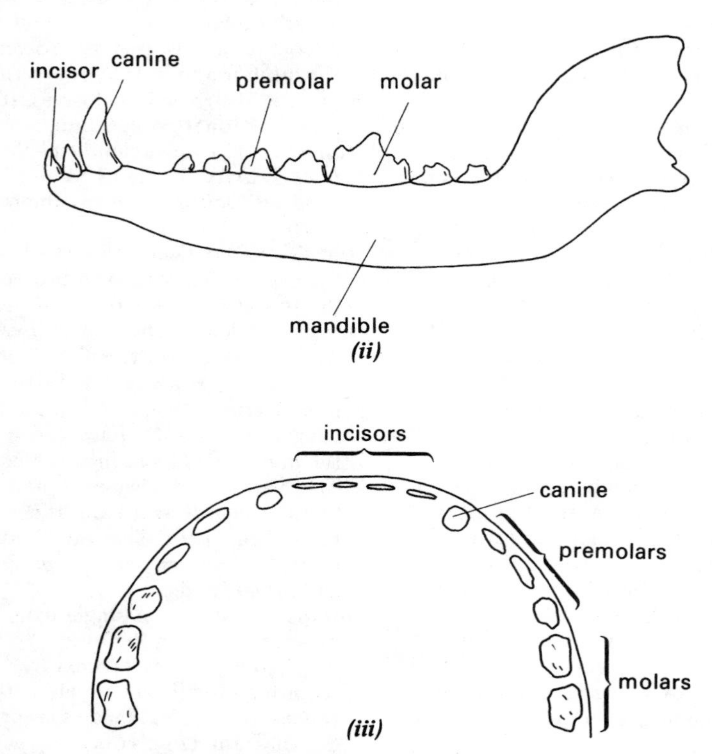

Fig. 159(i) Basic Structure of Human Teeth
(ii) Teeth in Lower Jaw of Dog
(iii) Plan of Human Dentition

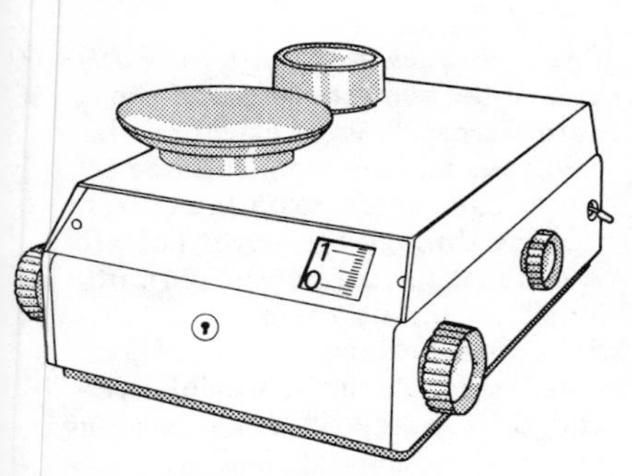

Fig. 160 Top-pan Balance

torso The part of the body other than head or limbs. The trunk.
total internal reflection When light is travelling through glass, water or perspex (etc.) towards air and the angle of *incidence* is greater than the *critical angle*, all the light is reflected (*reflection*) at the surface. See *critical ray* (Fig. 47).
touch A sense found in the *skin*. The touch *cells* are just below the surface. They are sensitive to *pressure* and pain.
toxic Poisonous.
toxin A poisonous substance made by *bacteria* in the body. See *anti-toxin*.
trace element A chemical needed by plants and animals in very small amounts.
tracer A *radioactive element* or *compound* used along with the normal substance in a *reaction*, the course of which can be followed by a Geiger counter.
trachea The air tube which goes to the *lungs* (Fig. 109i)—the *windpipe*. Also one of the many air tubes which go through the body of an insect.
tracheid A water tube in the wood of a plant. It is shorter than a *vessel*.
trajectory The path of an object thrown through the air.
transformer A *component* which will change the *potential difference* of an *alternating current* supply very efficiently. It contains two coils wound on the same framework or core, made of a special alloy. The alternating current flowing in the first (primary) coil magnetises the core and the rapidly changing *magnetic field* induces (*induction*) an alternating current in the second (secondary) coil. It is nearly true that,

$$\frac{\text{Potential difference across primary}}{\text{potential difference across secondary}} = \frac{\text{number of primary turns}}{\text{number of secondary turns}}$$

If the potential difference across the secondary is greater than that across the primary then it is a step-up transformer. If that across the primary is greater, then it is a step-down transformer.
transfusion The giving of *blood* from one person to another. It must be the right kind of *blood group*.
transistor A *component*, made of *semiconductors*, which has largely replaced the *triode* as a way of increasing the size of (amplifying) a small electric current.
transition element A series of *metal elements* in which the *electrons* in the inner *shells* are being built-up (i.e. these shells are being enlarged) while those in the outermost shell vary only slightly, e.g. iron, cobalt, manganese.
translocation Moving liquids through a plant.
translucent Allowing the passage of light, but cannot clearly be seen through, e.g. frosted glass, oiled paper, some thin plastic sheets.
transmutation Changing one *element* into another, e.g. a *base metal* such as lead into silver or gold—an old *theory* of the *alchemists*, later considered impossible. We now know that this process goes on continuously in *radioactive* elements and can be brought about by certain *nuclear* reactions. Radium is slowly but continually changing into lead.
transparent Something which lets

light through and which can be seen through clearly. The same idea can be used with reference to *radiations* other than light.

transpiration The loss of water from a plant. The liquid water is changed into *water vapour* by the heat of the air and passed out through the stomata (*stoma*). Transpiration helps to cool the plant and to make sure that more water is pulled up from the roots.

transverse processes The two pieces of *bone* which come out from the side of a *vertebra* (Fig. 166).

transverse wave A *wave* in which the to and fro movement of the particles (etc.) is at right angles to the direction in which the wave is travelling.

treble bond See triple bond.

tri- Prefix meaning 'three'.

triangle of forces If three *forces* which act at the same point can be represented, i.e. drawn in the same direction and with a length *proportional* to the force, as the three sides of a closed triangle, then the forces are in equilibrium.

triceps The *muscle* at the back of the upper arm. It works with the *biceps* in straightening the arm. See joint (Fig. 97 i).

trichloromethane $CHCl_3$ A dense, colourless, *volatile* liquid with a *characteristic* sweet smell. It produces an *anaesthetic vapour* and is also used as an industrial solvent. It is a *substitution* product of *methane.* It is also called chloroform.

tricuspid valve The strong piece of skin between the right *auricle* and the right *ventricle* of the *heart* (Fig. 91 ii). It keeps the *blood* flowing in the correct direction.

tri-iron tetroxide See iron(II) iron(III) oxide.

tri-lead tetroxide See dilead(II) lead(IV) oxide.

triode A *thermionic valve* with a *cathode, anode* and another net-like *electrode,* the *grid,* between them. Small changes in the *potential difference* between the grid and the cathode have a great effect on the electrons moving from cathode to anode and therefore on the current flowing through the valve. Therefore it can make small electric currents larger or amplify them.

triple-beam balance A single-pan balance on which the weights are added by sliding fixed masses along any of three separate *beams.*

triple bond A *bond* indicating the sharing of three pairs of *electrons* by two *atoms*, each contributing the same number, e.g. nitrogen $N{\equiv}N$, N_2; acetylene, C_2H_2 or $HC{\equiv}CH$. *Compounds* in which triple bonds occur are usually very *reactive* and will readily join with hydrogen (if *hydrocarbons*) to become *saturated compounds,* e.g. acetylene (ethyne), C_2H_2 becomes ethane, C_2H_6.

triple superphosphate *Calcium phosphate* treated with *phosphoric acid* to convert it all into soluble calcium tetrahydrogendiphosphate, $Ca(H_2PO_4)_2$, for use as a *fertilizer.*

tripod An iron stand having three legs and used to support apparatus being heated.

tritium An *isotope* of hydrogen containing two *neutrons* and one *proton* in the *nucleus.* It is *radioactive* with a *half-life* of 12.5 years. See also *deuterium.*

trivalent Having a *valency* of three. Also called tervalent.

trochanter The lump at the top of the *femur* bone to which the *muscles* which turn the leg are joined.

tropism The growth of part of a plant towards or away from a particular *stimulus*, e.g. *geotropism* (gravity), *phototropism* (light), *heliotropism* (sun).

trough See *pneumatic trough.*

trypsin An *enzyme* used to break up *proteins* in *digestion.* It is made in the *pancreas.*

tube feet The suckers on the arms of a starfish.

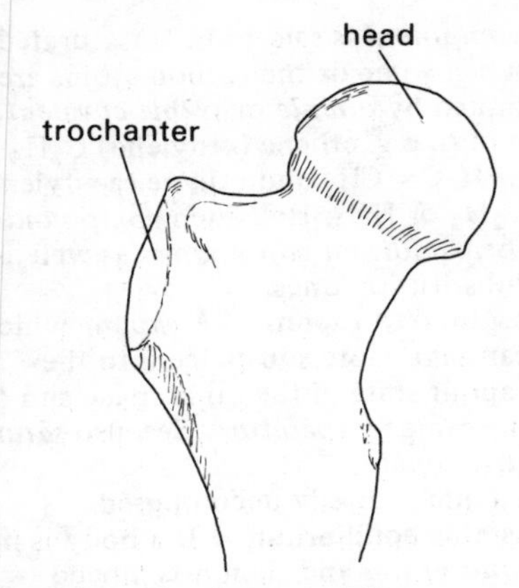

Fig. 161 Trochanter of Femur

tuber A swollen underground part of a plant. It may be a *stem* or a *root.* See *vegatative reproduction* (Fig. 164).

tuberculosis A disease of the *lungs* caused by *bacteria* which make small swellings (tubercles) in the lungs.

tubular florets The small *flowers* in the middle of a daisy or sunflower. The tubular florets are surrounded by *ray florets.* Also called disc florets.

tungsten W At. No. 74 R.A.M. 184 A hard grey *metal* which melts at a very high temperature. Used in alloys having a high resistence to wear and for electric lamp *filaments.* Also known as wolfram.

tuning fork A metal instrument in the shape of a fork with two long arms (prongs), which gives out a pure note when struck.

turbid Not clear–said of *liquids* containing particles in *suspension.*

turgor The condition of a *cell* when it is swollen up with water. Also called turgidity.

turgor pressure The pressure on the wall when the *cell* is turgid (*turgor*).

turpentine $C_{10}H_{16}$ An oily, liquid *hydrocarbon* having a pine-like odour. It burns with a sooty flame (being an *unsaturated* compound) and is used in lacquers and polishes and paints, as an industrial solvent and for cleaning paint brushes.

twig The end part of a branch of a tree. It bears the *buds.*

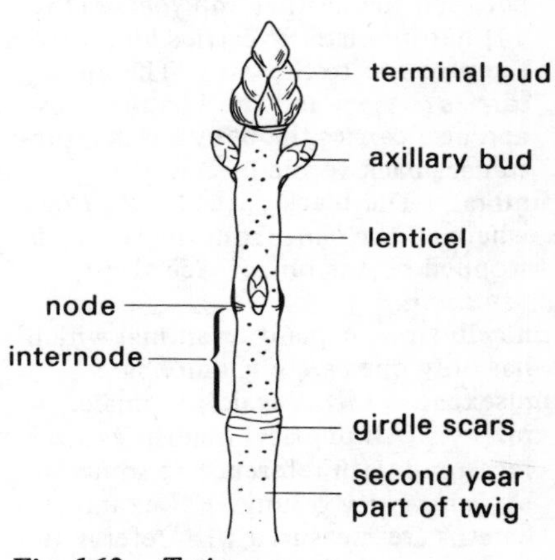

Fig. 162 Twig

twins Two babies which are born at the same time. They may or may not be *identical twins.*

tympanum The *ear drum.* See *ear* (Fig. 59).

typhoid fever A *disease* caused by *bacteria* found in bad water or milk.

typhus A *disease* caused by a *germ* carried by a louse.

ulna Part of the *skeleton* (Fig. 148). One of the two *bones* in the lower part of the arm.

ultra-violet radiation *Radiation* which is part of the *electromagnetic spectrum* but has a shorter *wavelength* than light. It cannot be seen by the eye but can be studied with suitable photographic film or other instruments. It is used in medicine and industry.

umbilicus The navel. The place on the *abdomen* where the baby was joined to the mother before birth.

umbilical cord The connection between the mother and *foetus* (Fig. 77) before birth. It carries *blood* from the *placenta* to the baby. The blood carries *oxygen* and food to the baby and also carries the baby's waste substances back to the placenta.

umbra The black part of a *shadow* where all the *light* from the source is stopped by the object. See also *penumbra.*

unicellular A plant or animal which has only one *cell,* e.g. amoeba.

unisexual Either male or female.

unit 1. All physical quantities are measured with reference to some basic quantity or unit. For example lengths are measured with reference to the *metre,* time with reference to the second, etc. Only *SI units* should now be used in science although many other units have been used in the past (e.g. the British set of units, based on the pound, the foot and the second.) 2. (Electrical energy) The unit of electrical *energy* used in commerce, which is equal to the *kilowatt-hour.* See also *cost of electricity.*

univalent Having a *valency* of one, e.g. sodium, hydrogen, chlorine, the ammonium or nitrate *ion.* It is sometimes called monovalent.

univalve A *mollusc* with one shell, e.g. snail.

universal indicator A mixture of *indicators* used for determining the approximate *pH* of a solution by the colour developed. It is also called a wide-range indicator or pH paper.

unsaturated compound An *organic compound* is said to be unsaturated when some of the carbon atoms are linked by *double* or *treble covalent bonds,* e.g. ethene (ethylene) C_2H_4 or $H_2C{=}CH_2$ and ethyne (acetylene) C_2H_2 or $HC{\equiv}HC$. Such *compounds* form *addition compounds* as well as substitution ones.

unsaturated vapour A *vapour* which can take more substance into the vapour state in the same space and at the same *temperature.* See also *saturated vapour.*

unstable Easily decomposed.

unstable equilibrium If a body is in *equilibrium* and then it is moved slightly to one side it may go on moving away from its first position. If it does so it was in unstable equilibrium, e.g. a ball on top of a *convex* surface. See also *neutral* and *stable equilibrium.*

upthrust The upward *force* pushing on a body when it is placed in a *fluid.* See also *Archimedes' principle* and the *principle of flotation.*

uranium U 92 At. No. 92 R.A.M. 238 A hard, white *metal* which undergoes *nuclear fission* when bombarded by *neutrons.* Naturally-occurring uranium is a mixture of three *isotopes* of atomic masses 234, 235 and 238 (99.3%) and the 235 isotope is the one used in nuclear *reactors.*

urea $CO(NH_2)_2$ A white crystalline compound, soluble in water which is excreted (*excretion*) in the *urine.* It is made in the body by the breakdown of proteins. It is used as a *nitrogenous fertilizer* and for making urea-formaldehyde *plastics.*

ureter The tube from the *kidney* to the *urinary bladder.* It carries the *urine.* See *excretion* (Fig. 67).

urethra The tube from the *urinary bladder* to the outside. It also carries *sperms* in the male. See *excretion* (Fig. 67).

uric acid The excretory (*excretion*) product of insects and birds.

urinary bladder The place where the *urine* is stored before being removed from the body. See *excretion* (Fig. 67).

urinary tubule A part of the *kidney* through which the *urine* passes to the *ureter.*

urine The watery liquid made in the *kidneys.* It is used to carry waste substances out of the body and to maintain correct water balance in the body.

urinogenital system The combined system for *sexual reproduction* and *excretion.*

utriculus Part of the inner *ear* (Fig. 59) which bears the *semicircular canals.*

u-tube A piece of apparatus used for the drying of gases. It can be filled with with *anhydrous* calcium chloride or *silica gel* to remove any water present in the gas.

vaccination Scratching the arm and putting on *vaccine* in order to guard against smallpox.

vaccine A liquid containing weak or dead *viruses* or *bacteria.* Vaccines are used to protect people from more serious *diseases*, e.g. the vaccine of *cow pox* can prevent smallpox in humans.

vacuole A space inside a *cell*, filled with liquid.

vacuum A space from which as much substance as possible has been taken away.

vacuum flask See *Thermos flask.*

vagina The opening of the female *reproductive organs* (Fig. 141 i) through which the baby is born.

vagus nerve A large *nerve* which comes from the back of the *brain.* It carries *nerve impulses* to the *heart* and *stomach.* It controls the heart beat and the breathing rate.

valency 1. A term expressing the capacity of an *atom, ion* or *group* for entering into combination with another one. 2. The combining power of an element, ion or group with reference to hydrogen or chlorine which have a valency of one. 3. The number of atoms of hydrogen or chlorine which one atom of an element will combine with or displace, e.g. iron has a valency of 3 in the reaction with chlorine, $2Fe + 3Cl_2 \rightarrow 2FeCl_3$ and of 2 when it reacts with a dilute acid; $Fe + 2HCl \rightarrow FeCl_2 + H_2$; 4. The valency of an ion is equal to its charge, e.g. the ammonium ion, NH_4^+, has a valency of 1 (+ 1) and the sulphate ion, SO_4^{2-}, has a valency of 2 (– 2). See also *electrovalency, covalency.*

valve 1. A part of a *pump* etc., which only lets a *fluid* flow one way. 2. See *thermionic valve.* 3. A structure made of *skin* or *muscle* which allows liquids, e.g. *blood*, to be carried in one direction. There is a muscle valve at the end of the *stomach* and skin valves inside the *veins* and *heart.*

Van de Graaff generator An *electrostatic* machine for producing very high *potentials.* A moving belt carries *electric charge* to a hollow metal ball (sphere), on an insulating (*insulator*) stand, where it is stored.

vanadium(V) oxide V_2O_5 A purple powder used as a catalyst in the *contact process* for the manufacture of sulphuric acid. Also called vanadium pentoxide.

vane The flat part of a *feather.*

vapour A gas which can be turned to a liquid by increasing the *pressure* alone. For every gas there is a *temperature,* the critical temperature, below which it is a vapour but above which it cannot be turned to a liquid by any increase in pressure. See also *saturated* and *unsaturated vapour.*

vapour density See *relative vapour density.*

vapour pressure The *pressure* of a *vapour* which reaches its highest value at a given *temperature* when the vapour becomes *saturated.* The pressure is then called the *saturation vapour pressure.* If a liquid is heated its vapour pressure increases with the temperature and when this equals the atmospheric pressure the liquid then boils.

variation The differences between the young produced by the same parents. The theory of *Evolution* is based upon such differences. The variation makes it possible for some *organisms* to continue to live under changing conditions.

varnish A solution of a *resin* in linseed oil or turpentine which is painted over a wood or metal surface. The *solvent evaporates* leaving a shiny layer which protects the surface.

vas deferens The main tube which carries the *sperms* from the *testis* to the *urethra.* See *reproductive organs* (Fig. 141 ii).

vas efferens One of the many small tubes which carry the *sperms* to the *vas deferens* via the *epididymis.*

vascular Having to do with carrying liquids around a body, e.g. *blood* system, the *vascular bundles* in a plant.

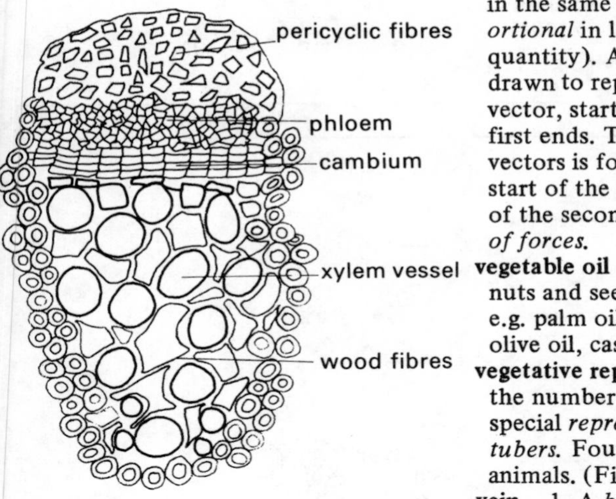

Fig. 163 Vascular Bundle in cross-section

vascular bundle Tubes which pass through a plant. There are two kinds of tubes, the *vessels* which carry liquids from the *root* to the *leaves* and the *sieve tubes* which carry food liquids from the leaves to all parts of the plant.

vaseline Trade name for petroleum jelly.

vaso-constriction The tightening of the *blood vessels* in the *skin* in cold weather to stop body heat loss.

vaso-dilation The loosening of the *blood vessels* in the *skin* to allow body heat loss and so to cool the body.

vector 1. A quantity which has both size and direction. See also *scalar.* 2. Any animal which carries a *disease* causing *germ*, e.g. a mosquito is a vector of *malaria.*

vector addition If two, or more, *vectors* are to be added together then it is important to consider the directions in which they act. The addition is done most easily by drawing. A line is drawn to represent the first vector (i.e. in the same direction as, and *proportional* in length to, the vector quantity). A second line is then drawn to represent the second vector, starting this line where the first ends. The total of the two vectors is found by joining the start of the first line to the end of the second. See also *parallelogram of forces.*

vegetable oil An oil obtained from nuts and seeds of trees and plants, e.g. palm oil, coconut oil, maize oil, olive oil, castor oil.

vegetative reproduction Increasing the number of young without having special *reproductive cells*, e.g. *runners, tubers.* Found more in plants than in animals. (Fig. 164.)

vein 1. A *blood vessel* which carries *blood* towards the *heart.* The blood is usually *deoxygenated,* but the veins from the lungs carry *oxygenated* blood. Each vein contains *valves* which make sure that the blood travels in the right direction. 2. A set of tubes in a leaf which carries water and food and helps to support the leaf.

velocity *Speed* in a given direction, and therefore a *vector* quantity. The *SI unit* is metres per second and the symbol is u, v or w.

velocity ratio In a *machine* the velocity ratio equals the distance moved by the *effort* divided by the distance moved by the *load.* It is also equal to the *velocity* at which the effort moves divided by the velocity at which the load moves.

vena cava A main *vein* of the body, entering the *heart* (Fig. 91 ii).

venation The arrangement of the *veins* in the body or in a *leaf.*

ventricle One of the two large hollow spaces in the *heart* (Fig. 91 ii). *Blood* is pumped from the ventricles into the main *arteries.*

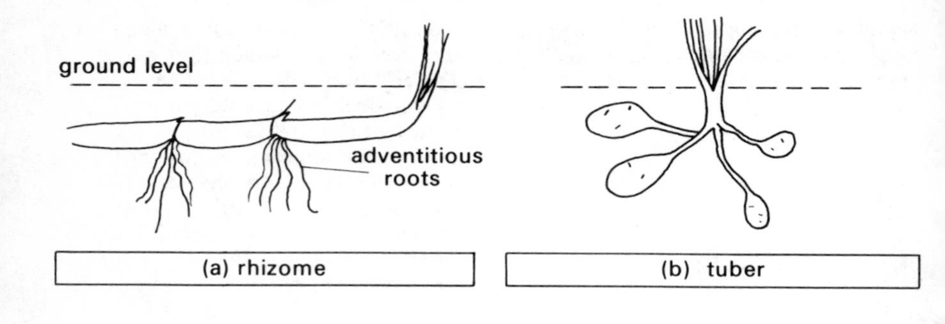

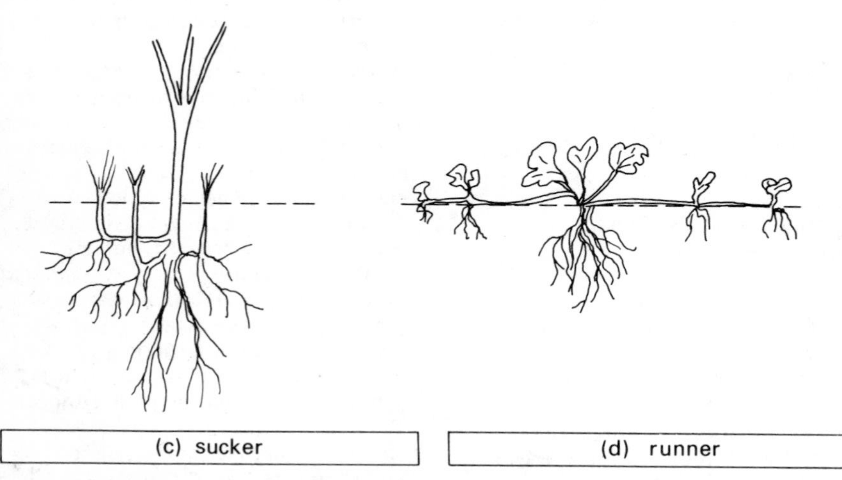

Fig. 164 Vegetative Reproduction in Different Plants

Also used for the spaces inside the *brain.*

vernier A part of many instruments which helps in the accurate measurement of length. It is an extra scale made to move over the main scale of an instrument with its zero mark showing the length to be measured. The movable or vernier scale is marked in units which are

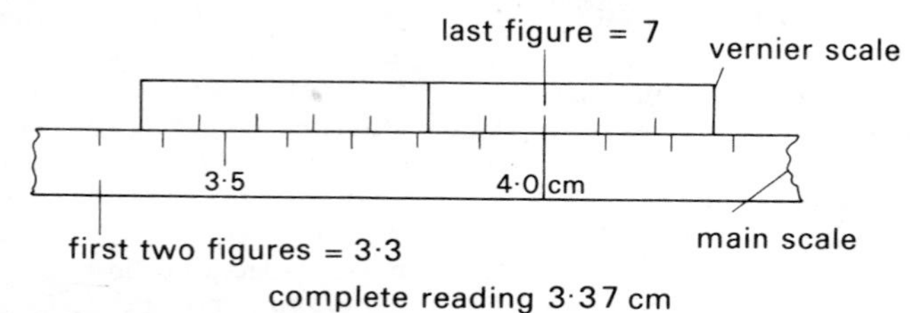

Fig. 165 The Use of a Vernier Scale

only $\frac{9}{10}$ (or sometimes $\frac{19}{20}$ etc.) as long as those on the main scale. In the drawing (which shows the scales bigger than usual) the length is between 3.3 and 3.4 cm but the next figure has to be guessed. If the second, movable scale is used, the next number in the length measurement is on this scale where the marks on the two scales come together, in this case 7. Thus the length is 3.37 cm.

vertebra One of the bones of the *spinal column.* The *spinal cord* goes through the vertebral canal. *Muscles* are joined to the bony parts.

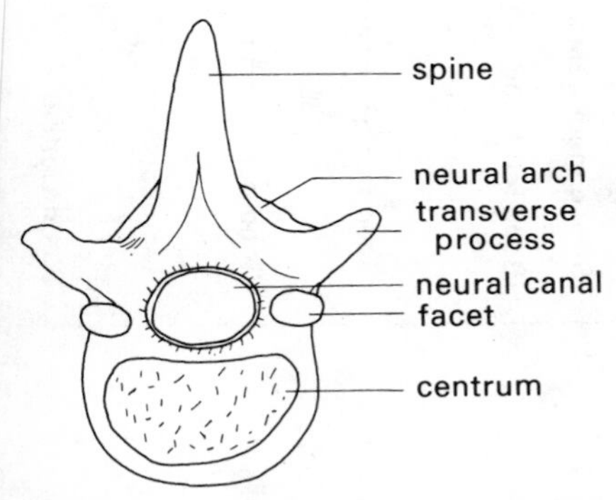

Fig. 166 Anterior view of a Vertebra

vessel Tubes in the *wood* of a plant or tubes carrying *blood* or *lymph* in an animal.

vestibule The space of the *inner ear.*

vestigial organ Part of the body for which no use is known, e.g. *appendix* in man.

villus (*pl.* villi) A small hair-like tube on the wall of the *intestine.* Food is taken into the *blood* or *lymph* inside the villus. The large number of villi makes a better surface for food *absorption.*

vinegar A very *dilute solution* of *acetic acid* (3–6%) obtained by the action of *bacteria* on wine or beer.

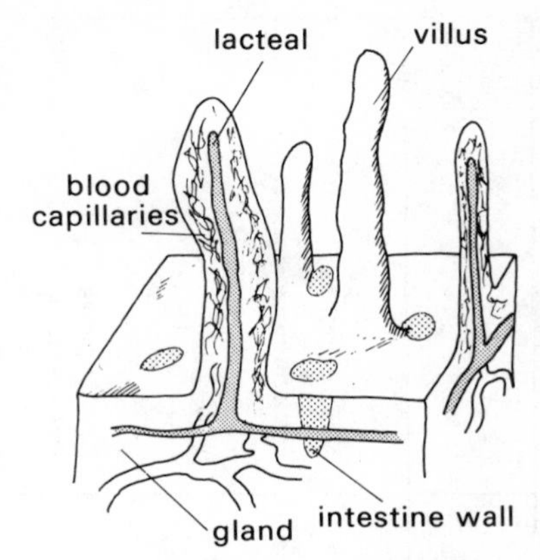

Fig. 167 Intestinal Wall in cross-section section

vinous Having a wine-like smell.

vinyl group The *unsaturated* $CH_2{=}CH-$ group present in such *compounds* as vinyl choride (chloroethene), CH_2CHCl. These compounds usually undergo *polymerization* fairly easily.

virtual image A type of image produced by a *lens* or *mirror* where the light only seems to pass through the *image* position. A virtual image cannot, therefore, be seen on a screen if one is placed at the image position. See also *real image.*

virus A *disease* causing organism, smaller than *bacteria.* Different viruses cause different diseases, e.g. smallpox, influenza. A virus can only live actively in a living *cell.*

viscera The name given to the main soft parts inside the body, e.g. *heart, stomach.*

viscous Thick and syrupy–like molasses or treacle.

vision The sense of sight. The field of vision is how widely an animal can see when it looks straight in front.

visual purple A chemical substance formed in the *rods* of the eye *retina.* It makes it possible for animals to see in weak light. It is related to *Vitamin A.*

Name	Where found	Use in body	Disease if lacking
Vitamin A	Liver, eggs, vegetables	Good sight Strong membranes Good growth	Night blindness Eye and nose trouble
Vitamin B group			
(1) Thiamine	Yeast, unpolished rice	Good digestion and nerves	Poor appetite Beri–beri
(2) Riboflavin	Yeast, meat, milk	Good growth	Skin and nerve troubles
(3) Niacin	Yeast, meat, milk	Strong membranes	Pellagra–intestine troubles
(4) Folic acid	Yeast, green vegetables	Good red blood	Anaemia
Vitamin C (ascorbic acid)	Fresh fruit and vegetables	Good skin and bones	Scurvy
Vitamin D (calciferol)	Liver, eggs, butter, milk	Good bones and teeth	Rickets
Vitamin E	Green vegetables	Reproduction	Sterility
Vitamin K	Green vegetables	Blood clotting	Much bleeding

vitamins Substances in small quantities in the food which are needed for good health. They are usually named by letters, e.g. Vitamin A, B, C. If they are absent from food, *disease* may occur, lack of Vitamin A causes night blindness, of Vitamin B *beri-beri*, of Vitamin C *scurvy*, of Vitamin D *rickets.*

vitelline membrane The skin around the *yolk* of a bird's egg.

vitreous humor The thick liquid inside the eye ball. It keeps the *eye* (Fig. 68) firm and allows light to pass through to the *retina.*

vitriol (oil of) The old common (trivial) name for sulphuric acid.

viviparous Giving birth to young, not laying eggs.

vocal cord Tight pieces of skin in the throat. They move when air passes through them and different sounds are made.

volatile Easily *vaporized*; having a high *vapour pressure.*

volt The *SI unit* of *potential* and *potential difference.* The symbol is V. If the potential difference between two points is one volt then one *joule* of electrical energy will be changed into other forms in one second if a current of one *ampere* is flowing or the total energy change is one joule for each *coulomb* of electricity that passes.

voltage A common, rather than a scientific word, for *potential difference* but one that is much used.

voltaic cell Another name for a *simple cell.*

voltameter A container made especially for the study of *electrolysis.*

voltmeter An instrument for measuring a *potential difference* in *volts.* It can be of the *moving coil* or *moving iron* type. A voltmeter is always connected across that part of the circuit where the potential difference is to be measured. The voltmeter contains a high *resistance* so that only a very small current flows. Changing this resistance will change the range of the voltmeter.

volume The space taken up by an object. The *SI unit* is metres cubed (although centimetres cubed and *litres* are in common use) and the symbol is V.

volumetric analysis A method of determining the *concentration* of a *solution* or the percentage of a particular *element* or group present in a compound by reactions involving the comparison of volumes of solutions which react together. One solution of known concentration—a *standard solution*—is usually added from a burette to a definite volume of the other and an *indicator* may show the completion of the reaction.

voluntary action A movement of the body made on purpose as a result of a message from the *brain* to *muscles*, e.g. kicking a football.

vulcanization Heating rubber with *sulphur* in order to make it less sticky and to bring about changes in its properties. The sulphur is usually mixed with zinc oxide and an accelerator.

vulva The opening of the female *reproductive system.* It leads into the *vagina.*

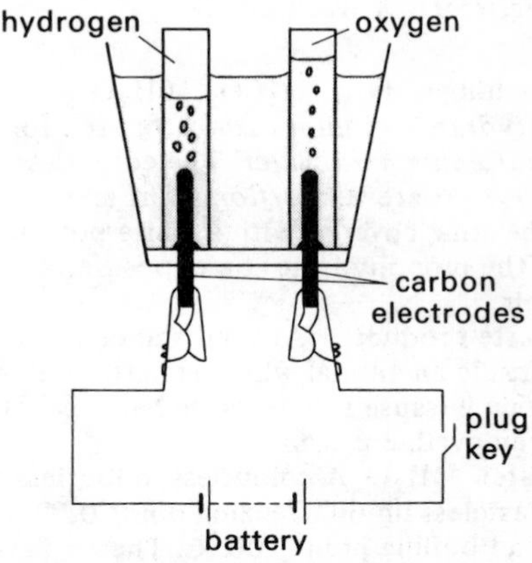

Fig. 169 Simple Voltameter

wall pressure The pressure of a *cell* wall which makes it resist swelling due to *turgor pressure.* It prevents the cell from bursting.

warm blooded The name given to those animals, birds and mammals, whose *blood* stays the same temperature however hot or cold it may be outside—*homoiothermic.*

wash bottle A glass or plastic container filled with distilled water and provided with a jet so that a fine stream of water may be directed where required, e.g. for washing a precipitate (*precipitation*) on a filter paper.

Fig. 170 A Wash Bottle

washing soda $Na_2CO_3.10H_2O$ *Hydrated sodium carbonate* used for *softening hard water.* The colourless *crystals* are very *efflorescent* and become covered with a white powder (the monohydrate) on exposure to air.

waste product Any substance made inside an animal which is better outside because it may cause harm if not removed, e.g. *urine.*

water H_2O A colourless, odourless, tasteless liquid, freezing point 0 °C and boiling point 100 °C. The common *oxide* or *hydroxide* of *hydrogen,* being produced when that gas burns in air. It is present in all living things and occurs naturally in abundance, though rarely in the pure state because of its property of dissolving many substances. It has a maximum density of 1 gram per cm^3 at 4 °C.

water-bath A beaker or vessel in which water can be heated. A solution to be *evaporated* slowly (for the formation of *crystals*) is placed in a dish on top of the water bath and is heated by the steam. Objects may also be placed in the bath when they need to be heated at a temperature below the boiling point of the water. For higher temperatures oil or liquid paraffin may be used in place of the water.

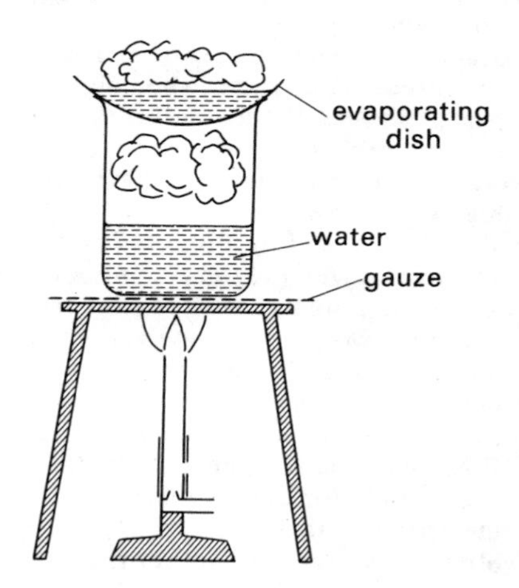

Fig. 171 Evaporation on a Water-bath

water culture solution A liquid of mixed chemicals which will give all the *mineral salts* a plant needs.

water cycle The natural cycle in which water is *evaporated* from the sea and eventually is returned as rain. The water *is vaporized* and rises into the air. It may be carried by winds over the land until it meets some obstacle, such as a mountain range, or is cooled down. It then *condenses* and falls as rain. This finds its way to the rivers and so back to the sea. It is the heat of the sun which

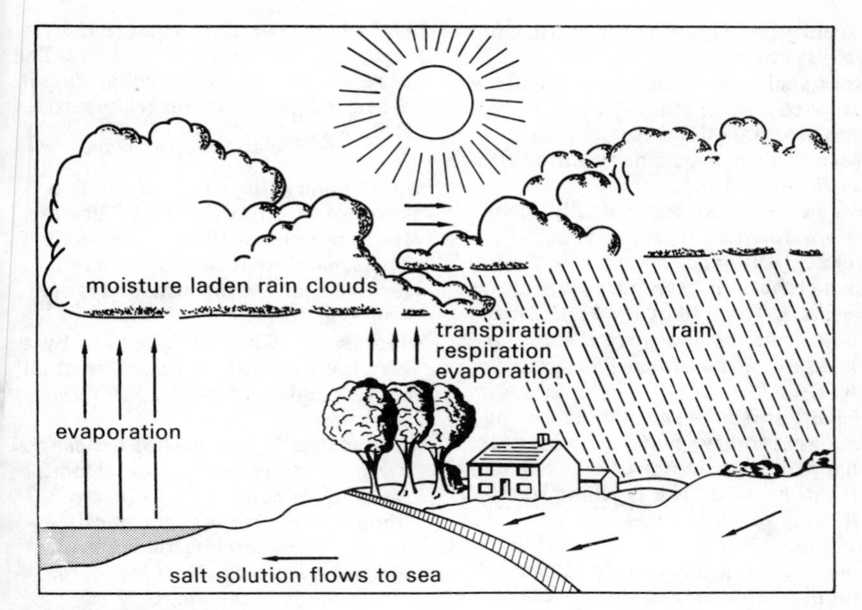

Fig. 172 The Water Cycle

supplies the energy necessary for the evaporation.

water gas See *blue water gas, carburetted water gas.*

water intake Water going into the plant through its *roots.*

water of crystallization Water that is taken up when some *crystals* are formed and which plays a part in giving them their shape and colour, e.g. cobalt chloride crystals are pink, but the *anhydrous* chloride is blue. The water of crystallization can usually be easily removed by heating. A definite small number of *moles* of water are present in each mole of the compound, as in $CuSO_4.5H_2O$ and $Na_2CO_3.10H_2O$.

water output The amount of water lost from the plant by *transpiration.*

water sac The *amnion.*

water softener See *ion-exchange.*

water uptake The amount of water passing from the *roots* to the leaves.

water-glass See *sodium silicate.*

watt The *SI unit* of *power.* The symbol is W. It is the power used when one *joule* of *work* is done per second.

wave A repeated to and fro motion that carries *energy* from a source. For example, sound, *light* and all members of the *electromagnetic spectrum.* Waves may be of several types, e.g. *longitudinal, progressive, standing, stationary* and *transverse* waves.

wave front A line across a *wave* where the particles (etc.) are all in step with each other.

wavelength The distance between any point on a *wave* and the next place where the same thing happens

Fig. 173 Wavelength

again. The *SI unit* is the metre and the symbol is λ.

weak acid An acid which is ionized only to a small extent in water, e.g. ethanoic (acetic) acid.

wean To change a baby's food from *milk* to solid food.

weathering Rocks gradually getting worn down by frost and rain.

weed Any plant in cultivated ground which has not been purposely grown.

weight The pull of the earth on an object. As this is a *force* it must be measured in *newtons.* The symbol is *W, G,* or *P.*

weightlessness The state of having or seeming to have no *weight.* This can happen when an object is a long way from the earth, if it is falling freely or moving freely in orbit.

welding The joining of *metals* by heating them sufficiently to cause them to *melt* and so *fuse* together.

wet test A test carried out in solution for the purpose of identifying the presence of a particular *element, ion* or group, e.g. if addition of aqueous silver nitrate produces a white precipitrate (*precipitation*) insoluble in dilute nitric acid, but easily soluble in ammonia solution, we know that the original solution contains a *chloride.*

wheatstone network An arrangement of four *resistors,* a *galvanometer* and a *cell* which is used to measure the *resistance* of one of the resistors. The *components* are connected as shown. When the galvanometer reads zero, $\frac{R_1}{R_2} = \frac{R_3}{R_4}$. See also *metre bridge.*

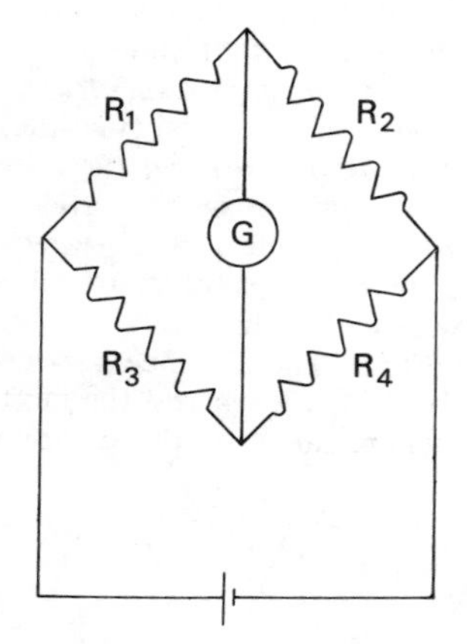

Fig. 174 The Wheatsone Network

white blood cells One of the two kinds of *cells* in the *blood.* They are fewer in number than the red cells. Their main use is to destroy any *germs* which get into the body. See *cell* (Fig. 33ii).

white light The *light* given out by a very hot object. It is a mixture of all *colours* and can form a *continuous spectrum.*

white matter The part of the *central nervous system* (Fig. 34 ii) which is made up of neuron fibres (*nerve fibres*). There are no neuron *cell bodies.* The white matter is outside the *grey matter* in the *spinal cord,* but inside.it in the *brain.*

Wilson cloud chamber See *cloud chamber.*

wilting A state of a plant when it is suffering from lack of water. The *stem* and *leaves* hang down.

wind dispersal The scattering of *seeds* by the wind. The seeds may have wings or hairs so that the wind can carry them a long distance. The seeds are normally very light.

wind pipe The *trachea* leading through the throat to the *lungs.*

wind pollination The scattering of *pollen* by the wind. Plants which are wind pollinated make a lot of very light pollen.

wing The *organ* of flight in birds and insects. In birds the wing bears *feathers* to give strength and support in flying. The wing of birds is the *fore limb.* In insects the wing is made of tough skin folded out from the body.

wire gauze A mesh of wires, woven into a square, and used for supporting glass vessels being heated.

womb The *uterus,* where the young *mammal* is developed before birth.

wood fibres The very strong parts of the wood of a plant. They are long, thin strips which are nearly solid.

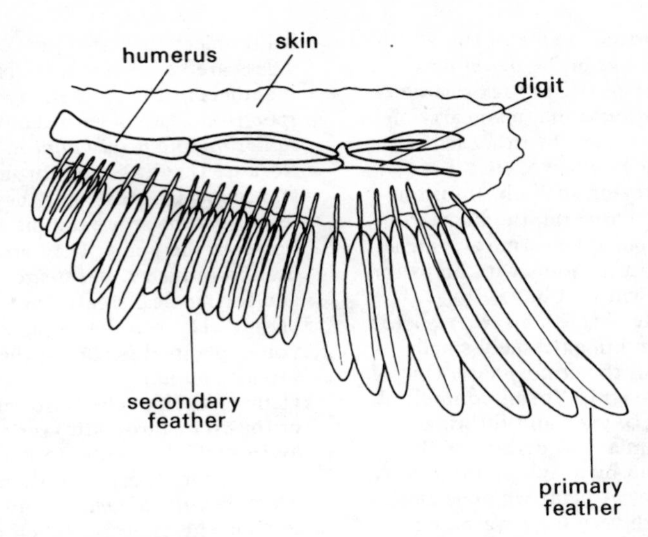

Fig. 175 Structure of a Wing

wood vessels The hollow tubes in wood. They are used to carry water through the plant.

wool The long curly hair of a sheep. It prevents the heat of the body from being lost.

work When a *force* moves in the direction in which it is acting it does work. The work is measured by the force multiplied by the distance moved. The *SI unit* is the *joule* and the symbol is *W*.

Woulfe bottle A bottle used for washing, absorbing or drying gases. It usually contains water, aqueous sodium hydroxide or concentrated sulphuric acid.

wrist A *joint* made of several bones. The *carpus*. The movement allows the hand to be turned.

wrought iron Pig-iron (*cast-iron*) which has been treated to remove all the *carbon*. The purest form of commercial iron, it is tough and can be welded. It is used for making chains, hooks, bars and various kinds of agricultural instruments.

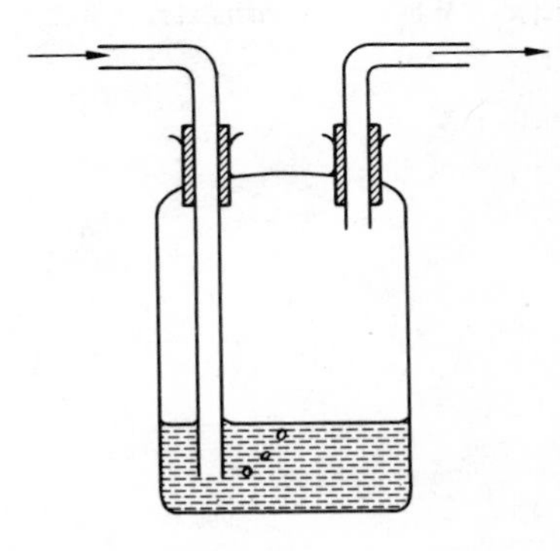

Fig. 176 Woulfe Bottle

Wulf electroscope An *electroscope* in which a thin aluminium strip, fixed at one end and joined to a quartz fibre (really a delicate spring) at the other, forms a bow as the *potential* rises. This continues until it touches an *earthed* metal plate when it springs back. The number of times this happens per second shows the size of the (very small) current which is then flowing.

X-chromosome One of the 23 *chromosomes* in the *ovum* of a woman. It carries the *sex character.* Half the *sperms* in a man carry an X-chromosome, the other half carrying a *Y-chromosome.* If a sperm carrying an X-chromosome joins with an ovum, the baby will be a girl, but if a sperm carrying a Y-chromosome joins with an ovum, the baby will be a boy.

xenon Xe At. No. 54 R.A.M. 131 A gas found in extremely small amounts in the atmosphere. One of the *argonons.* It forms compounds only with oxygen and fluorine.

xerophthalmia A disease of the eyes caused by a lack of *Vitamin* A.

xerophyte A plant which is able to live with very little water, e.g. cactus.

x-rays When *electrons* strike a solid object they give out X-rays. These are very short *wavelength* members of the *electromagnetic spectrum* coming between *gamma radiation* and *ultra-violet radiation.* Because they can pass through substances which stop *light* they are used to study, for example, broken bones in the body. They are also used in industry, in research and for medical treatment.

xylem The wood of a plant. In young plants it is part of the *vascular bundle.*

xylene C_8H_{10} The third member of the *arene* (aromatic) series of *hydrocarbons;* similar to benzene and toluene. It is a colourless liquid which occurs in *coal tar* and is used as a solvent. It is also called dimethylbenzene.

Y-chromosome One of the 23 chromosomes found in half the *sperms* of a man (See *X-chromosome*.

yeast A one-celled *fungus* much used in biology because it can carry out *fermentation.*

yellow 1. The *colour* of light in a *spectrum* which comes between orange and green (spectral yellow). 2. A mixture of red and green light (compound yellow). Both of these have similar effects on the human eye.

yellow spot The *fovea.* The most sensitive part of the *retina* of the *eye* (Fig. 68). It carries many *cone cells.*

yolk The food substance of the *egg* (Fig. 61) of animals. Best seen in birds' eggs, where the yellow yolk is in the middle of the egg. Yolk is made mostly of *oil* and *proteins.*

yolk sac The bag which contains the *yolk* in a very young *embryo.*

Young's double slits Two close parallel slits placed in front of a single slit and used to study the *interference* of light.

zeolites Compounds occurring naturally as a complex sodium aluminium *silicate.* They are able to remove calcium *ions* from *hard water* and to replace them with sodium ions. The term is also used for artificial substances (permutits) also used for *softening* water by the *ion-exchange* method.

zinc Zn At. No. 30 R.A.M. 65.4 A bluish-white *metal* which is fairly *reactive*, especially when slightly impure. Occurs as *calamine* (zinc carbonate) and *zinc blende* (the sulphide). Used in *alloys* such as brass, as an *electrode* in certain *cells* and for *galvanizing* iron sheets to protect them from *corrosion.*

zinc blende Zinc sulphide, ZnS, one of the *ores* of zinc.

zone A region, often heated, in which a reaction takes place.

zoology The study of animals.

zooplankton The very small animals which float in water. See also *phytoplankton.*

zoospore An *asexual reproductive cell* which is able to swim. Found in many *algae.*

zygospore A thick walled resting *spore* formed by the joining together of two *isogametes.* It is found in many fungi (*fungus*).

zygote A *cell* formed by the joining together of a male and female *gamete.* The zygote will grow into a new *organism* if conditions are suitable.

zymase A group of *enzymes* made by *yeast*. They help yeast in *fermentation.*

The Elements Arranged in Alphabetical Order of their Chemical symbol

Symbol	Name	Atomic number	Approx. relative atomic mass	Symbol	Name	Atomic number	Approx. relative atomic mass
Ar	Argon	18	40	Mg	Magnesium	12	24
Ac	Actinium	89	227	Mn	Manganese	25	55
Ag	Silver	47	108	Mo	Molybdenum	42	96
Al	Aluminium	13	27	N	Nitrogen	7	14
Am	Americium	95	243	Na	Sodium	11	23
As	Arsenic	33	75	Nb	Niobium	41	93
At	Astatine	85	210	Nd	Neodymium	60	144
Au	Gold	79	197	Ne	Neon	10	20
B	Boron	5	11	Ni	Nickel	28	58.5
Ba	Barium	56	137	No	Nobelium	102	253
Be	Beryllium	4	9	Np	Neptunium	93	237
Bi	Bismuth	83	209	O	Oxygen	8	16
Bk	Berkelium	97	249	Os	Osmium	76	190
Br	Bromine	35	80	P	Phosphorus	15	31
C	Carbon	6	12	Pa	Protoactinium	91	231
Ca	Calcium	20	40	Pb	Lead	82	207
Cd	Cadmium	48	112.5	Pd	Palladium	46	106.5
Ce	Cerium	58	140	Pm	Promethium	61	147
Cf	Californium	98	251	Po	Polonium	84	210
Cl	Chlorine	17	35.5	Pr	Praseodymium	59	141
Cm	Curium	96	247	Pt	Platinum	78	195
Co	Cobalt	27	59	Pu	Plutonium	94	242
Cr	Chromium	24	52	Ra	Radium	88	226
Cs	Caesium	55	133	Rb	Rubidium	37	85.5
Cu	Copper	29	63.5	Re	Rhenium	75	186
Dy	Dysprosium	66	162.5	Rh	Rhodium	45	103
Er	Erbium	68	167	Rn	Radon	86	222
Es	Einsteinium	99	254	Ru	Ruthenium	44	101
Eu	Europium	63	152	S	Sulphur	16	32
F	Fluorine	9	19	Sb	Antimony	51	122
Fe	Iron	26	56	Sc	Scandium	21	45
Fm	Fermium	100	253	Se	Selenium	34	79
Fr	Francium	87	223	Si	Silicon	14	28
Ga	Gallium	31	69.5	Sm	Samarium	62	150.5
Gd	Gadolinium	64	157	Sn	Tin	50	118.5
Ge	Germanium	32	72.5	Sr	Strontium	38	87.5
H	Hydrogen	1	1	Ta	Tantalum	73	181
He	Helium	2	4	Tb	Terbium	65	159
Hf	Hafnium	72	178.5	Tc	Technetium	43	99
Hg	Mercury	80	200.5	Te	Tellurium	52	127.5
Ho	Holmium	67	165	Th	Thorium	90	232
I	Iodine	53	127	Ti	Titanium	22	48
In	Indium	49	115	Tl	Thallium	81	204.5
Ir	Iridium	77	192	Tm	Thulium	69	169
K	Potassium	19	39	U	Uranium	92	238
Kr	Krypton	36	84	V	Vanadium	23	51
Ku	Kurchatovium	104		W	Tungsten	74	184
La	Lanthanum	57	139	Xe	Xenon	54	131
Li	Lithium	3	7	Y	Yttrium	39	89
Lr	Lawrencium	103	257	Yb	Ytterbium	70	173
Lu	Lutetium	71	175	Zn	Zinc	30	65.5
Md	Mendeleevium	101	256	Zr	Zirconium	40	91

Table 1 *List of Elements*

Properties of Common Materials*									
	Density (kg m^{-3})	Melting point (°C)	Boiling point (°C)	Specific heat capacity (J kg^{-1} K^{-1})	Specific latent heat of fusion (J kg^{-1})	Specific latent heat of vaporization (J kg^{-1})	Refractive index	Linear expansivity (K^{-1})	Resistivity (ohm metre)
Aluminium	2700	660	2400	910	40×10^{4}	–	–	23×10^{-6}	2.6×10^{-8}
Brass	8500	1030	–	370	–	–	–	18×10^{-6}	8×10^{-8}
Bronze	8800	1030	–	360	–	–	–	18×10^{-6}	30×10^{-8}
Carbon (graphite)	2300	3500	4600	710	–	–	–	7.9×10^{-6}	–
Carbon (diamond)	3500	–	–	–	–	–	2.42	–	–
Carbon Dioxide	1.97	(–79	sublime)	840	–	36×10^{4}	1.0005	–	–
Constantan	8900	1080	–	410	–	–	–	16×10^{-6}	48×10^{-8}
Copper	8900	1083	2600	390	21×10^{4}	–	–	17×10^{-6}	1.7×10^{-8}
Glass (crown)	2600	1100	–	670	–	–	~ 1.5	8×10^{-6}	–
Glass (flint)	4000	1200	–	500	–	–	~ 1.6	8×10^{-6}	–
Glycerol	1260	20	290	2400	17×10^{4}	83×10^{4}	1.47	–	–
Gold	19300	1063	3000	130	7×10^{4}	–	–	14×10^{-6}	2.4×10^{-8}
Hydrogen	0.84	–259	–253	14000	–	45×10^{4}	1.0001	–	–
Invar	8000	1530	–	500	–	–	–	1×10^{-6}	81×10^{-8}
Iron	7900	1535	3000	110	27×10^{4}	–	–	12×10^{-6}	10×10^{-8}
Lead	11300	327	1740	130	2.5×10^{4}	–	–	29×10^{-6}	21×10^{-8}
Manganin	8500	–	–	400	41×10^{4}	–	–	18×10^{-6}	44×10^{-8}
Mercury	13600	–39	357	140	1.2×10^{4}	29×10^{4}	–	–	96×10^{-8}
Naphthalene	1140	80	218	1310	–	–	–	–	–
Nitrogen	1.17	–210	–196	1040	2.6×10^{4}	20×10^{4}	1.0003	–	–
Oxygen	1.33	–219	–183	910	1.4×10^{4}	24×10^{4}	1.0003	–	–
Paraffin oil	800	–	–	2200	–	–	1.43	–	–
Perspex	1190	80	–	1500	–	–	1.50	80×10^{-6}	–
Polythene	920	140	–	2200	–	–	1.52	200×10^{-6}	–
Silver	10500	961	2200	240	10×10^{4}	–	–	19×10^{-6}	1.6×10^{-8}
Steel (mild)	7800	1400	–	420	–	–	–	15×10^{-6}	15×10^{-8}
Tungsten	19300	3380	5900	–	–	–	–	4×10^{-6}	5.5×10^{-8}
Water	1000	0	100	4200	33×10^{4}	226×10^{4}	1.33	–	–
Zinc	7100	420	910	390	10×10^{4}	–	–	30×10^{-6}	6×10^{-8}

* All values are approximate and given for room temperature (about 20° C) where appropriate

Table 2 Properties of Common Materials